CAMBRIDGE LIBRARY COLLECTION

Books of enduring scholarly value

Life Sciences

Until the nineteenth century, the various subjects now known as the life sciences were regarded either as arcane studies which had little impact on ordinary daily life, or as a genteel hobby for the leisured classes. The increasing academic rigour and systematisation brought to the study of botany, zoology and other disciplines, and their adoption in university curricula, are reflected in the books reissued in this series.

Fauna Boreali-Americana; or The Zoology of the Northern Parts of British America

Sir John Richardson (1787–1865), surgeon, naturalist and Arctic explorer, went on Sir John Franklin's first two Arctic expeditions as ship's doctor and naturalist, and made observations and collected a large number of plant and animal specimens from the Canadian Arctic. On his return to England after the second expedition he began to write this four-volume work of natural history, first published between 1829 and 1837. A volume is dedicated to each of the classes of mammal, bird, fish and insect, which are found in the Canadian Arctic. This work is an interesting example of pre-Darwinian natural history, full of detailed descriptions of the appearance, anatomy and behaviour of the different species. Volume 1, first published in 1829, focuses on mammals. Descriptions of the species sometimes include details of interactions between humans and that species; for example, unfortunate encounters between sailors and polar bears.

Cambridge University Press has long been a pioneer in the reissuing of out-of-print titles from its own backlist, producing digital reprints of books that are still sought after by scholars and students but could not be reprinted economically using traditional technology. The Cambridge Library Collection extends this activity to a wider range of books which are still of importance to researchers and professionals, either for the source material they contain, or as landmarks in the history of their academic discipline.

Drawing from the world-renowned collections in the Cambridge University Library and other partner libraries, and guided by the advice of experts in each subject area, Cambridge University Press is using state-of-the-art scanning machines in its own Printing House to capture the content of each book selected for inclusion. The files are processed to give a consistently clear, crisp image, and the books finished to the high quality standard for which the Press is recognised around the world. The latest print-on-demand technology ensures that the books will remain available indefinitely, and that orders for single or multiple copies can quickly be supplied.

The Cambridge Library Collection brings back to life books of enduring scholarly value (including out-of-copyright works originally issued by other publishers) across a wide range of disciplines in the humanities and social sciences and in science and technology.

Fauna Boreali-Americana; or The Zoology of the Northern Parts of British America

VOLUME 1: THE MAMMALS

JOHN RICHARDSON
ASSISTED BY WILLIAM SWAINSON
AND WILLIAM KIRBY

CAMBRIDGE
UNIVERSITY PRESS

CAMBRIDGE UNIVERSITY PRESS

Cambridge, New York, Melbourne, Madrid, Cape Town,
Singapore, São Paolo, Delhi, Mexico City

Published in the United States of America by Cambridge University Press, New York

www.cambridge.org
Information on this title: www.cambridge.org/9781108041676

This edition first published 1829
This digitally printed version 2012

ISBN 978-1-108-04167-6 Paperback

FAUNA
BOREALI-AMERICANA;

OR THE

ZOOLOGY

OF THE

NORTHERN PARTS

OF

BRITISH AMERICA:

CONTAINING DESCRIPTIONS OF THE OBJECTS OF NATURAL HISTORY COLLECTED ON THE LATE NORTHERN LAND
EXPEDITIONS, UNDER COMMAND OF CAPTAIN SIR JOHN FRANKLIN, R.N.

BY

JOHN RICHARDSON, M.D., F.R.S., F.L.S.

MEMBER OF THE WERNERIAN NATURAL HISTORY SOCIETY OF EDINBURGH, AND
FOREIGN MEMBER OF THE GEOGRAPHICAL SOCIETY OF PARIS,

SURGEON AND NATURALIST TO THE EXPEDITIONS.

ASSISTED BY

WILLIAM SWAINSON, Esq., F.R.S., F.L.S., &c.

AND

The Reverend WILLIAM KIRBY, M.A., F.R.S., F.L.S., &c.

ILLUSTRATED BY NUMEROUS PLATES.

PUBLISHED UNDER THE AUTHORITY OF THE RIGHT HONOURABLE THE SECRETARY OF
STATE FOR COLONIAL AFFAIRS.

LONDON:

JOHN MURRAY, ALBEMARLE-STREET.

MDCCCXXIX.

TO

THE RIGHT HONOURABLE

LORD VISCOUNT GODERICH,

THE FOLLOWING WORK,

UNDERTAKEN UNDER THE PATRONAGE OF HIS LORDSHIP,

IS,

BY PERMISSION,

INSCRIBED WITH THE UTMOST RESPECT,

BY

HIS MOST OBEDIENT SERVANT,

JOHN RICHARDSON.

FAUNA
BOREALI-AMERICANA,

BY

Dr. RICHARDSON.

a

INTRODUCTION.

THE objects of Natural History collected by the last Overland Expedition to the Polar Sea, under the command of Captain Sir John Franklin, to which I was attached as Surgeon and Naturalist, being too numerous for a detailed account of them to be comprised within the ordinary limits of an Appendix to the narrative of the proceedings of the journey, I was desirous of making them known to the world in a separate work. As it was necessary, however, in order to render such a publication useful, that many of the subjects, particularly in the Ornithological and Botanical parts, should be illustrated by figures, the expense would have been an insurmountable difficulty, had not His Majesty's Government, actuated by a most laudable desire of encouraging science, lent a liberal aid to the undertaking. On an application, which had the approval of the Secretary of State for Colonial Affairs, the Treasury granted one thousand pounds, to be applied solely towards defraying the expense of the engravings. A moiety of that sum has been allotted to the illustration of the Quadrupeds and Birds, and the remainder to the Fishes, Insects, and Plants; and care has been taken, by employing only the first artists, to render the plates worthy of the high patronage the work has received; while their number will demonstrate the rigid economy with which the funds for their execution have been distributed*.

---

* There are twenty-eight plates in this part; and fifty admirable coloured ones, of birds, have also been executed. The botanical plates will likewise be numerous, and many of them are already finished.

b

Having neither leisure nor ability to do justice to the different departments of such a work, without assistance, I have gladly availed myself of the aid of several kind friends and able naturalists,—the First Part, relating to the QUADRUPEDS, being the only one for which I am solely accountable. William Swainson, Esq., the able illustrator of the Ornithology of the Brasils, undertook to arrange and make drawings of the BIRDS, elucidate the Synonyms, furnish Remarks on the natural groups, and, in fact, to charge himself with the principal part of the Ornithology. The Reverend Mr. Kirby agreed to arrange and describe the INSECTS; and Dr. Hooker, Professor of Botany at Glasgow, relieved me entirely from the charge of describing the PLANTS. The number of specimens of these requiring that Dr. Hooker's part should extend to about two volumes of letter-press, it has been judged better to publish the Zoology and Botany in separate works,—the latter edited solely by Dr. Hooker, and as similar to the former in paper and type as possible*. The following introductory remarks are, therefore, drawn up principally with a view to the Zoological specimens.

First, with regard to the geographical limits of the country, whose ferine inhabitants are to be described.

The Expedition landed at New York, proceeded up the Hudson to Albany; from thence westward by the ridge-road to Niagara; then, after a short visit to the stupendous falls on that river, it crossed Lake Ontario to York, the capital of Upper Canada; and, passing by Lake Simcoe and the river Nattawasaga, it arrived at Penetanguishene, on the north-east arm of Lake Huron, in the beginning of April. Up to this place, owing to the early period of the year, and the mode of travelling, which was, for the greater part of our route, in carriages at a rapid rate, our collections were small, consisting, in Zoology, only of a few insects and serpents, and in Botany, principally of lichens

* Dr. Hooker is far advanced with his work, which will come out in parts; and Mr. Drummond has already, under his inspection, published two volumes of dried American mosses, containing two hundred and eighty-six species, collected by the Expedition.

and mosses. With these slight exceptions, the specimens brought to England were entirely collected to the north of the Great Canada Lakes, beyond the settled parts of Upper Canada, and, in fact, in a widely extended territory, wherein the scattered trading posts of the Hudson's Bay Company furnish the only vestiges of civilisation. The following work may, therefore, be termed a *Fauna;* or, more properly, *Contributions to a Fauna* of the *British American Fur Countries;* or it may be considered, in a general view, as comprising what is known of the Zoology of that part of America, which lies to the north of the 49th parallel of latitude, and which, to the east of the Rocky Mountains, at least, is exclusively British. I have, however, included in it descriptions of a few specimens obtained a degree or two to the southward of that latitude on Lake Huron and on the River Columbia, in both of which quarters there are several fur-posts of the Hudson's Bay Company. After having travelled through the Fur Countries lying to the eastward of the Rocky Mountains, for seven summers, and passed five winters at widely distant posts, it will scarcely be thought that I arrogate too much in saying, that almost all the quadrupeds that are objects of chase or interest to the natives, and a very great proportion of the birds, either came within the scope of my own observation, or were mentioned in the many conversations I had with the white residents and native hunters, on the natural productions of the country. But, although my opportunities of ascertaining the number of species actually inhabiting the northern parts of America were so great, I must confess, that a journey like ours, in which natural history was only a subordinate object, and at many periods of which the shortest delay beyond that absolutely necessary for refreshment and repose, was inadmissible, did not afford much opportunity for studying the manners and habits of the animals with the attention I could have wished to have devoted to that subject. The present work, therefore, though fuller than any preceding one, is to be considered only in the light of a sketch, in which many omissions remain to be supplied and inaccuracies to be corrected by future observers. To render the list as complete as possible, I have included those animals mentioned by preceding writers, which did not come under the notice of the Expedition; always carefully acknowledging the source of my quotations.

Sir John Franklin's narratives of his two journeys contain full information respecting the districts through which the Expedition travelled; but, to save reference, and to enable the reader of this work the more readily to discover the particular habitats, and to trace the geographical distribution of the species described in it, I have thought it proper to give a summary account of our route, followed by some compendious topographical notices.

The *First* of the two NORTHERN LAND EXPEDITIONS disembarked in the month of August, 1819, at York Factory, in Hudson's Bay, which is 90° of longitude east of the meridian of Greenwich. From thence, travelling between the 57th and 53d parallels of latitude, by Hayes' River, Lake Winipeg, and the Saskatchewan, it proceeded to Cumberland-house, situated beyond the 102d meridian, where it arrived towards the end of October. Early in January, 1820, the Commanding Officer, accompanied by Mr. (now Captain) Back, set out, to travel on snow-shoes up the Saskatchewan, nearly west-south-west to Carlton-house, in the 106th degree of longitude; and from thence, on a northerly and somewhat westerly course, by Green Lake, the Beaver River, Isle à la Crosse, and Buffalo lakes, across the Methy portage, and down the Elk River, to Fort Chepewyan, on the Atha-pescow or Athabascow Lake, or Lake of the Hills, as it is named by Sir Alexander Mackenzie. The other two officers of the Expedition (Lieutenant Hood and myself) stayed, during the remainder of the winter, at Cumberland-house; and after I had paid a visit in May to the plains of Carlton, and collected all the specimens of plants and animals I could procure at that season, set out in the month of June, to travel in canoe to Fort Chepewyan by the route of Beaver Lake, Missinippi or English River, Black-bear Island Lake, Isle à la Crosse, Buffalo Lake and Elk River. Having rejoined our companions, the whole party left Fort Chepewyan on the 18th of July, 1820; and, descending the Slave River, crossed Great Slave Lake, and ascended the Yellow-knife River, to the banks of Winter Lake, situated in latitude 64½°, and in the 113th degree of longitude, which it reached on the 19th day of August. A winter of nine months' duration was spent at this place in a log building, which was named Fort Enterprise; and in the beginning of June, 1821, while the snow was still lying on the ground, and the ice covering the river, the Expedition resumed its march. After the baggage and canoes had been dragged over ice and snow for one hundred and twenty miles to the north end of Point Lake, we embarked on the Coppermine River on the 1st of July, and on the 21st of the same month reached the Arctic Sea, when, turning to the eastward, we performed a coasting voyage of six hundred and twenty-six statute miles, to Point Turnagain, which is, owing to the deep indentations of the coast, only six degrees and a half of longitude to the eastward of the mouth of the Coppermine River. The rapid approach of winter now rendered it necessary to abandon the further pursuit of the enterprise; and on the 22d of August we retraced our course as far as Hood's River, which we ascended for a short way, and then set out to travel overland to Point Lake, on our way back to Fort Enterprise. Winter, clothed with all the terrors of an arctic climate, overtook the party early in September: it suffered dreadfully from famine, no supplies were obtained at Fort Enterprise, the majority of the party perished, and the survivors were on the verge of the grave, when the Indians brought supplies of provision, and conducted them to Fort Providence, the nearest of the Hudson Bay Company's posts. The want of the means of carriage, even at the most flattering periods of this disastrous journey, prevented us from attempting to preserve any bulky objects of natural history; but all the plants gathered previous to our reaching the mouth of the Coppermine River were saved, having been given in charge to five of the party who were sent back from thence. Those collected on the sea-coast, after having been carried for many days through the snow, were at length, on our strength being completely exhausted, reluctantly abandoned. The winter of 1821-22 was passed at Fort Resolution, on the south side of Great Slave Lake; and the summer of 1822 was consumed in returning by the route we had before travelled to York Factory, where we embarked for England in the month of September. The most interesting of the quadrupeds and birds collected on this Expedition were described by Joseph Sabine, Esq., in the Appendix to Sir John Franklin's narrative, and I published a list of the plants in the same work.

The *Second* or *Last* NORTHERN LAND EXPEDITION commenced, as far as regards the objects of natural history described in this work, at Penetanguishene, on St. George's day, the 23d of April, 1825, and having performed a coasting voyage along the northern sides of Lakes Huron and Superior, arrived at Fort William, a post of the Hudson's Bay Company, situated in Thunder Bay of the last-mentioned lake. From thence it ascended the Kamenistiguia to Dog Lake, and crossing a height of land of no great elevation at the source of the Dog River, and only between twenty and thirty miles from the shores of Lake Superior, it descended by a series of rocky rivers, interrupted by numerous cascades and portages, to Rainy Lake, the Lake of the Woods, and Lake Winipeg. On entering the Saskatchewan River, which falls into the last-mentioned lake, on its east side, the Second Expedition came upon the route of the first one already described, which it kept till its arrival at Fort Resolution, on Great Slave Lake. At Cumberland-house, Mr. Drummond, the Assistant Naturalist, was detached up the Saskatchewan to examine the plains of Carlton, and the eastern declivity of the Rocky Mountains, near the sources of the Peace River. His labours will be more particularly mentioned hereafter: at present I proceed to trace the progress of the Expedition, which, on its arrival at Fort Resolution, instead of directing its course across Great Slave Lake, as on the first journey, turned to the westward, along the south shore of the lake, and entered the Mackenzie, by far the largest of all the American rivers which fall into the Polar Sea, and which originating in the same elevated part of the Rocky

Mountain chain with the Columbia, the Missouri, and the Saskatchewan, or Nelson Rivers, flows under the names of Elk, Slave, or
Mackenzie River, on a north-north-west general course, through
fifteen degrees of latitude, until it discharges itself into the sea by a
mouth extending from the 133d to the 137th degree of longitude.
When the Expedition reached the 65th degree of latitude in its
descent of the Mackenzie, it turned to the eastward for seventy miles
up a river to Great Bear Lake, where a winter residence was erected,
on which the appellation of Fort Franklin was bestowed. Excursions
were made down the Mackenzie and along Bear Lake while the
navigation continued open, but the whole party were assembled at
their winter-quarters on the 5th of September. The extent of country
examined this first season may be judged of by the length of the route
of the Expedition, from its leaving Penetanguishene in the month of
April till its assembling at Great Bear Lake in September, which,
including Mr. Drummond's journey to the Rocky Mountains, Sir
John Franklin's voyage down the Mackenzie to the sea, and a voyage
round Great Bear Lake by myself, exceeded six thousand miles.
Towards the end of the month of June 1826, the Expedition left its
winter-quarters, and proceeded down the Mackenzie to the sea;
and the Commanding Officer, turning to the westward, sailed
along the coast until he attained the $70\frac{1}{2}°$ of latitude, and nearly
the 150th degree of longitude, when the lateness of the season
prohibiting a further advance, he retraced his way to Great Bear
Lake. In the mean time, a detachment under my charge had sailed
from the mouth of the Mackenzie eastward, round Cape Bathurst, in
latitude 71° 36′ north, to the mouth of the Coppermine River, whence
it travelled on foot to the north-east end of Great Bear Lake, and
from thence, in a canoe, to Fort Franklin. The extent of sea-coast
examined by the two branches of the Expedition exceeded twelve
hundred miles, and the whole distance travelled by them from the
time of their departure from Fort Franklin till their return to it again,
was upwards of four thousand miles. A collection of plants formed
by Captain Back, who accompanied Sir John Franklin, is peculiarly

interesting, as having been made principally on a coast skirting the northern termination of the Rocky Mountains. The Expedition returned to England the following summer; one division of it by way of Canada and New York, and the other by Hudson's Bay. I passed the early part of the winter at Great Slave Lake, where I obtained specimens of all the fur-bearing animals of that quarter, and afterwards travelled on the snow to Carlton-house on the Saskatchewan, where, with the assistance of Mr. Drummond, who joined me there, specimens of the greatest part of the birds frequenting that district were procured in the spring. I met Sir John Franklin at Cumberland-house in June, 1827, and accompanied him to Canada by the same route by which we came out, except that we went by the east side of Lake Winipeg, thus completing the circuit of that lake, and that instead of crossing Lake Ontario, on our way to New York, we gained the Uttawas from Lake Huron, by the route of the French River, and descended it to Montreal, whence we travelled to New York by way of Lake Champlain.

Having thus given in detail the routes of the other branches of the Expedition, it remains that I should mention the one pursued by Mr. Drummond, the Assistant Naturalist, to whose unrivalled skill in collecting, and indefatigable zeal, we are indebted for most of the insects, the greater part of the specimens of plants, and a considerable number of the quadrupeds and birds. This gentleman remained at Cumberland-house in the year 1825, after the rest of the party had gone to the north, collecting plants during the month of July, and then ascended the Saskatchewan for six hundred and sixty miles, to Edmonton-house, performing much of the journey on foot, and amassing objects of natural history by the way. Leaving Edmonton-house on the 22d of September, he crossed a swampy and thickly wooded country to Red Deer River, one of the branches of the Elk or Athapescow River, and along whose banks he travelled until he reached the Rocky Mountains, the ground being then covered with snow. Having explored the portage-road across the mountains to the Columbia River, for fifty miles, he hired an Indian hunter, with whom he returned to the head of the Elk River, on which he passed the winter making collections, under privations which would have effectually quenched the zeal of a less hardy naturalist.   In the month of April, 1826, he revisited the Columbia portage-road, and remained in that neighbourhood until the 10th of August, when he made a journey to the head waters of the Peace River, during which he suffered severely from famine.   Nothing daunted, however, he hastened back as soon as he obtained a supply of provisions, to the Columbia portage, with the view of crossing to that river, and botanizing for a season on its banks.   He had reached the west end of the portage, when he was overtaken by letters from Sir John Franklin, acquainting him that it was necessary to be at York Factory in 1827.   This rendered it necessary for him speedily to commence his return, which he did with great regret, for the view of the Columbia, whose banks are rich in natural productions, had stimulated his desire to explore them, and he remarks,—" The snow covered the ground too deeply to permit me to add much to my collections in this hasty trip over the mountains ; but it was impossible to avoid noticing the great superiority of the climate on the western side of that lofty range.   From the instant the descent towards the Pacific commences, there is a visible improvement in the growth of timber, and the variety of forest trees greatly increases.   The few mosses that I gleaned in the excursion were so fine, that I could not but deeply regret that I was unable to pass a season or two in that interesting region."   He now bade adieu to the mountains and returned to Edmonton-house, where he stayed some time, and then joined me at Carlton-house, as has been already mentioned.   His collections on the mountains and plains of the Saskatchewan amounted to about " fifteen hundred species of plants, one hundred and fifty birds, fifty quadrupeds, and a considerable number of insects."   He remained for six weeks at Carlton-house after I left that place, and then descended to Cumberland-house, where he met Captain Back, whom he accompanied to York Factory ; but he had previously the pleasure of seeing Mr. David Douglas, who, after collecting specimens

c

of plants for the Horticultural Society, for three years, on the banks of the Columbia and in North California, crossed the Rocky Mountains at the head of the Elk River, by the same portage-road that Mr. Drummond had previously travelled, and having spent a short time in visiting the Red River of Lake Winipeg, returned to England with that gentleman by way of Hudson's Bay. Thus, a zone of at least two degrees of latitude in width, and reaching entirely across the continent, from the mouth of the Columbia to that of the Nelson River of Hudson's Bay, has been explored by two of the ablest and most zealous collectors that England has ever sent forth; while a zone of similar width, extending at right angles with the other from Canada to the Polar Sea, has been more cursorily examined by the Expeditions.

Through the liberality of the Horticultural Society, and the influence of their learned Secretary, Joseph Sabine, Esq., ever readily exerted for the advancement of science, I have been permitted to examine and describe the specimens of quadrupeds collected by Mr. Douglas, and this gentleman, with a readiness to communicate the information he has acquired, that does him great credit, has kindly furnished me with some valuable notices of the habits of the animals which have been incorporated in this work. I have also had an opportunity of inspecting the specimens of quadrupeds obtained on the American coast of Behring's Straits, by Captain Beechey, on his late voyage in the Blossom; and the notes respecting them, made on the spot by Mr. Collie, Surgeon of that ship, by whom principally they were collected, have been submitted to my perusal. Previous to our setting out on the Second Expedition, Sir John Franklin addressed letters to many of the resident chief factors and traders of the Hudson's Bay Company, requesting their co-operation with our endeavours to procure specimens of Natural History, and their ready acquiescence with his desire was productive of much advantage to us. Not only were great facilities for the advancement of our pursuits afforded to us by Mr. John Haldane, Mr. James Leith, Mr. Alexander Stewart, Mr. John Prudens, Mr. Robert M‘Vicar, and other gentlemen, whose posts lay on our line of route; but a collection of birds and quadrupeds, of much interest, made at Fort Nelson on the River of the Mountains, a branch of the Mackenzie, was forwarded to us by Mr. Macpherson, together with some valuable specimens obtained in the same quarter by Mr. Smith, chief factor of that district. Mr. Isbister also had the kindness to prepare for us a copious collection of birds at Cumberland-house. These were not, however, the only channels through which the specimens described in the following pages were obtained. I have had ample opportunities for studying the specimens brought home by Sir Edward Parry, on his several expeditions; and much information was likewise derived from frequent visits to the museum of the Hudson's Bay Company, and from repeated examinations of the specimens imported by that Company from their posts on James's Bay, on the Columbia, and in New Caledonia, and presented by them to the Zoological Society and British Museum.

After this brief exposition of the various sources from whence the specimens were derived, I proceed to give *a concise general view of the nature of the different tracts of the country*, whose ferine inhabitants form the subject of the following pages. The most remarkable physical feature of the northern parts of America, is the great Mountain Ridge, which is continued under the appellation of the *Rocky Mountains**, in a north-north-west direction from New Mexico, to the 70th degree of latitude, where it terminates within view of the Arctic Sea, to the westward of the mouth of the Mackenzie River. The course of this chain is tolerably straight, and its altitude, though various in different places, is everywhere far superior to that of any other mountains existing in the same parallel of the American continent. Like the Andes, of which they seem to be a prolongation, the Rocky Mountains lie much nearer to the Pacific coast than to the eastern shore of America, and they give rise to several very large rivers. Over an elevated portion of the chain, extending from the 40th to the 55th degree of latitude, are spread the upper branches and sources of the *Columbia*, which falls into the Pacific in the 46th parallel. If the principal arms of this river had not a very circuitous course, the nar-

---

* Pennant names them the "Shining Mountains."

rowness of the stripe of country which intervenes between the summit of the ridge and the coast would have caused it to be little better than a mountain torrent. As it is, its arms spread far and wide, and it carries a great body of water to the sea. The head waters of the *Missouri* interlock with those of the southern branches of the Columbia; but that river, precipitating itself down the eastern declivity of the mountains, takes a devious course to the south-east, receiving in its way several great tributaries, and joining the *Mississippi*, which rises at the west end of Lake Superior, in a comparatively low, but hilly country. Their united streams traverse the whole of Louisiana, and fall into the Gulf of Mexico, after a course of four thousand and five hundred miles, reckoned from the head of the Missouri. The *Saskatchewan* is the third great river which issues from the same elevated part of the mountains, its feeding streams spreading from the 47th to the 54th parallel of latitude, and the more southern ones being interposed betwixt the head waters of the two preceding rivers. The upper streams of the Saskatchewan, after descending from the mountains, form two principal arms, which flow through comparatively naked, sandy plains, under the names of the North and South Branches, and then unite a short way below Carlton-house. From thence the river, continuing its course through a well-wooded country, passes by Cumberland-house, where it receives a considerable tributary that originates on the immediate banks of the Missinippi, a parallel river, and afterwards, flowing through Lake Winipeg, changes its name to Nelson River, and falls into Hudson's Bay, near Cape Tatnam. The whole course of the Saskatchewan or Nelson River, from the mountains to the sea, may be estimated, windings inclusive, at one thousand six hundred miles. Lake Winipeg, besides other large streams, receives the River Winipeg, which rises on a ridge of land bordering closely on Lake Superior, and also the Red River, whose eastern branch has its sources on the same heights with the Mississippi, and whose western branch originates close to the banks of the Missouri, some distance above where that river begins to turn to the southward. By means of short portages, then, one may pass from the respective branches of the Nelson, by the

Columbia, to the Pacific; by the Missouri or Mississippi to the Gulf of Mexico; by the St. Lawrence to the Atlantic, and also by the Elk or Mackenzie River, whose upper streams approach the north branches of the Saskatchewan to the Arctic Sea. The fourth great river which takes its rise from the same quarter of the Rocky Mountain range is the one just mentioned,—the *Mackenzie*, which is the third of the North American rivers in respect of size, being inferior only to the Missouri and St. Lawrence. The two principal arms of the Mackenzie are the Elk and Peace rivers. One of the main streams of the former, the Red Deer River, issues from the vicinity of the northern sources of the Columbia and Saskatchewan, whilst other feeders interlock with the head waters of the Beaver, Missinippi, or Churchill river. Having passed through the Athapescow Lake, the Elk River is joined by the Peace River, which, originating somewhat further north in the mountains within three hundred yards of the source of the Tacootchtessè or Frazer's River, affords a canoe route to all parts of New Caledonia. It is a singular fact, that the Peace River actually rises on the west side of the Rocky Mountain ridge, and is a large stream navigable for boats at the place where it makes its way through a narrow gorge bounded by lofty mountains, which are covered with eternal snows. Nearer the source of the river, and between it and the Tacootchtessè, the mountains are less lofty and more distant, and the country has there much of the character of elevated table-land. After its union with the Peace River the Elk River assumes the name of Slave River, which, on passing through Great Slave Lake, becomes the Mackenzie. At a considerable distance below the last-mentioned lake, and where the Mackenzie makes its first near approach to the Rocky Mountains, it is joined by a large stream, which rising a little to the northward of the Peace River, flows along the eastern base of the mountains. It obtained the name of the River of the Mountains from Sir Alexander Mackenzie; but its magnitude has since gained it the appellation of the South branch of the Mackenzie from the traders. The Mackenzie receives several other large streams on its way to the sea, and among others Great Bear Lake River, whose head-waters rise on the banks of the Coppermine River and Peel's River, which issues from the Rocky Mountains, in latitude 67°. Immediately after the junction of Peel's River the Mackenzie separates into numerous branches, which flow to the sea through a great delta, composed of alluvial mud. Here from the richness of the soil, and from the river bursting its icy chains, comparatively very early in the season, and irrigating the low delta with the warmer waters brought from countries ten or twelve degrees further to the southward, trees flourish, and a more luxuriant vegetation exists than in any place in the same parallel on the American continent. In latitude 68° there are many groves of handsome white spruce firs, and in latitude 69°, on the shores of the sea, lofty and dense willow-thickets cover the flat islands ; while currants and gooseberries grow on the drier hummocks, accompanied by some showy epilobiums and perennial lupins. The moose-deer, American hare, and beaver, accompany this display of vegetation to its limits. The whole course of the Mackenzie from the source of the Elk River to the sea, is about two thousand miles in length.

These are the principal rivers of the fur countries, but there are three others of shorter course, upon which some part of the collections of specimens were obtained, viz. *Hayes River*, which rises near Lake Winipeg, and holding an almost parallel course to Nelson's River, falls into the same part of Hudson's Bay. York Factory, which will be often mentioned in the following pages, stands on the low alluvial point that separates the mouths of these two rivers. The next river which I have to mention is the *Missinippi*, or, as it is occasionally named, the English River, which falls into Hudson's Bay at Churchill. Its upper stream, named the Beaver River, rises in a small ridge of hills, which separates the north branch of the Saskatchewan from a bend of the Elk River. The *Coppermine* is the last river which requires a particular notice. It has its origin not far from the east end of Great Slave Lake, and, taking a northerly course, flows through the Barren grounds to the Arctic Sea. It is a stream of no great magnitude in comparison with some of the branches of the Mackenzie : there are few alluvial deposits on its banks, and there is not, consequently, that richness of vegetation, which on the Mackenzie attracts certain quadrupeds to very high latitudes.

The Rocky Mountains have been crossed in four several places. First, by Sir Alexander Mackenzie, in the year 1793, at the head of the Peace River, between latitudes 55° and 56°. His route was followed, in 1806, by a party of the North-west Company, sent to make a settlement in New Caledonia, and is still occasionally used by the Hudson's Bay Company. Lewis and Clark, in the year 1805, crossed the Mountains in latitude 47°, at the head of the Missouri, in their way to the mouth of the Columbia River. For several years subsequent to that period, the North-west Company were in the habit of crossing in latitude 52½°, at the head of the North branch of the Saskatchewan, between which and one of the feeding streams of the Columbia there is a short portage; but of late years, owing to the hostility of the Indians, that route has been deserted, and the Hudson's Bay Company, who now have the whole of the Fur Trade of that country, use a portage of considerable length between the northern branch of the Columbia and the Red Deer River, one of the branches of the Elk or Mackenzie River. Some attempts have very recently been made to effect a passage in the 62nd parallel of latitude; but although several ridges of the mountains were crossed, it does not appear that any stream flowing towards the Pacific was reached.

The whole of the country lying to the eastward of the Rocky Mountains, and north of the Missouri and Great Lakes, is settled, or more or less frequently visited by the Hudson Bay Company's traders, and is well known to them, with the exception of the vicinity of the Polar Sea, and a corner bounded to the westward by the Coppermine River, Great Slave, Athapescow, Wollaston, and Deer Lakes, to the southward by the Churchill or Missinippi River, and to the northward and eastward by the sea. This *north-eastern corner* of the American continent is often mentioned in the following pages by the appellation of the *Barren-grounds*, which it has obtained from the traders on account of its being destitute of wood, except on the banks of some of the larger rivers that traverse it. The prevailing rocks in the district are primitive, and in one or two places only do they rise so as to deserve the name of a mountain-ridge, their general form being that of an assemblage of low hills with rounded summits, and more or less precipitous sides separated by narrow valleys. The soil of the latter is sometimes an imperfect peat earth, and in that case it nourishes a few stunted willows, glandular dwarf-birches, black spruce-trees, or larches; but more generally the soil consists of the debris of the rocks, which is a dry coarse quartzose sand, unfit to support any thing but lichens. All the larger valleys have a lake of very transparent water, often of great depth in their centre, and occasionally these lakes are perfectly land-locked, though they all contain fish. More generally one lake discharges its waters into another, through a narrow gorge, by a rapid and turbulent stream, and most of the rivers which flow through the Barren-grounds are little more than a chain of narrow lakes connected in this manner. The small caribou or rein-deer, and the musk-ox, are the principal and characteristic inhabitants of these lands, and the description by Linnæus, of the Lapland deserts frequented by the rein-deer, applies with perfect accuracy to this corner of America. " Nullum vegetabile in tota Lapponia tanta in copia reperitur ac hæc Lichenis species, (*Cenomyce rangiferina*) et quidem primario in sylvis, ubi campi steriles arenosi vel glareosi, paucis Pinis consiti; ibi enim non modo videbis campos per spatium unius horæ, sed sæpe duorum triumve milliarium *, nivis instar albos, solo fere hocce lichene obductos." " Hi Lichene obsiti campi, quos *terram damnatam* diceret peregrinus, hi sunt Lapponum agri, hæc prata eorum fertilissima, adeo ut felicem se prædicet possessor provinciæ talis sterilissimæ, atque lichene obsitæ." Being destitute of fur-bearing animals, no settlements have been formed within the Barren-grounds by the traders, and a few wretched families of Chepewyans, termed, from their mode of subsistence, " Caribou eaters," are the only human beings who reside constantly upon them. Were any one to penetrate into their lands, they might address him with propriety in the words used by the

---

* The Swedish mile is 5½ English miles.

Lapland woman to Linnæus, when he reached her hut, exhausted by hunger and the fatigue of travelling through interminable marshes. "O thou poor man, what hard destiny can have brought thee hither, to a place never visited by any one before! This is the first time I ever beheld a stranger. Thou miserable creature! how didst thou come, and whither wilt thou go*?" Parties of Indians occasionally cross these wilds in going from the Athapescow to Fort Churchill, but they almost always experience great privations, and very often lose some of their number by famine. Hearne, in his first and second journeys, traversed them in two directions; Sir John Franklin, in his first journey, travelled within their western limits; and Sir Edward Parry, in his second voyage, obtained specimens of the animals of Melville peninsula, which forms the North-east corner of the Barren-grounds. The Chepewyans, Copper Indians, Dog-ribs, Hare-Indians, and Esquimaux visit them annually for a short period of the summer season, in quest of caribou.

The following quadrupeds are known to inhabit the Barren-grounds:

Ursus arctos? Americanus.
  ,,   maritimus.
Gulo luscus.
Mustela (Putorius) erminea.
    ,,   ,,   vison.
Lutra Canadensis.
Canis lupus, et varietates ejus variæ.
    ,, (Vulpes) lagopus.
    ,,   ,,   ,, var. fuliginosa.

> More or less carnivorous or piscivorous. They prey much on the animals in the following section.

Fiber zibethicus.
Arvicola xanthognathus.
    ,, Pennsylvanicus.
    ,, borealis.
    ,, (Georychus) trimucronatus.
    ,,   ,,   Hudsonius.
    ,,   ,,   Grœnlandicus.
Arctomys (Spermophilus) Parryi.

> herbivorous.

* *Lachesis Lapponica,* p. 145.

d

Lepus glacialis.	Principal food the dwarf-birch.
Cervus tarandus, var. arctica.	Graminivorous, or more commonly
Ovibos moschatus.	lichenivorous.

A belt of low *primitive rocks* extends from the Barren-grounds to the northern shores of Lake Superior. It is about two hundred miles wide, and as it becomes more southerly, it recedes from the Rocky Mountains, and differs from the Barren-grounds, principally in being clothed with wood. It is bounded to the eastward by a narrow stripe of limestone, and beyond that there is a flat, swampy, partly alluvial district, which forms the western shores of Hudson's Bay. As far as regards the distribution of animals, the whole tract, from the western border of the low primitive rocks to the coast of Hudson's Bay, may be considered as one district, with the exception that the sea-bear seldom goes further inland than the swampy land which skirts the coast. The whole may be named the *Eastern district*, and the following animals inhabit it :—

Vespertiliones, species duo vel tres ignotæ.
Sorex palustris.
   ,,   Forsteri.
Scalops, species ignota.
Ursus Americanus.
   ,,   maritimus. { (Does not go further from the sea-shore than one hundred miles.)
Meles ?
Gulo luscus.
Mustela (Putorius) vulgaris.
   ,,   ,,   erminea.
   ,,   ,,   vison.
   ,,   martes.
   ,,   Canadensis.
Mephitis Americana, var. Hudsonica.
Lutra Canadensis.
Canis lupus, varietates variæ.
   ,, (Vulpes) lagopus.
   ,,   ,,   fulvus.
   ,,   ,,   ,,   var. decussata.
   ,,   ,,   ,,   ,,   argentata.
Felis Canadensis.

Castor fiber, Americanus et ejus varietates.
Fiber zibethicus et ejus varietates.
Arvicola xanthognathus.
    „   Pennsylvanicus.
    „   (Georychus) Hudsonius.
Mus leucopus.
Meriones Labradorius.
Arctomys empetra.
Sciurus (Tamias) Lysteri.
    „   Hudsonius.
Pteromys Sabrinus.
Lepus Americanus.
Cervus alces.
    „   tarandus, var. sylvestris.

The district just mentioned is bounded to the westward by a very flat limestone deposit, and the line of junction of the two formations is marked by a remarkable chain of rivers and lakes, among which are the Lake of the Woods, Lake Winipeg, Beaver Lake, and the middle portion of the Churchill or Missinippi River, all to the southward of the Methy portage; and the Elk River, Athapescow Lake, Slave River, Great Slave Lake, and Martin Lake, to the northward of it. The whole of this district is well wooded; it yields the fur-bearing animals most abundantly; and a variety of the bison, termed from the circumstance the wood bison, comes within its western border, in the more northern quarter. This animal has even extended its range to a particular corner, named Slave Point, on the north side of Great Slave Lake, which is also composed of limestone. The following animals may be found in the *limestone tract* :—

Vespertilio pruinosus.
Sorex palustris.
    „ Forsteri.
Condylura longicaudata. (Southern parts only.)
Ursus Americanus.
Gulo luscus.
Mustela (Putorius) vulgaris.
    „   „   erminea.
    „   „   vison.

Mustela martes.
    ,,   Canadensis.
Mephitis Americana, Hudsonica.
Lutra Canadensis.
Canis lupus occidentalis, var. grisea.
    ,,   ,,   ,,   atra.
    ,,   ,,   ,,   nubila.
    ,,   ,,   ,,   Sticte.
    ,,   (Vulpes) fulvus.
    ,,   ,,   ,,   var. decussata.
    ,,   ,,   ,,   argentata.
Felis Canadensis.
Castor fiber, Americanus et varietates ejus nigræ, variæ, et albæ.
Fiber zibethicus, colore interdum varians.
Arvicola xanthognathus.
    ,,   Pennsylvanicus.
Mus leucopus.
Meriones Labradorius.
Arctomys empetra.
    ,,   (Spermophilus) Hoodii (in the south-western limits of the district.)
Sciurus (Tamias) Lysteri (in the southern part of the district.)
    ,,   ,,   quadrivittatus (middle parts of the district.)
    ,,   Hudsonius.
    ,,   niger (southern border of the district.)
Hystrix pilosus.
Lepus Americanus.
Cervus alces.
    ,,   tarandus, sylvestris (only in a few spots.)
Bos Americanus.

Between this limestone district and the foot of the Rocky Mountains, there is an extensive tract of what is termed Prairie land. It is in general level, the slight inequalities of surface being imperceptible when viewed from a distance, and the traveller in crossing it must direct his course by the compass or the heavenly bodies, in the same way as if he were journeying over the deserts of Arabia. The soil is mostly dry and sandy, but tolerably fertile, and it supports a pretty thick sward of grass, which furnishes food to immense herds of the bison. Plains of a similar character, but still more extensive, have been described by the American writers as existing on the Arkansaw and Missouri Rivers. They gradually become narrower to the north-ward, and in the southern part of the fur countries they occupy about fifteen degrees of longitude, extending from Maneetobaw or Maneeto-woopoo, and Winepegoos Lakes to the foot of the Rocky Mountains. They are partially intersected by some low ridges of hills, and also by several streams, the banks of which are wooded, and towards the outskirts of the plain there are many detached clumps of wood and picturesque pieces of water, disposed in so pleasing a manner as to give the country the appearance of a highly cultivated English park. In the central parts of the plains, however, there is so little wood that the hunters are under the necessity of taking fuel with them on their journeys, or in dry weather of making their fires of the dung of the bison. To the northward of the Saskatchewan, the country is more broken, and intersected by woody hills; and on the banks of the Peace River, the plains are of comparatively small extent, and are detached from each other by woody tracts; they terminate altogether in the angle between the River of the Mountains and Great Slave Lake. The abundance of pasture renders these plains the favourite resort of various ruminating animals. They are frequented throughout their whole extent by buffalo and wapiti. The prong-horned antelope is common on the Assinaboyn or Red River, and south branch of the Saskatchewan, and extends its range in the summer to the north branch of the latter river. The black-tailed deer, the long-tailed deer, and the grisly bear, are also inhabitants of the plains, but do not wander further to the eastward.

The following list will shew the peculiarity of the group of ferine animals which frequent the district :—

Ursus ferox.
Canis latrans.
    ,, (Vulpes) cinereo-argentatus.
Arctomys (Spermophilus ?) Ludovicianus.
   ,,   ,,   Richardsonii.
   ,,   ,,   Franklinii.
   ,,   ,,   Hoodii.

Geomys? talpoides.
Diplostoma?
Lepus Virginianus.
Equus caballus.
Cervus alces.
   ,,   strongyloceros.
   ,,   macrotis.
   ,,   leucurus.
Antilope furcifer.
Bos Americanus.

The fur-bearing animals also exist in the belts of wood which skirt the rivers that flow through the plains; and the wolverene wanders over them as it does through every part of the northern extremity of America. The mephitis Americana Hudsonica breeds freely there; and the raccoon is found on the banks of the Red River, which is its most northern limit.

The following animals are found on the Rocky Mountains :—

Vespertilio subulatus.
Sorex palustris.
Ursus Americanus.
   ,,   ferox.
Gulo luscus.
Mustela (Putorius) erminea.
   ,,   ,,   vison.
   ,,   martes.
   ,,   Canadensis.
Mephitis   ?
Lutra Canadensis.
Canis lupus et ejus varietates.
   ,, (Vulpes) fulvus et ejus varietates.
Felis Canadensis.
Castor fiber, Americanus.
Fiber zibethicus.
Arvicola riparius.
   ,,   xanthognathus.
   ,,   Novoboracensis.
   ,,   (Georychus) helvolus.
Neotoma Drummondii.
Mus leucopus.

Arctomys empetra.
,,   ? pruinosus.
,,   (Spermophilus) Parryi, var. erythrogluteia.
,,   ,,   ,,   phæognatha.
,,   ,,   guttatus ?
,,   ,,   lateralis.
Sciurus (Tamias) quadrivittatus.
,,   Hudsonius.
Pteromys Sabrinus, var. alpina.
Hystrix pilosus.
Lepus Americanus.
,,   glacialis.
Lepus (Lagomys) princeps.
Cervus alces.
,,   tarandus? { (A large kind of caribou is said to frequent the mountains, but I have seen no specimens either of the animal or of its horns.)
,,   macrotis.
Capra Americana (on the highest ridges.)
Ovis montana (on the eastern side of the ridge.)
Bos Americanus (in particular passes only.)

The country lying between the Rocky Mountains and the Pacific is in general more hilly than that to the eastward; but there are some wide plains on the upper arms of the Columbia which have much of the character of the plains of the Missouri and Saskatchewan, and are inhabited by the same kind of animals. In particular the ursus ferox, canis latrans, canis cinereo-argentatus, the braro (perhaps meles Labradoria), cervus macrotis var. β. Columbiana, cervus leucurus, and aplodontia leporina, are enumerated by Lewis and Clark. Mr. Douglas also observed the condylura macroura, and several species of Felis and of Geomys and Diplostoma in that quarter. The sea-coast at the mouth of the Columbia is frequented by a species of fox very like the European one, or the red-fox of the Atlantic states of America. The Arctomys brachyurus and the Arctomys Douglasii also inhabit the banks of the Columbia; and the Arctomys Beecheyi, a species nearly allied to the latter, is found in the adjoining parts of California. The bison are supposed to have found their way across the mountains only very recently, and they are still comparatively few in numbers, and confined to certain spots.

The following brief description of New Caledonia, another district on the west of the Rocky Mountains, is extracted from Mr. Harmon's journal :—

" New Caledonia was first settled by the North-West Fur Company in 1806, and may extend from north to south about five hundred miles, and from east to west, three hundred and fifty or four hundred. The post at Stuart's Lake is nearly in the centre of it, and lies in 54½° north latitude, and 125° west longitude. In this large extent of country, there are not more than five thousand Indians, including men, women, and children. It is mountainous, but between its elevated parts there are pretty extensive valleys, along which pass innumerable small rivers and brooks. It contains a great number of lakes, one of which, Stuart's Lake, is about four hundred miles in circumference ; and another, Nateotain Lake, is nearly twice as large. I am of opinion that about one-sixth part of New Caledonia is covered with water. There are but two large rivers. One of these, Frazer's River, is sixty or seventy rods wide, rises in the Rocky Mountains within a short distance of the source of the Peace River, and is the river which Sir Alexander Mackenzie followed for a considerable distance when he went to the Pacific Ocean in 1793, and which he took to be the Columbia. The other large river of New Caledonia is Simpson's River, which takes its origin in Webster's or Bear Lake, and, after passing through several considerable lakes, falls into Observatory Inlet. The mountains of New Caledonia are not to be compared, in point of elevation, with those that skirt the Peace River between Finlay's Branch and the Rocky Mountain portage, though there are some which are pretty lofty, and on the summits of one in particular, which is visible from Stuart's Lake, the snow lies during the whole year.

" The weather is not severely cold, except for a few days in the winter, when the mercury is sometimes as low as 32° below zero of Fahrenheit's thermometer. The remainder of the season is much milder than it is on the other side of the mountains in the same latitude. The summer is never very warm in the day-time; and the nights are generally cool. In every month in the year, there are frosts. Snow generally falls about the 15th of November, and is all dissolved by the 15th of May. About M'Leod's Lake the snow is sometimes five feet deep, and I imagine that this is the reason that none of the large animals, except a few solitary ones, are to be met with.

" There are a few moose; and the natives occasionally kill a black bear. Caribou are also found at some seasons. Smaller animals like-wise occur, though they are not numerous. They consist of beavers, otters, lynxes, fishers, martins, minks, wolverines, foxes of different kinds, badgers, polecats, hares, and a few wolves. The fowls are, swans, bustards (*anas Canadensis*), geese, cranes, ducks of several kinds, par-tridges, &c. All the lakes and rivers are well furnished with excellent fish. They are, sturgeon, white-fish, trout, sucker, and many of a smaller kind. Salmon also visit the streams in very considerable num-bers in autumn. The natives of New Caledonia we denominate Carriers; but they call themselves Tâ-cullies, which signifies people who go upon water."

Captain Cook, in his third voyage, saw raccoons, foxes, martins, and squirrels, alive, on the coast of New Caledonia, and obtained skins of the following animals :—

Black-bear, brown-bear, glutton, grey wolf, arctic or stone fox, black fox, foxes of a yellow colour with a black tip to the tail, foxes of a deep reddish yellow intermixed with black, raccoon, land-otter, sea-otter, ermine, martins of three kinds: the common one, the pine-martin, and a larger one with coarser hair (*mustela Canadensis?*), lynx, spotted marmot, hares, and skin of an animal named *wanshee* by the natives. In addition to this list, Meares mentions moose-deer skins, and the skin of a very small species of deer, as among the articles of trade in possession of the natives at Nootka Sound.

To the north of New Caledonia there is a large projecting corner, which belongs to Russia, and has been traversed by the servants of the

Fur Company of that nation ; but of which no account has been given to the world, except of the coast, respecting which some information may be obtained from the narratives of Captain Cook, Kotzebue, and other voyagers. The few Indians of Mackenzie River, who have crossed the Rocky Mountains, report that, on their western side, there is a tract of barren grounds frequented by caribou and musk oxen ; and the furs procured by the Russian Company indicate that woody regions, similar to those to the eastward of the mountains, also exist there.

Langsdorff gives the following list of skins contained in the principal magazine of the Russian Fur Company, on the island of Kodiak, most of them collected on the peninsula of Alaska, Cook's River, and other parts of thecontinent.

Brown and red bears, black bears, foxes black and silver-gray, (the stone fox, *canis lagopus*, is not found to the southward of Oonalaska), glutton, sea, river, and marsh otters, lynx, beaver, zizel marmot, common marmot, hairy hedge-hog (*erinaceus ecaudatus*), rein-deer, American wool-bearing animal.

The quadrupeds which inhabit the shores of the Polar Sea, are the same that are comprised in the list of the animals of the Barren Grounds. On the remote North Georgian Islands, in latitude 75°, there are nine different species of mammiferous animals, of which five are carnivorous, and four herbivorous. The following is Captain Sabine's list of them :—

Ursus maritimus.
Gulo luscus.
Mustela erminea.
Canis lupus.
Canis lagopus.
Lemmus Hudsonius.
Lepus glacialis.
Bos moschatus   . ⎰ These two animals are only summer visitors.
Cervus Tarandus . ⎱ They arrive on Melville Island towards the
                      middle of May, and quit it on their return
                      to the South in the end of September.

I have not enumerated the seals, moose, or whales, in any of the lists ; nor have I attempted to give a description of any of them in the text, because my opportunities of examining them were too limited to enable me to record any new facts; neither had I the means of correctly ascertaining the species.

I have, in the text, described the different species of animals, from nature, as correctly as I could; and I have chosen rather to subject myself to the charge of proxility than to become obscure by aiming at too great conciseness, because, in the course of my researches, I have felt the difficulty of ascertaining the species, from the brief characters assigned to them by the old writers. I have for the same reason in many instances repeated some of the generic characters in the account of a species, particularly in cases where any doubt respecting the genus or sub-division of the genus existed. In the account of the manners of the animals, I have borrowed freely from preceding writers; and from none more frequently or more copiously than from Captain Lyons, whose "Private Journal" contains a great fund of information respecting the northern animals. I wish it to be understood, however, that in all cases, unless where a doubt is actually expressed, or where I state that I have had no opportunity of personal observation, the remarks I have quoted are sanctioned by the information I collected on the spot. The nomenclature of colours, made use of in the description, is a modification of Werner's, contained in Mr. Syme's useful little work*.

Before closing this introductory chapter, I have to discharge the agreeable duty of expressing my obligations to many gentlemen who have fostered the progress of the work. To the Right Honourable Lord Viscount Goderich my gratitude is especially due. To his attachment to the sciences I am indebted for that patronage and aid, which his high situation in his Majesty's Government enabled him to bestow, and without which this work could not have appeared. To the Right Honourable Thomas Frankland Lewis, also, I am under great obligations for the interest he has shewn in the advancement of the work, and for his kindness in forwarding my views. My gratitude is not less owing to the present Treasury Board, for the readiness with which they made the grant of money available; and to the late and

---

* *Werner's Nomenclature of Colours, with Additions.* By PATRICK SYME, Flower Painter. *Edinburgh,* 1821.

present Secretaries of State for Colonial Affairs, for their kindness in forwarding my applications through their department. I have next to express my best thanks to the Governor and Committee of the Hudson's Bay Company, for granting me free access to their museum, and to the manuscript accounts of the Fur Countries, in their possession, and for the strong recommendations they transmitted to the resident Chief Factors and Chief Traders, to forward the views of the Expedition, with respect to Natural History. To Mr. Garry, the Deputy Governor of that Company, I have to offer my thanks in an especial manner, not only for his general kindness and good offices, but for the free use of his valuable library, particularly rich in the works of the early travellers in America. I have also to mention my deep sense of the kindness of the Council of the Horticultural Society, and of Joseph Sabine, Esq., Secretary to that Institution, for the opportunity of examining and describing Mr. Douglas's specimens. To Charles Koenig, Esq., of the British Museum, I am under much obligation, for the facility he afforded me of examining the specimens in that collection; and I am equally indebted to N. A. Vigors, Esq., of the Zoological Society, for his aid in the consultation of the museum under his charge. I have, lastly, to express my gratitude to Sir John Franklin, and to the Officers associated with me under his command. To the former, for the kindness with which he embraced every opportunity during the progress of the Expedition, of forwarding my views with respect to that branch of its objects, which was more particularly intrusted to me; and to Captain Back, Lieutenant Kendall, and Mr. Dease, for their active assistance in the collection of specimens. Indeed, I may, with propriety, embrace this opportunity of saying, that I had the happiness of being placed under an Officer, who was endowed with the rare union of devoted attention to the duties of his profession, and of the most sincere attachment to the interests of general science,—and that, in him, and in the Officers under his command, I met with kind friends, whose agreeable society beguiled the tedium of a lengthened residence in the Arctic wilds.

EXPLANATION

OF THE

REFERENCES TO AUTHORS.

BARTON . . . . Medical and Physical Journal, edited by Professor Barton, Philadelphia. (This work is quoted after M. Say.)

BEWICK . . . . Bewick's History of Quadrupeds. 1st and 2nd editions, with wood cuts.

BILLINGS . . . . . . Expedition to the Northern Parts of Russia, by Commodore Joseph Billings, 1785 to 1794; narrated by Martin Sauer. London, 1802.

BLAINVILLE . . . . . Bulletin des Sciences par la Société Philomatique, 1791 et seq. Paris.

BRISSON . . . . Le Règne Animal Divisé en ix. Classes. 1 vol. in 4to. Paris, 1756.

BUFFON . . . . Histoire Naturelle, Generale et Particuliere, avec la Description du Cabinet du Roi. Paris, 1749. 36 vols. in 4to.

CARTWRIGHT . . . . Journal of Sixteen Years' Residence in Labrador, by G. Cartwright. 1 vol. 8vo. London.

CARVER . . . . Travels in North America, by J. Carver, Esq., in the Years 1766, 1767 and 1768. London, 1778.

CATESBY . . . . . . The Natural History of Carolina, Florida, and the Bahama Islands, by Mark Catesby. 2 vols. fol. with App. London, 1731 and 1743.

C. HAMILTON SMITH . . *Vide* SMITH.

CHAMPLAIN . . . . . Voyages du Sieur de Champlain Xaintongeois. 1613.

CHARLEVOIX . . . . Histoire de la Nouvelle France, avec le Journal d'un Voyage dans l'Amerique, Septentrionale, par le P. Pierre Francois Xavier de Charlevoix, a Paris an 1777. 12mo. tom. 5.

CLERK of the CALIFORNIA *Vide* SMITH and DRAGE.

CLINTON . . . . . . Transactions of the Literary and Philosophical Society of New York, instituted in the Year 1814. Introductory Discourse by the Hon. De Witt Clinton, LL.D., &c. 4to. New York, 1815.

COOK . . . . Voyage to the Pacific Ocean, in 1776—1780, performed under the Direction of Captain Cook. London, 1784. 4to. 3 vols.

COXE . . . . Account of Russian Discoveries between Asia and America, by William Coxe, A.M., F.R.S. London, 1787.

CUVIER, or CUVIER, Baron. Tableau Elementaire de l'Histoire Naturelle des Animaux. 1 vol. in 8vo. Paris,

„ „ „ 1798. Leçons d'Anatomie Comparée, Recueilles et Publiées, par MM. Dumeril et Duvernay. 5 vols. in 8vo. Paris, 1803—1805.

„ „ „ Recherches sur les Ossemens Fossiles de Quadrupedes. 4 vols. in 4to. Paris, 1812.

„ „ „ Le Règne Animal Distribue d'après sur Organisation, par M. Le Chr. Cuvier. 4 vols. 8vo. Paris, 1817.

CUVIER, FRED. . . . Histoire Naturelles des Mammifères. En folio.

DE MONTS . . . . . *Vide* MONTS.

DENYS . . . . Histoire de l'Amerique. (Quoted from Pennant.)

DESMAREST . . . . . Mammalogie en Description des Especes des Mammifères, par M. A. G. Desmarest. 4to. Paris, 1820.

DIXON . . . . A Voyage round the World, in the Years 1785, 1786, 1787, and 1788, by Captain George Dixon. London, 1789.

DOBBS . . . . An Account of Hudson's Bay, by Arthur Dobbs, Esq. London, 1744

DRAGE . . . . *Vide* SMITH and CLERK of the CALIFORNIA.

DUDLEY . . . . Philosophical Transactions, January, 1727. Of the Moose-deer in America, by Paul Dudley, Esq.

DU PRATZ . . . . . *Vide* PRATZ.

EDWARDS . . . . . . Natural History of Birds and other rare undescribed Animals, by George Edwards. 7 vols. 4to. London, 1743.

ELLIS . . . . Voyage to Hudson's Bay in the Dobbs and California, by Henry Ellis, in the year 1746 and 1747. London, 1748. 8vo.

ERXLEBEIN . . . . . Systema Regni Animalis. 8vo. Leipzick, 1777.

FABRICIUS . . . . . Fauna Groenlandica Othonis Fabricii. 1 vol. 8vo. Hafniæ et Lipsiæ, 1780.

FERNANDEZ . . . . . Historia Animalium, auctore Francisco Fernandez, Phillippi Secundi Primario Medico. 1 vol. 4to. An. 1651, Roma.

FLEMING . . . . . . The Philosophy of Zoology, by John Fleming, D.D. 2 vols. 8vo. Edinburgh, 1822.

FORSTER . . . . . . Philosophical Transactions, vol. 62. An. 1777. Descriptions of Specimens of Animals brought from Hudson's Bay, by J. Reinhold Forster.

FRANKLIN . . . . . Narrative of an Expedition to the Shores of the Polar Sea, in the Years 1819, 1820, and 1821, by John Franklin, Capt. Royal Navy. 1 vol. 4to. London, 1822.

„ . . . . . Narrative of a Second Expedition to the Shores of the Polar Sea in the Years 1825, 1826, and 1827, by John Franklin, F.R.S., Captain Royal Navy. 1 vol. 4to. London, 1828.

GASS . . . . Journal of the Travels of a Corps of Discovery under Captain Lewis and Captain Clarke to the Pacific Ocean, in the Years 1804, 1805, and 1806. 8vo. By Patrick Gass. 1 vol. 8vo. London, 1808.

GEOFFREY . . . . . Geoffroy St. Hilaire, Annales du Museum d'Histoire Naturelle de Paris. 20 vols. in 4to. De 1822 a 1823.

GMELIN . . . . . . Systema Naturæ Linnei, ed. 13. An. 1790. J. F. Gmelin.

GODMAN . . . . American Natural History, by John D. Godman, M.D. 3 vols. 8vo. Philadelphia, 1826.

GRAHAM . . . . *Vide* HUTCHINS.

GRIEVE . . . . . . History of Kamskatcha, translated from the Russian of Krascheninikoff, by James Grieve, M.D. Gloucester, 1764.

GRIFFITH . . . . . The Animal Kingdom, by Baron Cuvier, translated by Edward Griffith, and Others. 8vo. London, An. 1827 et seq.

GULDENSTED . . . . Novi Commentarii Petropolitani, 1749—1775. 20 vols.

HAMILTON SMITH . . *Vide* SMITH.

HARLAN . . . . . . Fauna Americana, being a Description of the Mammiferous Animals inhabiting North America, by Richard Harlan, M.D. 8vo. Philadelphia, 1825.

HARMON . . . . . . A Journal of Voyages and Travels in the Interior of North America, between the 47th and 58th Degrees of Latitude, by Daniel William Harmon, a Partner in the North West Company. Andover, 1820.

HEARNE . . . . . . Journey to the Northern Ocean, by Samuel Hearne, in the Years 1769, 1770, 1771, and 1772. London, 1807.

HENNEPIN . . . . . Nouvelle Decouverte d'un Tres grand Pays situè dans l'Amerique, par R. P. Louis de Hennepin. Amsterdam, 1698.

HENRY . . . . Travels and Adventures in Canada and the Indian Territories, by Alexander Henry, in the Years 1760—1776. New York, 1809.

HERIOT . . . . . . Travels through Canada, by George Heriot, Esq. London, 1807.

HERNANDEZ . . . . . Rerum Medicarum Novæ Hispaniæ Thesaurus Francisci Hernandez, Reccho Editore. Roma, 1651.

HISTOIRE DE L'AMERIQUE. Histoire de l'Amerique Septentrionale. Tom. 2, 12mo. Amsterdam, 1723.

HONTAN . . . . . . *Vide* LAHONTAN.

HUTCHINS . . . . . MS. Account of Hudson's Bay, written about the year 1780. Mr. Hutchins furnished much intelligence to Pennant respecting the Zoology of Hudson's Bay. In a few first sheets of this work Mr. Graham is through mistake quoted as the author of these manuscript notices.

JAMES . . . . The dangerous Voyage of Captain Thomas James, for the Discovery of a North-West Passage. London, 1633, reprinted 1740.

JAMES . . . . Expedition to the Rocky Mountains, under the Command of Major Long, by Edwin James. 3 vols. London, 1823. The American edition is also quoted occasionally.

JAMESON . . . . . . Transactions of the Wernerian Society of Edinburgh, vol. iii. p. 306. Account of the Rocky Mountain Sheep, by Professor Jameson.

JEREMIE . . . . . . Voyage au Nord. (Quoted from Pennant.)

JOSELYN . . . . . . New England. (Quoted from Pennant.)

JOUTEL . . . . Voyage to Mexico, by Mr. Joutel, translated from the French. London, 1719.

KALM . . . . Peter Kalm's Travels in North America, translated by J. R. Foster. The abridgement in Pinkerton's collection of voyages is also quoted.

KLEIN . . . . Isaac Theodore Klein, Quadrupedum Dispositio. 4to. Lipsiæ, 1751.

KRASHENINIKOFF . . . *Vide* GRIEVE.

LAHONTAN . . . . . Voyages dans l'Amerique de M. La Baron de la Hontan. Vol. 2 en 12mo. A la Haye, 1703.

LANGSDORFF . . . . Voyages and Travels to various Parts of the World, in the Years 1803, 1804, 1805, 1806, and 1807, by G. H. von Langsdorff. 2 vols. London, 1813.

LAWSON . . . . . . History of Carolina. (Quoted from Pennant.)

LEACH . . . . Leach, W. Elford. Zoological Miscellany.

„   „   „ Appendix to Ross's Voyage to Baffin's Bay. 1819.

LESSON . . . . . . Manuel de Mammalogie, par Réne Primeverre Lesson. 12mo. Paris, 1827.

LEWIS and CLARKE . . Travels to the Pacific Ocean in 1804, 1805, and 1806, by Captains Lewis and Clarke. 3 vols. 8vo. London, 1807.

LICHTENSTEIN . . . . Voyage a Boukhara, par M. Le Baron Georges de Meyendorff, en 1820. Paris, 1826. Description, par M. Lichtenstein des Animaux Recueilles dans le Voyage, par M. Eversman.

LINN. . . . . Systema Naturæ, Carolo a Linnè. Ed. xii. 1766.

„   „   „ Fauna Suecica. 8vo. 1746.

LINN. GMELIN . . . . Systema Naturæ Linnei. Ed. xiii. Cura Gmelini, Leipsig, 1788.

LONG'S JOURNEY . . . *Vide* JAMES.

LYON . . . . Private Journal of Captain G. F. Lyon during a Voyage of Discovery under Captain Parry. 8vo. London, 1824.

MC GILLIVRAY . . . . New York Medical Repository, vol. vi. p. 238. Account of the Mountain Ram, by William Mc Gillivray. 1803.

MACKENZIE (SIR ALEX.) Travels to the Polar Sea and to the Pacific Ocean, in the Years 1789—1791, by Alexander Mackenzie. London.

MACKENZIE (SIR GEORGE) Travels in Iceland.

MARTEN . . . . . . Voyage to Spitzbergen and Greenland, by F. Marten. 8vo. London, 1711.

MEARS . . . . . . Voyages to the North-West Coast of America in 1788 and 1789, by John Meares, Esq. 4to. London, 1790.

MEYENDORFF . . . . *Vide* LICHTENSTEIN.

MITCHILL . . . . . Medical Repository of New York. An. 1821. (Quoted from M. Say.)

MONTS, DE . . . . . Nova Francia. The three last voyages of Monsieur de Monts, of M. Pontgrave, and of M. De Poutrincourt, into La Cadia. London.

ORD . . . . Guthrie's Geography, American Edition. Philadelphia. (Quoted from Harlan.)

„ . . . . Journal of the Academy of Sciences of Philadelphia. Vol. iv. p. 305.

PALISOT DE BEAUVAIS . Bulletin des Sciences par la Société Philomatique depuis, 1791. Paris.

PALLAS . . . . . . Novæ Species Quadrupedum e Glirium Ordine. Erlang. In 4to.

„ . . . . . . Spicelegia Zoologica. Berolini, 1767—1780.

„ . . . . . . Voyage dans Plusieurs Provinces de l'Empire de Russie. 8 vols. in 8vo. Paris.

PARRY , ... Voyage for the Discovery of a North-West Passage, performed in the Years 1819, 1820, in His Majesty's Ships the Hecla and Griper, by William Edward Parry, R.N., F.R.S. 4to. London, 1821.

,, ... Second Voyage for the Discovery of a North-West Passage in the Years 1821, 1822, 1823, in the Fury and Hecla, by Captain William Edward Parry, R.N. F.R.S. London, 1824.

PENNANT . . . . . . History of Quadrupeds. 3d Edition. 2 vols. 4to. London, 1793.

,, ... Arctic Zoology. 2 vols. 4to. 1784.

PIKE . . . . Travels on the Missouri and Arkansaw, by Lieutenant Pike, in 1805 and 1806. Edited by T. Rees, Esq. London, 1811.

PRATZ, DU . . . . . Voyage de Louisiana. (Quoted from Pennant.)

RAFINESQUE, or RAFINESQUE-SMALTZ. }Annals of Nature. (Quoted from Desmarest.)

,, ,, American Monthly Magazine, (Ditto).

,, ,, Precis, les Decouvertes Somiologiques. En 18mo. Palerme, 1814.

RAY . . . . Raii Synopsis Methodica Animalium. 8vo. Londini, 1693.

RICHARDSON . . . . . Appendix to Captain Parry's Second Voyage. London, 1824.

,, ... Zoological Journal. 1828, 1829. London.

SABINE, (JOSEPH) . . . Franklin's First Journey. Zool. Appendix. London, 1822.

,, ,, . . . Linnean Transactions, vol. xiii.

SABINE, (Capt. EDWARD) Supplement to the Appendix of Captain Parry's First Voyage in 1819, 20. London, 1824.

SAGARD-THEODAT . . *Vide* THEODAT.

SAUER . . . . . . , . *Vide* BILLINGS.

SAY . . . . His Zoological Notices, in the Notes to Long's Expedition to the Rocky Mountains, are quoted. *Vide* JAMES.

SCHOOLCRAFT . . . . Travels to the Sources of the Mississippi River, by H. R. Schoolcraft. Albany, 1821.

SCHREBER . . . . . Histoire des Mammifères. In 4to. Erlangen, 1775, et suiv.

SHAW . . . . General Zoology, by George Shaw, M.D., F.R.S. 16 vols. 8vo. London, 1800—1812.

SMITH, (CAPTAIN) . . Voyage by Hudson's Straights, in the California, by Captain Francis Smith; by the Clerk of the California, in 1746 and 1747. (The Clerk's name was Drage. Ellis, the Agent for the Proprietors in the Dobbs, the consort of the California, gives another account of the voyage, but less full on points of Natural History.) *Vide* Ellis.

SMITH, (C. H.) . . . . His papers in the Linnean Transactions, and in Griffith's Translation of Cuvier, are quoted.

STELLER . . . . . . Acta Petropolitana.

TEMMINCK . . . . . Monographies de Mammalogie et Tableau Methodique des Mammifères. 4to. Paris, 1827.

f

THEODAT . . . . . . Histoire du Canada, par le F. Gabriel Sagard-Theodat. 12mo. Paris, 1636.

TRAILL . . . . . . Voyage to Greenland, by J. Scoresby. Appendix.

UMFREVILLE . . . . Present State of Hudson's Bay, by Edward Umfreville. London, 1790. 8vo.

ULLOA . . . . Voyage. (Quoted from Pennant.)

WARDEN . . . . . . Account of the United States of North America. Edinburgh, 1819.

VOYAGE DE L'AMERIQUE Voyage de l'Amerique dans le Vaisseau Pelican. En 1697. Amsterdam, 1723.

SYSTEMATIC LIST OF THE SPECIES.

f 2

NORTHERN ZOOLOGY.

DIRECTIONS FOR PLACING THE PLATES.

ERRATA.

Page xxvi. line 17, for " duo," read " duæ."
—— xxxviii. line 10, for " en," read " ou."
—— xxxviii. line 3d from the bottom, for " Geoffrey," read " Geoffroy."
Plate 18, B., page 206, for " Douglasii," read " bulbivorum."

NORTHERN ZOOLOGY.

PART I.

MAMMALIA.

[1.] 1. VESPERTILIO PRUINOSUS. (Say.) *Hoary Bat.*

GENUS. Vespertilio. LINN. *Sub-genus.* Vespertilio, GEOFFROY.
V. Pruinosus. SAY. *Long's Exped.*, vol. i. p. 167. American edition. (vol. i. p. 331, Engl. ed.)
HARLAN. *Fauna Amer.* p. 21.
Hoary Bat. GODMAN. *Nat. Hist.* vol. i. p. 68, and *fig.* t. No. 3.

This species of Bat was first noticed by Mr. Nuttall, at Council Bluffs, on the Missouri; and Mr. Say, in Long's Expedition, describes an individual captured in the same neighbourhood. Dr. Godman states, that it has been taken near Philadelphia. The specimen I have described below was caught at Cumberland-house on the Saskatchewan, in latitude 54°, and presented to me by Mr. Isbister, resident clerk at that post. This individual is larger than Mr. Say's, but there seems to be no other difference. Godman's figure does not represent the tail forming a small obtuse point to the interfemoral membrane, such as it exists in my specimen. After a minute examination, I could find no traces of more than two incisors in the upper jaw. Mr. Say found the same number; but it is possible, that some cutting-teeth may have dropped out in both specimens. The number of teeth would bring this species of Bat into the genus *Nycticeius* of Rafinesque; but the whole habit of the animal shews that it is properly classed in Geoffroy's genus *Vespertilio*, a subdivision of the great Linnæan genus.

DESCRIPTION.

Dental formula, incisors $\frac{2}{6}$, canines $\frac{1-1}{1-1}$, grinders $\frac{5-5}{6-6} = 34$.
The *superior incisors* are conical and sharp pointed, separated from each other by a wide naked space, and closely adjoining to the canine tooth on their respective sides. They are

B

slightly dilated exteriorly at their bases, but can scarcely be termed tuberculated.   In height, they equal the molar teeth.   The *inferior incisors* are arranged in contact with each other in a convex line, and are very short.   They have obtuse, slightly two-lobed crowns, which expand laterally beyond their roots.   The *upper canine teeth* are conical, obscurely three-sided and sharp pointed.   They stand twice as high as the molar teeth.   The *inferior canine teeth* are of the same size with the superior ones, and have each a minute and rather obtuse lobe at the base on the inner side.   The *molar teeth* have high, sharp, pyramidal points.

The *nostrils* are two lines apart, turned a little outwards, and have a raised obtuse, naked margin.   There is a depression between the nostrils superiorly, but no furrow on the margin of the lip, which is hoary within and without.   The *eyes* are surrounded by fur, but situated clear of the ear and its tragus.   The *ears* are shorter than the head, nearly circular, entirely covered with fur behind, except a small lobe, which projects anteriorly, and is overlapped by the tragus.   On the inside there are some detached patches of hair. The margins are entire, and the folds around the auditory opening have a resemblance to those of the human ear.   The *tragus* is scalene-triangular, fixed by one of its angles, and is well characterised by Mr. Say as very obtuse at the tip, and arquated.   It is thinly hairy exteriorly.   The margin of the mouth and the chin are black and hairy; and the crown of the head and throat are yellowish-brown.   The occiput, and the rest of the superior parts, are covered with a long and very fine *fur*, which is blackish-brown at the base, then shining yellowish-brown, followed by very dark umber-brown, and, lastly, tipped with white, producing a hoary and almost silvery colour on the back.   The fur of the under parts is also hoary, but has less lustre.   The *interfemoral membrane* is triangular, and at its apex there is a very slight smooth projection of the tail.   It is hairy above; its fur, towards the middle, being coloured like that of the back; but, near its margins, and particularly towards the apex, a reddish-brown tint prevails.   The *wing-membrane* presents some small hairy patches above the elbow-joint, and at the roots of the metacarpal-bones.   Underneath, it has a close coat of yellowish-brown fur on each side of the humerus; also a hairy patch beneath the brachial-bone, and others beneath the metacarpal-bones at their origin.   The first finger has one joint; the second, three; and the others, two each.   The thumb has one phalanx, which is much longer than its metacarpal-bone, and is armed with a short but strong, curved, black claw.   The *hind-feet* are covered with hoary fur above, and have short, curved claws, which are excavated underneath.

DIMENSIONS.

	Inches.	Lines.		Inches.	Lines.
Length of the head and body . .	4	0	Space betwixt the upper canine teeth .	0	3½
„ tail . . .	2	0	„ lower canine teeth .	0	2½
Spread of the wings	15	0	„ ears . . .	0	7
Length of head . . .	1	1	Length of thumb and claw . .	0	6
Space betwixt the nostrils nearly .	0	2	Diameter of the ear, (every way,) about .	0	6

[2.]   2. VESPERTILIO SUBULATUS.   (Say.)   *Say's Bat.*

Vespertilio Subulatus. SAY. *Long's Exped.* vol. ii. p. 65. (or vol. ii. p. 253, Eng. ed.)
Subulate Bat. GODMAN. *Nat. Hist.* vol. i. p. 71.

DESCRIPTION.

Dental formula, incisors $\frac{2-2}{6}$, canines $\frac{1-1}{1-1}$, grinders $\frac{6-6}{6-6}$ = 38.

The *upper incisors* are short, and are arranged in two distant pairs, each pair being close to the canine tooth of the same side. Each tooth has a small interior pointed lobe. The *lower incisors* are very short, and have two obtuse lobes. The *canine teeth* are a little longer than the grinders, nearly straight, subulate, and sharp pointed*. The two anterior *grinders* on each side, both above and below, are small, short, conical, and sharp pointed. The one adjoining to them, also simply conical, is higher than the three posterior grinders of each side, which, in the lower jaw, have a double row of acute points; and, in the upper jaw, a triple row; the inner row of the latter being much lower than the outer ones.

The *head* is short, broad, and flat: the nose blunt, with a small, flat, naked muzzle. The nostrils, situated at the two anterior corners of the muzzle, are small, roundish, naked, and scarcely one line apart. The tip of the lower jaw is rounded, and naked. *Eyes* concealed by the fur, and situated near the ears, but not covered by them. *Ears* about the length of the head, or a little longer, thin, membranous, ovate, obtuse; slightly undulated, but not notched posteriorly, and curving forwards at the base; slightly ventricose anteriorly, without folds. The ear is hairy at the base behind, and there are a very few scattered hairs on its inner surface. The *tragus* is thin, broadly subulate below, tapering to a point upwards, and ending in a small obtuse tip; it is attached by one corner at the base, is about two-thirds of the height of the ear, and is not curved or falciform.

The *back* has a shining yellowish-brown colour; the belly a yellowish-gray. The fur, soft and fine, is longest on the back (three lines), and both above and below is blackish at the roots. With the exception of the small naked space behind the nostrils, the head is covered with fur, but a little shorter than that on the back; towards the mouth it assumes a blackish colour; it is rather coarser on the lips, and there are a few longer hairs or whiskers, but they are not stiff nor very conspicuous.

The *interfemoral membrane* is broad, and tapers to a point along the tail, which it envelopes. It is thinly clothed at the base with fur similar to that on the back in colour, but shorter. It is also fringed with a few scattered hairs on its posterior, free margin, which is not undulated.

* The bifid point of one of the canine teeth in Mr. Say's specimen seems to have been an accidental circumstance.

B 2

The *tail* projects about a line beyond the membrane. The toes of the hind-feet are rather long, and have white, slender claws, not greatly curved, with a few long hairs projecting over them. The *wing-membrane* is naked, and the joints of the fingers correspond with those of the *vespertilio pruinosus*, and the rest of the genus, as restricted by Geoffroy. The thumb is about two lines and a half long, including its slender claw, which rather exceeds half a line.

DIMENSIONS.

	Inches.	Lines.		Inches.	Lines.
Length of body and head . . .	1	10	Height of the tragus	0	4½
„ tail . . .	1	6	Spread of wings from tip of the middle		
„ head	0	9	finger of the right wing to the tip of		
Height of ear	0	8	the corresponding finger of the left		
Breadth of ditto near the middle . .	0	4	wing	10	0
It is broader at the base.					

This Bat is the most common species near the eastern base of the Rocky Mountains on the upper branches of the Saskatchewan and Peace Rivers. Mr. Say's specimen was obtained near the head of the Arkansas, within sight of the mountains; and the description he gives of it corresponds so nearly with my specimens, that I have no hesitation in considering them to be the same. Say's Bat has a general resemblance to the *Vespertilio pipistrellus* of the British isles; but the latter has one grinder of a side fewer, weaker canine teeth, a smaller ear, and a shorter thumb and claw. Its fur is likewise shorter, and its back and belly do not exhibit such distinct shades of colour. It seems to approach near to the *Vespertilio emarginatus* of Geoffroy, as Mr. Say has remarked; but I have not been able to obtain a specimen of the latter with which I might compare it. The Carolina Bat differs in the shape of the tragus, which is semi-cordiform, but resembles this one nearly in the colour of the fur and in general form.

[3.] 1. SOREX PALUSTRIS. (Richardson.) *American Marsh-Shrew.*

GENUS. Sorex. LINN.

Sorex palustris. RICHARDSON. *Zoological Journal,* No. xii. *April,* 1828.

S. (palustris) caudâ corpus longitudine excedenti, auriculis subvestitis vellere latentibus, corpore cinerascenti-nigro ; subter cinereo.

Shrew, with the tail longer than the body, short hairy ears concealed by the fur, back somewhat hoary-black, belly ash-gray.

DESCRIPTION.

The *dimensions* of this animal are nearly the same with those of the *Musaraigne de Daubenton,* or *Water Shrew* of Pennant, and are considerably greater than those of the *S. constrictus,* with which it seems to have some relations.

Dental formula ; intermediary incisors $\frac{2}{2}$, lateral incisors $\frac{4}{4}$ grinders $\frac{4}{3}$ = 30.

The two posterior lateral incisors are smaller than the two anterior ones on the same side, and the latter are a little longer than the posterior lobes of the intermediary incisors. All the lateral incisors have small lobes on their inner sides. The tips of the teeth have a shining chestnut-brown colour.

Form.—The *muzzle* is shorter in proportion and broader than that of the *Sorex parvus.* The whole upper lip is bordered with *whiskers,* and the tips of the posterior ones, which are the longest, reach behind the ears. The extremity of the muzzle is naked and two-lobed. The *eyes* are visible. The *ear* is shorter than the fur ; its inferior margin is folded in ; there is a heart-shaped lobe covering the auditory opening, and a transverse fold above it. The ears, particularly the superior margins, are clothed with thick tufts of fur, like that on the rest of the head. The tail appears to be rounded, or slightly four-sided from its base, to near the tip, where it is compressed and terminated by a small pencil of hairs. It is covered by a close coat of short hair. The feet are clothed with rather coarse, short, adpressed hairs, those on the sides of the toes being arranged somewhat in a parallel manner, but not very distinctly.

The *fur* resembles that of the mole in softness, closeness, and lustre. On the superior or dorsal aspect it is black, with a slightly hoary appearance when turned to the light. On the ventral aspect it is ash-coloured. At the roots it is bluish-gray. The outside of the thighs and upper surface of the tail correspond in colour with the back, the under surface of the tail and inside of the thighs with the belly. The feet are paler than the back and a little hoary. The nails are whitish.

DIMENSIONS.

	Inches.	Lines.		Inches.	Lines.
Length from nose to origin of tail	3	6	Length of nose, from upper incisors, scarcely	0	2
„ of tail	2	7	Height of ear	0	3
„ of head	1	2	Length of hind-foot from heel to end of the		
„ from nose to eye	0	7	nails	0	9

This animal agrees with the *S. constrictus* in having two lateral incisors more in the upper jaw than some other species of the genus, but the *Sorex brevicaudus* of Say is described by Dr. Harlan as having five lateral incisors (" minute false molars ") on each side, and the same thing occurs in the following species. When compared with a specimen of the *water-shrew* in the British Museum, the colour of its fur appeared different, the points of the teeth darker, the ears smaller, and the tail longer than in the water-shrew. Several specimens of this animal were obtained, but the descriptions were drawn up from the prepared skins, and some uncertainty consequently exists as to the true shape of the tail. The *S. palustris* most probably lives in the summer on similar food with the water-shrew; but I am at a loss to imagine how it procures a subsistence during the six months of the year in which the countries it inhabits are covered with snow. It frequents borders of lakes, and Hearne tells us that it often takes up its abode in beaver houses.

[4.]   2. Sorex Forsteri. (Richardson.) *Forster's Shrew-Mouse.*

Shrew, No. 20. Forster. *Phil. Trans.* vol. lxii, p. 381.
Sorex Forsteri. Richardson. *Zool. Journ.* No. 12, *April*, 1828.
Sorex (Forsteri) caudá tetragoná longitudine corporis, auriculis brevibus vestitis, dorso xerampelino, ventre murino.
Forster's Shrew-mouse, with a square tail as long as the body, short furry ears, back of a clove-brown colour,
        belly pale yellowish-brown.

This little animal is common throughout the whole of the fur countries to the 67th degree of latitude, and its minute foot-prints are seen every where in the winter, when the snow is sufficiently fine to retain the impression. I have often traced its pathway to a stalk of grass, by which it appears to descend from the surface of the snow, but a search for its habitation by removing the snow was invariably fruitless. I was unable to procure a recent specimen, and the following description is drawn up from one prepared by Mr. Drummond. It is the smallest quadruped the Indians are acquainted with, and they preserve skins of it in their conjuring bags. The power of generating heat must be very great in this diminutive creature to preserve its slender limbs from freezing when the temperature sinks 40 or 50 degrees below zero. The Sorex Forsteri approaches the *S. tetragonurus* of Desmarest in dimensions, and agrees with it in some other points.

DESCRIPTION.

Dental formula, interm. incisors $\frac{2}{2}$, lateral incisors $\frac{5-5}{2-2}$, grinders $\frac{4-4}{3-3} = 32$.

The teeth are white, brightly tinged with chestnut brown on the points. The upper *intermediary incisors* have each a posterior obtuse lobe. The *lateral incisors* of the upper jaw are crowded and somewhat tiled; the four anterior ones of a side are broad and obtusely conical, the fifth is flattish on the crown. The first *grinder* is smaller than either of the two which succeed it; and the fourth is the smallest of all. In the *lower jaw* the intermediary incisors have two distinct obtuse posterior lobes, and a slight undulation producing the rudiment of a lobe towards their points: the lateral incisors have a central mammillary point; and the anterior grinder is a little larger than the other two. The *muzzle* is very slender, and has a naked and a deeply lobed tip. The *whiskers* reach to the occiput, and are composed of a few white hairs, intermixed with many black ones. The *ear* is as long as the fur of the head, and is clothed within and without, but particularly on its margins, and folds, with hairs of the same colour and length of those on the crown of the head. It is rounded, but from a small fold of its upper margin appears pointed. Its circumference is ample for the size of the animal. There is a semicircular lobe projecting from the inferior margin of the ear, and covering the auditory opening, and above it there is a transverse fold. The ear is not perceptible until the fur is blown aside. The *fur* forms a fine, short, close coat, which on the dorsal aspect of the animal has a grayish-brown or clove-brown colour, and on the ventral aspect a dull yellowish-brown. The *tail* is four-sided and tapers gradually from the root to its extremity, which is terminated by a pencil of hairs. It is covered with dark-brown hair above, and pale, yellowish-brown hair beneath. The *feet* are five-toed, and are clothed with short, adpressed, pale yellowish-brown hairs. The *nails* are slender and white.

DIMENSIONS.

	Inches.	Lines.		Inches.	Lines.
Length of head and body	2	3	Length from upper incisors to nostrils	0	2
,, of tail	1	3	Height of the ear	0	2
,, of head	0	9½			

[5.]   3. Sorex Parvus.   (Say?)   *Small Shrew-Mouse.*

Sorex parvus. Say. *Long's Expedition*, vol. i, p. 163 ?
Sorex, No. 89. Museum of the Zoological Society.

There is a specimen of a shrew-mouse in the Museum of the Zoological Society, which answers nearly to the description of the *Sorex parvus* by Say, except that its tail is considerably longer.   Not to add unnecessarily to the number of specific names, I have adopted Mr. Say's, until a comparison of authentic specimens shall determine whether it belongs to the same or a different species.   Forster, in the Philosophical Transactions, mentions the *Sorex araneus* as an inhabitant of Hudson's Bay.   The large naked ears of that species would distinguish it at once from the *S. parvus*.

Description of the specimen in the Zoological Museum.
Form.—Ears very short, and indicated only by a brownish tuft of hair, shorter than the rest of the fur. Muzzle more slender than that of *S. palustris,* but not so much so as that of *S. Forsteri*. The tail is apparently cylindrical the greater part of its length, pointed and perhaps slightly compressed at the tip. The *fur,* from its root to near the tip, has a dark blackish-gray colour, but from its closeness only the tips are seen, and on the back they have a brownish-black colour, on the head and sides brownish-gray, and on the belly ash-gray. The feet have a brownish tinge. The points of the teeth are dark reddish-brown.

DIMENSIONS.

		Inches.	Lines.
Length of head and body		2	9
„ tail		1	9
„ from nostrils to incisors		0	1½

Mr. Collie, surgeon of his Majesty's ship Blossom, caught a Shrew-mouse on the shores of Behring's Straits, which he describes as having a dark brownish-gray colour above, and a gray tint beneath. It measured, from the tip of the snout to the root of the tail, two inches and four lines, and its tail was one inch long. This specimen agrees still more nearly with Mr. Say's description than the one in the Zoological Museum does, and if it is allowed to be of the same species, it gives to the *Sorex parvus* a range of twenty-three degrees of latitude.

[6.]   1. SCALOPS CANADENSIS.   (Cuvier.)   *Shrew-Mole.**

GENUS. Scalops. CUVIER.
Brown Mole. PENNANT. *Arctic Zool.*, vol. i. p. 141.
Sorex aquaticus. *Lin. Syst.*
Musaraigne-taupe. CUVIER. *Tab. Elém.*
Scalope de Canada. CUVIER. *Règne An.*, vol. i. p. 134.
Shrew-Mole. GODMAN. *Nat. Hist.*, vol. i. p. 84, *t. v. fig.* 3.
Mole. LEWIS AND CLARKE. *Journey, &c.*, vol. iii. p. 42.

DESCRIPTION.

Dental formula, incisors $\frac{2}{2}$, grinders $\frac{10-10}{10-10} = 44$.

The two *upper* incisors have an exact resemblance, in shape and position, to the two middle incisors of man.   They occupy the end of the jaw, and are twice as broad, and somewhat higher than the grinders which immediately follow.   The four first grinders of a side are conical, and obscurely three-sided.   The fifth is a little compressed, and has a minute projection at its base posteriorly.   The sixth is still more compressed, and has a larger posterior projection.   These six anterior grinders (termed conical teeth or false grinders by some authors) are nearly equal to each other in height, and occupy the whole jaw between the incisors and posterior higher grinders.   They stand at equal but small distances from each other, and from the incisors, not exceeding the quarter of the breadth of a single tooth.   The four posterior grinders are larger, and rather exceed the incisors in height.   The first of them, or seventh grinder, does not differ much from the preceding one ; it is compressed, has an acute lobe posteriorly, and a minute one on the inside anteriorly.   The two next grinders are composed of two exterior triangular folds of enamel, and one interior one, producing, besides some subordinate points, three conspicuous sharp ones, of which the interior one is lower than the other two.   The tenth or last grinder is smaller than the two which precede it. In the *lower jaw*, there are two incisors, shaped like the upper ones, but much smaller and lower than the closely adjoining grinders.   They are succeeded on each side by seven small conical but rather obtuse grinders, which are flat on the inside.   These teeth are close to each other, but do not touch, and they have their points gently inclined forwards.   They increase gradually but slightly in height, in proportion as they are situated further from the incisors ; and the three which are farthest back have a minute projection at their bases posteriorly. The foremost of these conical teeth on each side, which is almost in contact with the incisors, closely resembles the two which follow it ; but it is by many considered as an incisor, and when one or both lower incisors have dropped out, it does indeed approach to its fellow, and then becomes more opposed to the upper incisors.   They stand, like the other grinders,

---

* The English trivial name of Shrew-mole is a translation of Pennant's epithet *Sorex talpæformis*, or of Cuvier's *Musaraigne-taupe*, and is adopted from Dr. Godman.

in the plane of the limbs of the jaw, or nearly at a right angle with the planes of the incisors. The three posterior lower grinders of each side resemble the upper ones reversed, but have no lobe corresponding to the interior one of the upper teeth. They rise more above the sockets than the upper grinders do, and they have, as Dr. Godman has observed, a considerable resemblance to the grinders of a Bat. In old individuals, all the teeth are worn down and have rounded crowns.

The Shrew-mole has a thick cylindrical body, like that of the Common Mole, without any distinct neck. Its limbs are very short, being concealed by the skin of the body nearly down to the wrist and ankle-joints. The fore-extremities are situated nearly under the auditory opening. The moveable snout is almost linear, and projects about four lines and a half beyond the incisors. It is naked towards its extremity, particularly above; below, it is thinly clothed with hairs for about two-thirds of its length next the incisors. There is a conspicuous furrow, extending nearly its whole length, on the upper surface; and, beneath, there is also a furrow, reaching half its length from the incisors. Beyond the latter, the snout is transversely wrinkled beneath; and its small, flat, or truncated extremity is smooth and callous. The small oblong nostrils open in an inclined space, immediately above this circular callous end. The *eyes* are concealed by the fur, and scarcely to be found in the dried specimen *. The auditory openings are covered by the fur, and there is no external ear. The *tail* is thickest about one-third from its root, and tapers from thence to its tip, which is acute. It is whitish, and is sparingly clothed with short hairs. Its vertebræ are equally four-sided. The fore-arm, rather slender, and projecting only about three lines from the body, is, consequently, concealed by the fur. The five fingers, extremely short, and united to the roots of the nails, form, with the wrist, a large, nearly circular palm. The nails of all the fingers are large, white, and have a semi-lanceolate form, with narrow, but rather obtuse points. They are nearly straight, convex above, and slightly hollowed beneath. The middle one is the largest, the others gradually diminish on each side of it, and the exterior one is the smallest of all. The palms are turned outwards and backwards, and the whole fore-foot bears a close resemblance to that of the Common Mole. The hind-feet are more slender than the fore ones, and the nails are one-half shorter, much more compressed, and sharper, and, in fact, nearly subulate. They have a slight curvature laterally corresponding with the direction of the toes inwards, and are somewhat arched, but cannot be said to be in any manner hooked. They are excavated underneath. The fore and hind feet are thinly clothed above with adpressed, pale hairs. The palms and soles are naked, but are bordered posteriorly with white hairs, which curve a little over them.

The *fur* has the same velvety appearance with that which clothes the Common Mole. It has considerable lustre on the surface; and, in most lights, exhibits a brownish-black tint. When blown aside, it shews a greyish-black colour, from the roots to near the tips. It has the same colour over the whole body, but there is a slight tinge of chestnut-brown on the forehead and about the base of the snout, and on the throat it is shorter and paler.

---

* Dr. Godman informs us, that the aperture in the skin is just big enough to admit an ordinary sized human hair.

DIMENSIONS.

	Inches.	Lines.		Inches.	Lines.
Length of head and body	7	8	Length from wrist joint to tip of the middle nail	1	0
„ tail	1	6	„ heel to end of middle claw	0	10
„ fore-palm	0	6	Greatest breadth of the hind-foot	0	3
Breadth of fore-palm	0	7	Distance from auditory opening to the end		
Length of middle fore-nail	0	6	of the snout	1	7

The animal described above inhabits the banks of the Columbia and the adjoining coasts of the Pacific in considerable numbers, and is, doubtless, the mole mentioned by Lewis and Clarke as resembling, in all respects, the mole of the United States. Sir Alexander Mackenzie saw many animals, which he terms "moles," on the banks of a small stream near the sources of the Columbia; but as we are led to infer, from the way in which he speaks of them, that they were in numbers above ground, I am inclined to think that they were *sewellels*, belonging to the genus *aplodontia*, and not Shrew-moles*. I did not obtain recent specimens of the Shrew-mole on the late expedition, and am unable to say what are the exact limits of its range to the northward. I do not think, however, that it can exist, at least on the east side of the Rocky Mountains, beyond the fiftieth degree of latitude, because the earth-worm on which the Scalops, like the Common Mole, principally feeds, is unknown in the Hudson's Bay countries. On the milder Pacific shore, it may, perhaps, reach a somewhat higher latitude. There are two specimens of the Shrew-mole from the Columbia preserved in the Museum of the Hudson's Bay Company, and Mr. David Douglas has kindly furnished me with others which he obtained in the same quarter. The Columbia animal seems to be of larger dimensions, and has a longer tail than the Shrew-moles of the United States; but I have not detected any other peculiarities by which it might be characterised as a distinct species. Authors, probably from their specimens being of different ages, have varied considerably in their descriptions of the dentition of the Scalops, and several of them have mentioned edentate spaces between the incisors and grinders. In the adult animal, from which my description was taken, no such spaces exist. In a large and apparently very old individual, the incisors, and all the small grinders, are so worn and rounded, as to appear like a row of small pearls set in the jaw. Baron Cuvier informs us, that the animals of the genus Scalops unite to the teeth of the Desmans (*mygale*); and the simply pointed muzzle of the Shrews, large hands, armed with strong nails, fitted for digging into the earth, and entirely similar to those of the Moles. It is evident, from my description of the teeth of the Columbia Shrew-mole, that

* MACKENZIE's *Voyage to the Pacific,* &c. p. 314.

C 2

they approach closely to those of the Desmans, there being merely some not very important variations in the shape, particularly of the upper incisors *.

The Shrew-mole resembles the Common European Mole in its habits, in leading a subterranean life, forming galleries, throwing up little mounds of earth, and in feeding principally on earth-worms and grubs. Dr. Godman has given a detailed and interesting account of their manners, particularly of one which was domesticated by Mr. Titian Peale. He mentions that they are most active early in the morning, at mid-day, and in the evening, and that they are well known in the country to have the remarkable custom of coming daily to the surface *exactly at noon*. They may then be taken alive by thrusting a spade beneath them and throwing them on the surface, but can scarcely be caught at any other period of the day. They burrow in a variety of soils, and in wet seasons are observed to retreat to the higher grounds. The captive one in possession of Mr. Peale ate considerable quantities of fresh meat, either cooked or raw, drank freely, and was remarkably lively and playful, following the hand of its feeder by the scent,—burrowing, for a short distance, in the loose earth, and, after making a small circle, returning for more food. When engaged in eating, he employed his flexible snout in a singular manner to thrust the food into his mouth, doubling it so as to force it directly backwards.

From the great resemblance of the Shrew-mole to the common one, they might be readily mistaken for each other by a casual observer; and Bartram and others, who have asserted the existence of a species of the genus *talpa* in America, are, on this account, supposed, by later writers, to have been mistaken. There are, however, several true moles in the Museum of the Zoological Society which were brought from America, and which differ from the ordinary European species in being of a smaller size, and in having a shorter and thicker snout. Their fur is brownish-black. I could not learn what district of America they came from.

---

* I have termed "first grinders" the teeth named "inferior lateral incisors" by Cuvier, because they have an exact resemblance to the other small grinders in form and size.

[7.] 1. CONDYLURA LONGICAUDATA. (Illiger.) *Long-tailed Star-nose.*

Long-tailed Mole. PENNANT. *Hist. Quadr.*, vol. ii. p. 232. *t.* 90. *f.* 2. *Arctic Zool.*, vol. i. p. 140.
Talpa longicaudata. ERXLEBEIN. *Syst.*, tom. i. p. 118.
Condylure à longue queue. DESMAREST. *Mamm.*, tom. i. p. 158.
Condylura longicaudata. HARLAN. *Faun.*, p. 38.
Naspass-kasic. CHIPPEWAYS, and SAULTEUR INDIANS.

The Zoological Society recently obtained several specimens of a Star-nose from Moose Factory, Hudson's Bay, which agree so closely with Pennant's description of his Long-tailed Mole, that I have had no hesitation in referring it to that species. They were not accompanied by any account of their habits, or notice of the exact locality where they were killed; but as the most southern fur posts depending upon Moose Factory are situated upon the borders of Lake Superior, it is probable that they came from that quarter. Pennant's specimen was received from New York. It is remarkable that M. Desmarest, who derives all his knowledge of this animal from Pennant, should make *"point des crêtes nasales"* part of its essential character. In the History of Quadrupeds, it is termed the " Long-tailed Mole with a radiated nose; " and in Arctic Zoology, it is said to have "the nose long, the end radiated with short tendrils." Perhaps M. Desmarest was misled by the miserable figure in the History of Quadrupeds.

DESCRIPTION.

The Long-tailed Star-nose has a thick body, with a long head, tapering towards the end of the nose, which is furnished with a cartilaginous fringe, having eighteen rays in the circumference, and two shorter bifid ones attached beneath the nostrils. The body is covered with a soft, short, velvety coat of fur of a brownish-black colour on the surface, and a bluish-black hue towards the roots. The nose is of the same colour with the body. The tail, slender and tapering, is covered with short hair, and is about one-third shorter than the body. Its vertebræ are equally four-sided. The extremities are short, and bear a resemblance to those of the Common Mole. The palms are not so broad as those of the Mole, but have a similar form. They are naked; and the back of the hand is covered with scales, with a few intermixed hairs. The claws are large, white, convex, linear, and obtuse. The hind extremities are longer than the fore ones. The legs, short and slender, are thinly covered with hair. The feet are longer and narrower than the hands, covered above as far as the ankle-joint with scales. The hind-claws are white, narrow, and sharp-pointed

DIMENSIONS.

		Inches.	Lines.
Length of head and body		4	9
,, tail		2	9
,, head		1	3

[8.]   1. Ursus Americanus.   (Pallas.)   *American Black-Bear.*

Black Bear. Pennant. *Arct. Zool.*, vol. i. p. 57, *and Introduction*, p. cxx. *Hist. Quad.*, vol. ii. p. 11.
          Warden. *United States*, vol. i. p. 195.
          Godman. *Nat. Hist.*, vol. i. p. 194.
Ursus Americanus. Pallas. *Spicel. Zool.*, vol. xiv. p. 6—26.
          Harlan. *Fauna*, p. 51.
Sass. Chepewyan Indians.
Musquaw (pl. musquawuck). Cree Indians, or, when reference is made to the black colour of the fur, it is termed
       cuskeeteh musquaw.   The cinnamon-coloured variety is named oosaw-wusquaw, the first letter of the proper
       name being altered *euphoniæ causâ.*
Mucquaw. Algonquins. Maconsh (a young bear.) Iidem.

The different species of Bears resemble each other so strongly in form ; and colour, when described by general and frequently indefined terms, affords so uncertain a mark of discrimination, that much doubt has arisen as to what are species and what merely varieties.   These doubts can be removed only by a rigid comparison of the skeletons of the different kinds, combined with careful observation of the habits of the animals in their native retreats, and a more attentive consideration of their geographical distribution than has hitherto been given.   Buffon, classing the American and European Bears together, distinguishes two species of land Bear differing from each in colour and manners*.   Naturalists of the present day are generally of opinion that there are two or more species of Bear in the northern parts of the New World, differing specifically from those of the old continent.   The Polar Bear is perhaps the only species which is common to both continents, but it may with justice be considered as a sea animal, inhabiting the ice floating between them.

The *Black Bear of America* was first described as a distinct species by Pallas, and with reason, although some late writers continue to confound it with the Black Bear of Europe †.   It has a milder disposition, and lives more on vegetable

---

* " Il faut distinguer," dit-il, " deux espèces dans les ours terrestres, *les bruns* et *les noirs,* lesquels n'ayant pas les mêmes appetits naturels, ne peuvent pas être considérés comme deux espèces distinctes et séparées.   De plus, *l'ours blanc terrestre* n'est qu'une variété de l'une ou de l'autre de ces espèces.   Nous comprenons ici sous la dénomination d'ours bruns, ceux qui sont bruns, fauves, roux, rougeâtres, et par celle d'ours noirs ceux qui sont noirâtres aussi bien que tout à fait noirs."—Buffon, *Hist. Nat.*, vol. viii. p. 248.

† Baron Cuvier, in his elaborate work *Sur les Ossemens Fossiles,* distinguishes the *ursus niger Europæus* from the European Brown Bear, or *ursus arctos* of authors.   The *ursus niger* has the frontal bone of its cranium flattened, especially transversely, and separated from the temporal depressions by well marked ridges, which unite behind at an acute angle to form an elevated sagittal crest.   Its fur is blackish, rough, and more or less woolly.   The well marked depressions and ridges of the cranium giving lodgment and origin to the strong muscles of the lower jaw, shew that the Black Bear of Europe is more decidedly a beast of prey than the Brown one, in which respect they differ from the bears of corresponding colours which inhabit the New World.

substances than the latter, and there are corresponding differences in the form of its cranium, which is shorter, with less convex zygomatic arches, and consequently a smaller space for lodging the crotaphite muscle. Its forehead is not flat like that of the Black Bear of Europe, but arched, although not so much so as the forehead of the Brown Bear. Its temporal ridges, however, are well marked, and unite to form a sagittal crest. Its nose is continued nearly on the same line with the forehead, and is rather arched, which produces the most striking peculiarity in the physiognomy of this species. Its ears are high, oval, rounded at the tips, and far apart. The palms and soles of the feet are short in comparison with those of the Brown Bear. The fur on the body is long, straight, shining and black, and the mæsial line of the nose is also black or very deep brown, but there is a large pale yellowish-brown patch on each side of the muzzle. The naked extremity of the nose is a little oblique, not being so directly truncated as that of the Brown Bear. The hair of the feet projects beyond the claws, which are black.

The *Cinnamon Bear* of the Fur Traders is considered by the Indians to be an accidental variety of this species, and they are borne out in this opinion by the quality of the fur, which is equally fine with that of the Black Bear. The *Yellow Bear of Carolina* is also referred by Cuvier to this species, as is likewise the *Ours Gulaire* of M. Geoffroy, which has a white throat. The white markings on the throat of the animal, mentioned by the latter author, are perhaps analogous to the white collar which many of the European Brown Bears exhibit when young. Captain Cartwright remarks that the cubs of the Black Bear, on the Labrador coast, are often marked with white rings round the neck*, and Pennant notices the same thing of the bears of Hudson's Bay.

The Black Bear is smaller than the other American bears which we have to describe, the total length of an adult seldom exceeding five feet. Its favourite food appears to be berries of various kinds, but when these are not to be procured, it preys upon roots, insects, fish, eggs, and such birds or quadrupeds as it can surprise. It does not eat animal food from choice; for when it has abundance of its favourite vegetable diet, it will pass the carcase of a deer without touching it. It is rather a timid animal, and will seldom face a man unless it is wounded, or has its retreat cut off, or is urged by affection to defend its young. In such cases its strength renders it a dangerous assailant. I have known the female confront her enemy boldly until she had seen her cubs attain the upper branches of a tree, when she made off, evidently considering them to be in safety, but in fact leaving

* CARTWRIGHT'S *Journal of a Residence in Labrador.*

them an easy prey to the hunter.   The speed of the Black Bear when in pursuit is said not to be very great, and I have been told that a man may escape from it, particularly if he runs into a willow grove or amongst long grass: for the caution of the Bear obliges it to stop frequently and rise on its hind legs for the purpose of reconnoitring.   I have, however, seen a Black Bear make off with a speed that would have baffled the fleetest runner, and ascend a nearly perpendicular cliff with a facility that a cat might envy.

This Bear, when resident in the fur countries, almost invariably hibernates, and about one thousand skins are annually procured by the Hudson's Bay Company, from Black Bears destroyed in their winter retreats.   It generally selects a spot for its den under a fallen tree, and having scratched away a portion of the soil, retires to it at the commencement of a snow-storm, when the snow soon furnishes it with a close, warm covering.   Its breath makes a small opening in the den, and the quantity of hoar frost which occasionally gathers round the aperture serves to betray its retreat to the hunter.   In more southern districts, where the timber is of a larger size, Bears often shelter themselves in hollow trees.   The Indians remark that a Bear never retires to its den for the winter until it has acquired a thick coat of fat, and it is remarkable that when it comes abroad in the spring it is equally fat, though in a few days thereafter it becomes very lean.   The period of the retreat of the Bears is generally about the time when the snow begins to lie on the ground, and they do not come abroad again until the greater part of the snow is gone.   At both these periods they can procure many kinds of berries in considerable abundance.   In latitude 65°, their winter repose lasts from the beginning of October to the first or second week of May; but on the northern shores of Lake Huron, the period is from two to three months shorter.   In very severe winters, great numbers of Bears have been observed to enter the United States from the northward.   On these occasions, they were very lean, and almost all males; the few females which accompanied them were not with young *.   The remark of the natives above-mentioned, that the fat Bears alone hibernate, explains the cause of these migrations.   The Black Bears in the northern districts couple in September, when they are in good condition from feeding on the berries then in maturity.   The females retire at once to their dens, and conceal themselves so carefully that even the lyncean eye of an Indian hunter very rarely detects them; but the males, exhausted by the pursuit of the female, require ten or twelve days to recover their lost fat.   An unusually early winter will, it is evident, operate

* PENNANT'S *Arctic Zoology*, vol. i. p. 60.

most severely on the males, by preventing them from fattening a second time; hence their migration at such times to more southerly districts. It is not, however, true that the Black Bears generally abandon the northern districts on the approach of winter, as has been asserted, the quantity of Bear skins procured during that season in all parts of the fur countries being a sufficient proof to the contrary. The females bring forth about the beginning of January, and it is probable that the period of their gestation is about fifteen or sixteen weeks, but I believe it has not been precisely ascertained. The number of cubs varies from one to five, probably with the age of the mother, and they begin to bear long before they attain their full size.

The Black Bear inhabits every wooded district of the American continent, from the Atlantic to the Pacific, and from Carolina to the shores of the Arctic sea. They are, however, more numerous inland than near the sea-coast. Langsdorff observes, that " the valuable Black Bear, the skins of which form part of the (Russian) Company's stock, are not the produce of the Aleutian islands, but of the continent of America, about Cook's river, Prince William's Sound, and other places *.

The strength and agility of the Bear, together with its tenacity of life, render an attack upon it hazardous, and its chace has been considered by the rude inhabitants of the northern regions as a matter of the highest importance. Many of the native tribes of America will not join the chace until they have propitiated the whole race of Bears by certain speeches and ceremonies, and when the animal is slain they treat it with the utmost respect, speak of it as of a relation, offer it a pipe to smoke, and seldom fail to make a speech in exculpation of the act of violence they have committed in slaying it, although the hunter at the same time glories in his prowess. This veneration for the Bear seems to have arisen from the ability and pertinacity with which it defends itself; and it is interesting to observe in how similar a manner the same feeling manifests itself in tribes speaking diverse languages, and widely separated from each other by geographical position. Thus, Regnard informs us that the chase of the Bear is the most solemn action of the Laplander, and the successful hunter may be known by and exults in the number of tufts of bear's hair he wears in his bonnet. When the retreat of a Bear is discovered, the ablest sorcerer of the tribe beats the *runic* drum † to discover the event of the chase, and the side on which the animal

* LANGSDORFF's *Voyages*, vol. ii. p. 74.

† The same kind of drum, shaped like a double-headed tambourine, and painted with arbitrary characters or rude representations of wild beasts and of the heavenly bodies, is common throughout all the various North American tribes.

D

ought to be assailed.   During the attack, the hunters join in a prescribed chorus, and beg earnestly of the Bear that he will do them no mischief.   When they have killed him, they put the body into a sledge to carry it home ; the rein-deer which has been employed to draw it, is exempted from labour during the rest of the year ; and means are also taken to prevent it from approaching any female.   A new hut is constructed expressly for the purpose of cooking the flesh ; and the huntsmen, joined by their wives, begin again their songs of joy, and of thanks to the animal for permitting them to return in safety *.   Leems also acquaints us, that the Laplanders never presume to call the Bear by its proper name of *Guourhja*, but term it "the old man in the fur cloak," because they esteem it to have the strength of ten men and the sense of twelve †.   It is also said that the Bear is the great master of the Kamskatkans in medicine, surgery, and the polite arts.   They observe the herbs he has recourse to when ill or wounded, and acknowledge him as their dancing-master, mimicking his attitudes and graces with great aptness ‡.   Bear-dances, in which the gestures of the animal are copied, are also common with the North American Indians.

The following extract § from the narrative of Mr. Alexander Henry, one of the first Englishmen who penetrated into the fur countries after the reduction of Canada under the British arms, will serve to contrast the manners of the Indians with those of the Laplanders, and it contains besides some remarks on the habits of the Bear peculiarly valuable as coming from an eye-witness worthy of all credit. " In the course of the month of January, (whilst on the banks of Lake Michigan,) I happened to observe that the trunk of a very large pine-tree was much torn by the claws of a bear, made both in going up and down.   On further examination, I saw that there was a large opening in the upper part, near which the smaller branches were broken.   From these marks, and from the additional circumstance that there were no tracks on the snow, there was reason to believe that a Bear lay concealed in the tree.   On returning to the lodge, I communicated my discovery ; and it was agreed that all the family should go together, in the morning, to assist in cutting down the tree, the girth of which was not less than three fathoms.   The women, at first, opposed the undertaking, because our axes being only of a pound and a half weight, were not well adapted to so heavy a labour ; but the hope of finding a large Bear, and obtaining from its fat a great quantity

* REGNARD's *Journ. to Lapland.* (PINKERTON's *Voy.* vol. i. p. 194.)
† LEEMS's *Danish Lapland.* (*Idem.* vol. i. p. 485.)
‡ *Arctic Zoology*, vol. i. p. 65. *Introd.*, p. cxx.
§ HENRY's *Travels*, p. 142.

of oil, an article at the time much wanted, at length prevailed.   Accordingly, in the morning, we surrounded the tree, both men and women, as many at a time as could conveniently work at it; and there we toiled, like beavers, till the sun went down.   This day's work carried us about half-way through the trunk; and the next morning we renewed the attack, continuing it till about two o'clock in the afternoon, when the tree fell to the ground.   For a few minutes every thing remained quiet, and I feared that all our expectations were disappointed; but as I advanced to the opening, there came out, to the great satisfaction of all our party, a Bear of extraordinary size, which, before she had proceeded many yards, I shot.

" The Bear being dead, all my assistants approached, and all, but more particularly my old mother, (as I was wont to call her,) took his head in their hands, stroking and kissing it several times; begging a thousand pardons for taking away her life; calling her their relation and grandmother; and requesting her not to lay the fault upon them, since it was truly an Englishman that had put her to death.   This ceremony was not of long duration; and if it was I that killed their grandmother, they were not themselves behind hand in what remained to be performed.   The skin being taken off, we found the fat in several places six inches deep.   This being divided into two parts loaded two persons; and the flesh parts were as much as four persons could carry.   In all, the carcase must have exceeded five hundred weight.   As soon as we reached the lodge, the Bear's head was adorned with all the trinkets in the possession of the family, such as silver armbands, and wrist-bands, and belts of wampum; and then laid upon a scaffold, set up for its reception, within the lodge.   Near the nose was placed a large quantity of tobacco.

" The next morning no sooner appeared, than preparations were made for a feast to the manes.   The lodge was cleaned and swept; and the head of the Bear lifted up, and a new stroud blanket, which had never been used before, spread under it.   The pipes were now lit; and Wawatam blew tobacco-smoke into the nostrils of the Bear, telling me to do the same, and thus appease the anger of the Bear, on account of my having killed her.   I endeavoured to persuade my benefactor and friendly adviser, that she no longer had any life, and assured him that I was under no apprehension from her displeasure; but the first proposition obtained no credit, and the second gave but little satisfaction.   At length the feast being ready, Wawatam made a speech, resembling, in many things, his address to the manes of his relations and departed companions; and we then all

D 2

ate heartily of the Bear's flesh. It is only the female Bear that makes her winter lodging in the upper parts of trees, a practice by which her young are secured from the attacks of wolves and other animals. She brings forth in the winter-season; and remains in her lodge till the cubs have gained some strength. The male always lodges in the ground, under the roots of trees. He takes to this habitation as soon as the snow falls, and remains there till it has disappeared. The Indians remark, that the Bear comes out in the spring with the same fat which he carries in in the autumn; but, after the exercise of only a few days, becomes lean. Excepting for a short part of the season, the male lives constantly alone."

La Hontan * has also given a very full account of the ceremonies attending a Bear-hunt by the Canadian Indians, which does not differ greatly from Mr. Henry's. The women of the Chepewyan and Dog-rib tribes will not touch a bear's skin, nor even step over it; so that one spread at the door of a tent is an effectual bar against female intruders. Even the men of some of the tribes refuse to eat bear's flesh, or pemmican which contains bear's grease. The Laplanders, also, prohibit their women from eating certain portions of a bear. The flesh of a bear, when in good condition, resembles greasy and rather flabby pork; and when the animal has been fed on the sea-coast, and by the banks of rivers, has also a fishy taste. The skin of a Black Bear, with the fur in prime order, and the claws appended, was, at one period, worth from twenty to forty guineas, and even more, but at present the demand for them is so small, from their being little used either for muffs or hammercloths, that the best, I believe, sell for less than forty shillings.

* LA HONTAN, *Journal de Voy.*, vol. v. p. 169, *et seq.* See also SCHOOLCROFT's *Narrative*, &c., p. 183.

[9.]   2. Ursus Arctos? Americanus.   *Barren-ground Bear.*

Grizzly Bear.   Hearne's *Journey, passim.*
Brown Bear, variety ♂, Grizzly.   Pennant's *Arct. Zool.* vol. i. p. 62.

The Brown Bears of America are, by some authors, supposed to be merely varieties of the Black Bear of the preceding article; whilst others have considered them to belong to a distinct species, whose identity, however, with the Brown Bear of Europe has not been ascertained; neither has any one given it a new specific appellation. The obscurity in which the subject is involved has been increased by the accounts received from the natives of another species, named the Grisly Bear (*Ursus ferox*) having been amalgamated with the descriptions that authors have given of their Brown Bear. Warden * mentions a Brown Bear under the appellation of the "Ranging Bear," and says that it has the general shape of the Black Bear, but that its body and legs are longer, and that it is more ferocious when wounded. It is said to be an inhabitant of the United States, particularly of the western districts; but it never came under our notice, and the remainder of this article has no relation to it. From the inquiries I made throughout the woody country from Lake Superior to Great Slave Lake, being ten degrees of latitude, I learnt that the natives of those districts are acquainted with only two species of Land Bear, *viz.*, the *Common Black Bear*, including the cinnamon-coloured and other varieties, and the *Grisly Bear*, which is confined to the lofty chain of the Rocky Mountains, and the extensive plains that skirt their bases. The barren lands, however, lying to the northward and eastward of Great Slave Lake, and extending to the Arctic Sea, are frequented by a species of Bear, which differs from the American Black Bear in its greater size, profile, physiognomy, longer soles, and tail; and from the Grisly Bear also, in colour and the comparative smallness of its claws. Its greatest affinity is with the Brown Bear of Norway; but its identity with that species has not been established by actual comparison. It frequents the sea-coast in the autumn in considerable numbers, for the purpose of feeding on fish.

The general colour of this Bear is a dusky-(or sometimes yellowish)-brown, but

* Warden's *United States.*

the shoulders and flanks are, in the summer season at least, covered with long hair, which is frequently very pale towards the tips. The Indians and interpreters, who are not very precise in their application of the few terms they have to express varieties of colour, often denominate them "White Bears." Hearne calls them "Grizzly Bears," and some confusion has been produced by late writers having applied the same name to Lewis and Clark's *Ursus ferox*. Pennant, who describes them as a variety of the American Black Bear, considers them at the same time to be of the same species with the "Silver Bear" that inhabits the north of Europe. It is, indeed, very probable, that the Brown Bear which Captain King informed Pennant was an inhabitant of Kamskatka, is of this species, which may, in fact, extend all along the north of the old continent; but this, in the present state of our knowledge, is mere matter of conjecture. Mention is made in the narrative of Cook's third voyage * of Bears of a brown or sooty colour inhabiting the American coast near Cook's river. Langsdorff also informs us that Brown and Red Bears are abundant on the Aleutian Islands, where the Black Bear does not exist †. These authors do not furnish us with any details whereby the species may be determined; but the Bears they mention live in similar districts with the Barren-ground Bear, and differ in that respect from the *Ursus ferox*, which exists principally, perhaps only, in the buffalo districts.

The Indians dread the Barren-ground Bears, and are careful to avoid burning bones in their hunting encampments, lest the smell should attract them. Keskarrah, an old Indian mentioned in the Narrative of Captain Franklin's first Journey, was seated at the door of his tent, pitched by a small stream not far from Fort Enterprise, when a large Bear came to the opposite bank, and remained for some time apparently surveying him. Keskarrah considering himself to be in great danger, and having no one to assist him but his aged wife, made a speech to the following effect: "Oh Bear! I never did you any harm; I have always had the highest respect for you and your relations, and never killed any of them except through necessity. Go away, good Bear, and let me alone, and I promise not to molest you." The Bear walked off; and the old man, fancying that he owed his safety to his eloquence, favoured us, on his arrival at the fort, with his speech at length. The Copper Indians often cautioned us against these "White Bears" of the barren lands, which they said would attack us if they saw us, but we received no such caution in travelling through the districts

* COOK's *Third Voyage*, vol. ii. p. 376.   † LANGSDORFF's *Travels*, vol. ii. p. 74.

frequented by the Black Bear. It does not, however, possess the boldness of the *Ursus ferox*, as all the individuals we saw fled at once. The Barren-ground Bear resorts to the coast of the Arctic Sea in the month of August, and preys indiscriminately upon animal and vegetable matters. In the stomach of one which I opened there were the remains of a seal, a marmot, a large quantity of the long sweet roots of some *astragali* and *hedysara*, together with some berries, and a little grass. Many long white worms adhered to the interior of the stomach which held this farrago. Hearne has given the name of Grizzle Bear Hill to an eminence which had been much ploughed up by the Bears in quest of the *Arctomys Parryi*, termed by him " Ground Hog." The appellation of " grizzly," first used by Hearne to designate this Bear, being also applied by the traders and American authors to the *Ursus ferox*, I have given this one the *ad interim* name of Barren-ground Bear, until its difference from, or identity with, the *Ursus arctos* of Linnæus be fully established *.

DESCRIPTION.

We saw several of these animals during Captain Franklin's first Expedition. An old and lean male, killed on the shores of the Arctic Sea on the 1st of August, 1821, was of a nearly uniform yellowish-brown colour, except on the forehead and back, where the tips of the fur were paler. The fur, which was straight, and of the fineness of coarse wool, was giving place to a thin coat of blackish hair. Its forehead was broad, and slightly convex, and the arch of the orbit rose conspicuously at the root of the nose, which was straight. The legs were long, and the claws, of an intermediate size between those of the Black and Grisly Bears, projected beyond the hairs, and were more pointed than the claws of the latter.

Dental formula, incisors $\frac{6}{6}$, canines $\frac{1-1}{1-1}$, spurious molars $\frac{2-2}{1-1}$, grinders $\frac{3-3}{4-4} = 36$.

The *incisors* were worn flat, except one on each side, which adjoined the canine-teeth, and which rose in a point above the others. The *canines* strong, conical, and slightly curved, projected an inch and a quarter above the gums. Two small and pointed spurious molar-teeth (*dents espacèes*) rose on each side of the upper jaw, and were succeeded by three tuberculated molars that increased in size from the first to the last. The first of these was pointed anteriorly, and had a lobe posteriorly which exhibited the section of two points. The other two were worn quite flat; the second, or carnivorous-tooth, presented the section of two pairs of points; and the last, and largest, the section of three pairs. In the lower jaw, one small spurious molar-tooth was situated close to the canines. The first of the true molars was pointed, without any flattened lobe; the remaining three differed little from each other in size, though the middle one was rather the largest; and their crowns were worn so smooth, that no vestige remained of the points they originally possessed.

* In the appendix to Captain Parry's second voyage, from a hasty consideration of the subject, I erroneously stated the Barren-ground Bear to be the brown variety of the American Black Bear.

The dimensions of this individual were as follows * : —

	Feet.	Inches.		Feet.	Inches.
Length from the muzzle to the root of the tail	5	2	Length from anterior angle of the eye to the centre of the auditory opening	0	10
„ of tail	0	6	Distance from the tip of one ear to the tip of the other	0	10
Height from the sole of the fore-foot to the top of the shoulder	2	9	Breadth of fore-foot, which was nearly circular	0	6
„ of hind-quarters	2	6			
Length of muzzle from the nostrils to the anterior angle of the eye	0	6	Length of the sole of the hind-foot	0	10

[10.]   3. URSUS FEROX.   (Lewis and Clark.)   *Grisly Bear.*

Grizzle Bear. UMFREVILLE's *Hudson's Bay*, p. 168. *An.* 1790.
Grisly Bear. MACKENZIE's *Voyage, &c.*, p. 160. *An.* 1801.
White, or brown-gray Bear. GASS' *Journal of Lewis and Clark's Expedition*, pp. 45, 116, 346. *An.* 1808.
Grisly, brown, white, and variegated Bear, (Ursus ferox) LEWIS and CLARK's *Voyages, &c.*, vol. i. pp. 284, 293, 343, 375 ; vol. iii. pp. 25, 268. *Anno* 1814. CLINTON, *Trans. Philos. and Liter. New York*, vol. i. pp. 56, 114. *An.* 1815.
Grizzly Bear. WARDEN's *United States*, vol. i. p. 197. *An.* 1819.
Grey Bear. HARMON's *Journey*, p. 417. *An.* 1820.
Ursus Cinereus. DESMAREST's *Mammal. No.* 253. *An.* 1820.
Ursus horribilis. SAY, *Long's Expedition*, vol. ii. p. 244, *note* 34. *An.* 1822.
Ursus Candescens. HAMILTON SMITH, *Griffith's An. Kingdom*, vol. ii. p. 229, *and* vol. v. *No.* 320. *An.* 1826.
Grizzly Bear. GODMAN's *Nat. Hist.*, vol. i. p. 131. *An.* 1826.
Meesheh musquaw. CREE INDIANS.
Hohhost. CHOPUNNISH INDIANS (Lewis and Clark).

PLATE I.

This animal has long been known to the Indians and fur traders as a distinct species, inferior to all the varieties of the Black Bear in the quality of its fur, and distinguished by its great strength and ferocity, its carnivorous disposition, the length of its claws, the breadth and length of its soles, and the shortness of its tail. It has attracted the attention of almost all travellers who have passed through the districts it inhabits, and is mentioned in several of the earlier French writers on America under the title of *Ours blanc,* not that it is ever seen of a white colour like the Polar Bear, but because the Canadian *Coureurs des bois* who were, and who remain to this day, almost the only interpreters of the Indian languages, translated

---

* Baron Cuvier describes the *Ursus Arctos,* or Brown Bear of Europe, as having the upper part of its cranium arched longitudinally and rounded laterally ; the forehead and occiput forming parts of the same curve, and there being no well-defined line of separation between the forehead, the middle portion of the parietal bones, and the temporal fosses. The sagittal suture beginning to be sensibly marked very near the occipital bones, and the nasal bones to be set in rather obliquely to the rounded forehead, producing the appearance of a depression at the root of the nose. The sole of the hind-foot is of moderate length.—*Ossemens Fossiles.*

URSUS FEROX.

Published by John Murray January 1829.

the terms used by the different tribes to signify hoary or light coloured, by the general epithet of *blanc*. Lewis and Clark, in their ably-executed Journey to the Shores of the Pacific, had numerous opportunities of observing its manners, and by their ample descriptions, first enabled naturalists to class it as a distinct species. It is true that Forster, long before, in his translation of Bossu's Travels, had intimated that the " White Bear of Louisiana " must be distinct from the Polar Bear, which it resembled in size, but the remark was suffered to pass unheeded. De Witt Clinton, in his discourse at the Institution of the New York Literary and Philosophical Society, is the first naturalist who, judging from Lewis and Clark's account, clearly asserted that this animal was specifically different from either the Polar or common American Bears. Since that time the various synonymes prefixed to this article, in the order of their publication, have been assigned to it. The English name of Grisly has been adopted in this work as being less liable to objection than one founded on colour alone ; and the Latin translation of it, *ferox*, which, as far as I have been able to ascertain, first occurs in Desmarest, and seems preferable to *cinereus*, is used for the specific appellation. Mr. Say, in the account of Major Long's Expedition, gives a description of the Grisly Bear, drawn up from male and female specimens, preserved in the Philadelphia Museum, and which, having been brought up in a state of confinement, were killed before they arrived at maturity. Figures of these specimens have been published in the American edition of Long's Expedition, and in Godman's Natural History. A young cub, caught on the Rocky Mountains, being brought to England by the Hudson's Bay Company about eight years ago, has been kept in the Tower ever since, and there is a spirited engraving of it by Landseer, in Griffith's Animal Kingdom. The etching, forming plate first of this work, is by the same able artist, the head being from that of an adult male, brought home by Mr. Drummond, and the form of the body and attitudes from the individual in the Tower. I was present at the death of a young Grisly Bear, killed at Carltonhouse on the Saskatchewan. It was a male, in its second year, which being pursued by mounted hunters, was overtaken after an hour's chase, through snow one foot deep. The hunters approached boldly, trusting in the fleetness of their horses ; although, from the size of its foot-prints, they were fully aware that it was a Grisly Bear, even before they saw it *. The skin and scull of this individual are now preserved in the Museum of the Edinburgh University, and a figure of it is given in the sixth number of the very excellent Illustrations of Zoology by Wilson.

---

* MACKENZIE mentions the foot-marks of a Grisly Bear as being nine inches long and proportionably wide. The foot-marks of the young one mentioned in the text were of equal dimensions.

E

DESCRIPTION.

The Grisly Bear has been well compared by Mr. Say with the Norwegian variety of the *Ursus Arctos*, to which it has a great resemblance in its general appearance. Its fur is long, and mostly of a dark brown colour, with paler tips, that on the flanks being generally lighter coloured in the summer season, and there is frequently a considerable admixture of gray hairs on the head. The whole muzzle is pale, without the dark central stripe which the Black Bear has. It is distinguished from the Brown and Black Bears, by shorter and more conical ears, placed further apart, and white, arched, and very long claws, compressed like the incisors of a squirrel, carrying their breadth on their upper surface, nearly to the tips, but chamfered away as it were beneath. They project far beyond the hair of the foot, and cut like a chisel when the animal strikes a blow with them. The forehead is broad, flattish, and continued nearly in a line with the nose, but there is in the older animals a distinct projection of the superciliary ridges of the frontal bone. The soles of its feet are longer and its heel is broader than those of the Brown Bear of Europe. Its tail is very small, so as to be hidden by the hair of the buttocks, and it is a standing joke among the Indian hunters, when they have killed a Grisly Bear, to desire any one unacquainted with the animal to take hold of its tail. The tail of the Black Bear is conspicuous enough, and that of the Barren-ground Bear is still longer*.

The strength and ferocity of the Grisly Bear are so great that the Indian hunters use much precaution in attacking them. They are reported to attain a weight exceeding eight hundred pounds, and Lewis and Clark mention one that measured nine feet from the nose to the tail, and say that they had seen a still larger one, but do not give its dimensions. This is far above the usual size of other Land Bears, and equals the larger specimens of the Polar Bear. Governor Clinton received an account of one fourteen feet long, from an Indian Trader, but even admitting that there was no inaccuracy in the measurement, it is probable that it was taken from the skin after it was removed from the body, when it is known to be capable of stretching several feet. The strength of this Bear may be estimated from its having been known to drag to a considerable distance the carcass of a

* " Two men visited the Indian village (on one of the upper branches of the Columbia) where they purchased a dressed bear skin, of an uniform pale reddish-brown colour, which the Indians called *yackah*, in contradistinction to *hohhost*, or the White Bear. This remark induced us to inquire more particularly into their opinions as to the several species of bears, and we therefore produced all the skins of that animal which we had killed at this place, and also one very nearly white, which we had purchased. The natives immediately classed the *white*, the deep and the pale *grizzly-red*, the *grizzly dark-brown*, in short, all those with the extremities of the hair of a white or frosty colour, without regard to the colour of the ground of the fur, under the name of *hohhost*. They assured us that they were all of the same species with the White Bear; that they associated together, had longer nails than the others, and never climbed trees. On the other hand, the *black skins*, those which were *black* with a number of entire *white hairs* intermixed, or with a *white breast*, the *uniform bay*, the *brown*, and *light reddish-brown*, were ranged under the class *yackah*, and were said to resemble each other in being smaller, and having shorter nails than the White Bear, in climbing trees, and being so little vicious, that they could be pursued with safety."—LEWIS AND CLARK, vol. iii. p. 215.

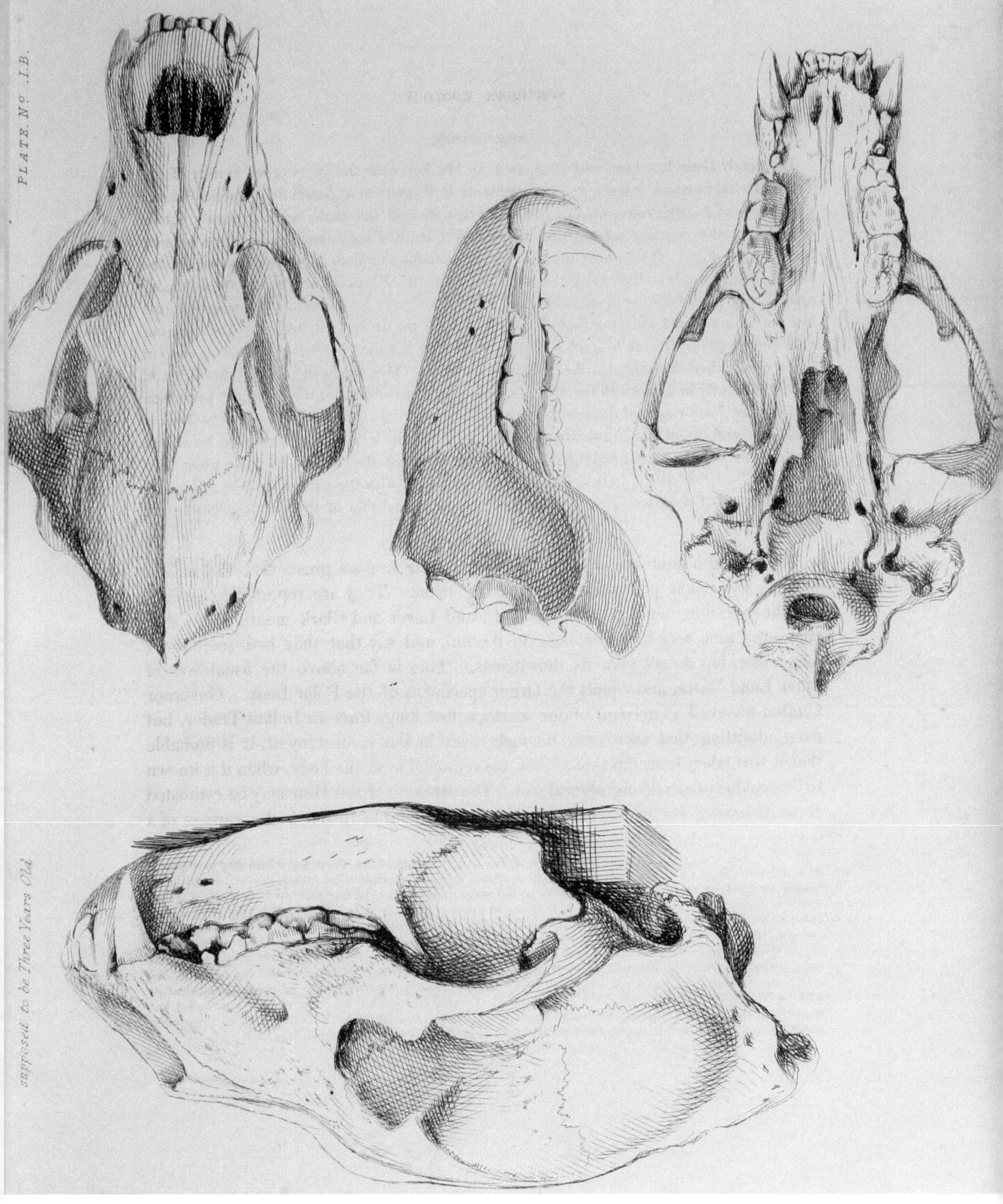

Buffalo, weighing about one thousand pounds. The following story is well authenticated. A party of voyagers, who had been employed all day in tracking a canoe up the Saskatchewan, had seated themselves in the twilight by a fire, and were busy in preparing their supper, when a large Grisly Bear sprung over their canoe that was tilted behind them, and seizing one of the party by the shoulder carried him off. The rest fled in terror with the exception of a Metif, named Bourasso, who, grasping his gun, followed the Bear as it was retreating leisurely with its prey. He called to his unfortunate comrade that he was afraid of hitting him if he fired at the Bear, but the latter entreated him to fire immediately, without hesitation, as the Bear was squeezing him to death. On this he took a deliberate aim, and discharged his piece into the body of the Bear, which instantly dropped its prey to pursue Bourasso. He escaped with difficulty, and the Bear ultimately retreated to a thicket, where it was supposed to have died; but the curiosity of the party not being a match for their fears, the fact of its decease was not ascertained. The man who was rescued had his arm fractured, and was otherwise severely bitten by the Bear, but finally recovered. I have seen Bourasso, and can add that the account which he gives is fully credited by the traders resident in that part of the country, who are best qualified to judge of its truth from their knowledge of the parties. I have been told that there is a man now living in the neighbourhood of Edmonton-house, who was attacked by a Grisly Bear, which sprung out of a thicket, and with one stroke of its paw completely scalped him, laying bare the scull, and bringing the skin of the forehead down over the eyes. Assistance coming up, the Bear made off without doing him further injury, but the scalp not being replaced, the poor man has lost his sight, although he thinks that his eyes are uninjured.

Mr. Drummond, in his excursions over the Rocky Mountains, had frequent opportunities of observing the manners of the Grisly Bears, and it often happened that in turning the point of a rock or sharp angle of a valley, he came suddenly upon one or more of them. On such occasions they reared on their hind legs and made a loud noise like a person breathing quick, but much harsher. He kept his ground without attempting to molest them, and they on their part, after attentively regarding him for some time, generally wheeled round and galloped off, though, from their known disposition, there is little doubt but he would have been torn in pieces had he lost his presence of mind and attempted to fly. When he discovered them from a distance, he generally frightened them away by beating on a large tin box, in which he carried his specimens of plants. He never saw more than four together, and two of these he supposes to have been cubs; he

E 2

more often met them singly or in pairs. He was only once attacked, and then by a female, for the purpose of allowing her cubs time to escape. His gun on this occasion missed fire, but he kept her at bay with the stock of it, until some gentlemen of the Hudson's Bay Company, with whom he was travelling at the time, came up and drove her off. In the latter end of June 1826, he observed a male caressing a female, and soon afterwards they both came towards him, but whether accidentally, or for the purpose of attacking him, he was uncertain. He ascended a tree, and as the female drew near, fired at and mortally wounded her. She uttered a few loud screams, which threw the male into a furious rage, and he reared up against the trunk of the tree in which Mr. Drummond was seated, but never attempted to ascend it. The female, in the meanwhile retiring to a short distance, lay down, and as the male was proceeding to join her, Mr. Drummond shot him also. From the size of their teeth and claws, he judged them to be about four years old. The cubs of the Grisly Bear can climb trees, but when the animal is fully grown it is unable to do so, as the Indians report, from the form of its claws. Two instances are related by Lewis and Clark, and I have heard of several others, where a hunter having sought shelter in a tree from the pursuit of a Grisly Bear, has been held a close prisoner for many hours, by the infuriated animal keeping watch below. The Black and Brown or even the Polar Bear ascend trees with facility. Some interesting anecdotes of contests with this Bear, selected from the narratives of Lewis and Clark, Major Long, and others, are related in Godman's Natural History, to which the reader is referred.

The Grisly Bears are carnivorous, but occasionally eat vegetables, and are observed to be particularly fond of the roots of some species of *psoralea* and *hedysarum*. They also eat the fruits of various shrubs, such as the bird-cherry, choke-cherry, and *hippophäe Canadensis*. The berries of the latter produce a powerful cathartic effect upon them. Few of the natives, even of the tribes, who are fond of the flesh of the Black Bear, will eat of the Grisly Bear, unless when pressed by hunger. Say and Gass mention a method which the Shoshonee or Snake Indians have of baking Bear's flesh in a pit filled with alternate layers of brush-wood and meat, and covered with earth *, which is nearly similar to the way in which the natives of the South-sea Islands prepare their dogs and hogs.

The Grisly Bear inhabits the Rocky Mountains and the plains lying to the eastward of them, as far as latitude 61°, and perhaps still farther north. Its southern range, according to Lieutenant Pike, extends to Mexico. There is a Brown

* Gass's *Journal, &c.*, p. 311.

Bear on the Andes of Peru, but whether it is of this species or not is not known *. Lewis and Clark could not ascertain that the Grisly Bear at all inhabited the country between the western declivity of the Rocky Monntains and the sea-coast, and remark that those which they saw about the great falls of the Columbia were more variegated in colour, and of a milder disposition than those near the sources of the Missouri, but certainly of the same species. Mr. Drummond observes that the Grisly Bears are most numerous in the woody country skirting the eastern base of the Rocky Mountains, particularly in districts which are interspersed with open prairies and grassy hills. They vary, he says, much in colour, from a very light gray to a dark chestnut. The latter variety is common about the sources of the Peace River, and, according to the Indians, is more ferocious than the gray one. The Black Bear, which inhabits the same districts, and frequently varies there to a cream-colour, never associates with the Grisly Bear.

The young Grisly Bears and gravid females hibernate, but the older males often come abroad in the winter in quest of food. Mackenzie mentions the den or winter retreat of a Grisly Bear, which was ten feet wide, five feet high, and six feet long. These dens are named *watee* by the Indians. As this Bear comes abroad before the snow disappears, its foot-marks are frequently seen in the spring, and when there is a crust on the snow, the weight of the animal often causes it to crack and sink for a yard or more round the spot trod upon. These impressions, somewhat obscured by a partial thaw, have been considered by the inexperienced as the vestiges of an enormously large quadruped, and the natives, although perfectly aware of the cause of the marks, are prone by their observations to heighten the wonder they perceive to be excited by them. Many reports of the existence of live Mammoths in the Rocky Mountain range, have, I doubt not, originated in this manner. Necklaces of the claws of a Grisly Bear are highly prized by the Indian warriors as proofs of their prowess.

---

* CONDAMINE's *Travels*, p. 82. ULLOA's *Voyage*, 461 (quoted from Arctic Zoology, p. clxx.)

[10.]   4. URSUS MARITIMUS. (Lin.) *Polar* or *Sea Bear*.

White Bear.   MARTEN's *Spitz. Trans.*, p. 107, t. O, *fig.* c.   An. 1675.
Ursus Maritimus.   LIN. *Syst.*
Ursus Albus.   BRISSON, *Règne Animal*, p. 260, sp. 2.   An. 1756.
L'ours Blanc.   BUFFON, vol. xv. p. 128.   An. 1767.
Ursus Marinus.   PALLAS's *It.* vol. iii. p. 691, *et* SPICEL. *Zool.* xiv. t. 1.   An. 1780.
Polar Bear.   PENNANT's *Arctic Zoology*, p. 53, and *Introd.* pp. lxxxix and cxciii.   An. 1784.
Ursus Albus.   Ross's *Voy.*, *App.* p. xliv. with a plate of the head, p. 199.   An. 1820.
Ursus Maritimus.   PARRY's *First Voy.*, *Supp.* p. clxxxiii.   FRANKLIN's *First Journey*, p. 648.
          PARRY's *Second Voy.*, *App.* p. 288.
Bear.   LYON's *Private Journal*, pp. 13 and 377.   An. 1824.
Wawpusk (pl. Wawpuskwuck).   CREE INDIANS.
Nannook.   ESQUIMAUX.   Nennook, GREENLANDERS.

Buffon had many doubts as to the Sea or Polar Bear being a distinct species from the Land Bear, of which there are white varieties in the northern countries. He acknowledges, however, that the distinctive characters which Marten, one of its earliest describers, has pointed out, would, if correct, establish it as a peculiar species.   A further acquaintance with the animal has fully confirmed Marten's observations.

DESCRIPTION.

The Polar Bear is distinguished from the other species by its narrow head and muzzle prolonged on a straight line with the flattened forehead; its short ears; long neck; the greater length of its body in proportion to its height; the soles of the hind-feet equalling one-sixth of the length of its body; and, lastly, the quality of its fur, which is very thick and long on the body, still more so on the limbs, and everywhere of a yellowish-white colour.   The naked extremity of the snout, the tongue, margins of the eyelids, and claws, are black; the lips purplish-black; the eyes dark brown, and the interior of the mouth pale violet.

I have met with no account of any Polar Bear, killed of late years, which exceeded nine feet in length, or four feet and a half in height.   It is possible that larger individuals may be occasionally found; but the greatness of the dimensions attributed to them by the older voyagers has, I doubt not, originated in the skin having been measured after being much stretched in the process of flaying. Marten, who seems to have been a correct observer, expressly states that the Polar Bear is of the same size with the German Bears.

The great power of the Polar Bear is portrayed in the account of a disastrous accident which befel the crew of Barentz's vessel on his second voyage to Waigats Straits.   "On the 6th of September, 1594, some sailors landed to search for a certain sort of stone, a species of diamond. During this search, two of the seamen lay down to sleep by one another, and a White Bear, very lean, approaching softly, seized one of them by the nape of the neck. The poor man, not knowing what it was, cried out, 'Who has seized me thus behind?' on which his companion, raising his head, said, 'Holloa, mate, it is a Bear,' and immediately ran away. The Bear having dreadfully mangled the unfortunate man's head, sucked the blood. The rest of the persons who were on shore, to the number of twenty, immediately ran with their matchlocks and pikes, and found the Bear devouring the body, which, on seeing them, ran upon them, and carrying another man away, tore him to pieces. This second misadventure so terrified them, that they all fled. They advanced again, however, with a reinforcement, and the two pilots having fired three times without hitting the animal, the purser approached a little nearer, and shot the Bear in the head, close by the eye. This did not cause him to quit his prey, for, holding the body, which he was devouring always by the neck, he carried it away as yet quite entire. Nevertheless, they then perceived that he began himself to totter, and the purser and a Scotchman going towards him, they gave him several sabre wounds, and cut him to pieces, without his abandoning his prey *."

In Barentz's third voyage, a story is told of two Bears coming to the carcass of a third one that had been shot, when one of them, taking it by the throat, carried it to a considerable distance, over the most rugged ice, where they both began to eat it. They were scared from their repast by the report of a musket, and a party of seamen going to the place, found that, in the little time they were about it, they had already devoured half the carcase, which was of such a size that four men had great difficulty in lifting the remainder †. In a manuscript account of Hudson's Bay, written about the year 1786, by Mr. Andrew Graham, one of Pennant's ablest correspondents, and preserved at the Hudson's Bay House, an anecdote of a different description occurs. "One of the Company's servants who was tenting abroad to procure rabbits (*Lepus Americanus*), having occasion to come to the factory for a few necessaries, on his return to the tent passed through a narrow thicket of willows, and found himself close to a White Bear lying asleep. As he had nothing wherewith to defend himself, he took the bag off his shoulder and held it before his breast, between the Bear and him. The animal arose on seeing the man, stretched himself and rubbed his nose, and having satisfied his curiosity by smelling at the bag, which contained a loaf of bread and a rundlet of strong beer, walked quietly away, thereby relieving the man from his very disagreeable situation."

The Polar Bears feed chiefly on animal substances, and as they swim and dive well, they hunt seals and other marine animals with great success. They are even said to wage war, though rather unequally, with the Walrus. They feed likewise on land animals, birds, and eggs, nor do they disdain to prey on carrion, or, in the absence of other food, to seek the shore in quest of berries and roots. They scent their prey from a great distance, and are often attracted to the whale vessels by the smell of burning *kreng*, or the refuse of the whale blubber. Captain Lyons thus describes the mode in which the Polar Bear surprises a seal. " The Bear, on seeing his intended prey, gets quietly into the water, and swims to leeward of him, from whence, by frequent short dives, he silently makes his approaches, and so arranges his distance, that, at the last dive, he comes to the spot where the seal is lying. If the poor animal attempts to escape by rolling into the water, he falls into the Bear's clutches, if, on the contrary, he lies still, his destroyer makes a powerful spring, kills him on the ice, and devours him at leisure." The same writer describes the pace of the Polar Bear, at full speed, as " a kind of shuffle, as quick as the sharp gallop of a horse."

The principal residence of the Polar Bear is on fields of ice, with which he frequently drives to a great distance from the land. In this way they are often carried from the coast of Greenland to Iceland, where they commit such ravages on the flocks, that the inhabitants rise in a body to destroy them. Captain Sabine mentions that he saw one about mid-way between the north and south shores of Barrow's Straits, which are forty miles apart, although there was no ice in sight to which he could resort to rest himself upon ; and Captain Lyons informs us, that the Polar Bears not only swim with rapidity, but are capable of making long springs in the water. They are not known to travel far inland. They have been found in higher latitudes than any other quadruped, having been seen by Captain Parry in his most adventurous boat-voyage beyond 82 degrees of north latitude. The limit of their incursions southward on the shores of Hudson's Bay, and of Labrador, may be stated to be about the 55th parallel. They are often seen about York Factory in the autumn, having most probably drifted from the northward on the ice during the summer. Pennant, who has collected, from good authorities, much information relative to their range, states, that they are frequent on all the Asiatic coasts of the Frozen Ocean, from the mouth of the Obi eastward, and abound in Nova Zembla, Cherry Island, Spitzbergen, Greenland, Labrador, and the coasts of Baffin's and Hudson's Bays. They were seen by

Captain Parry within Barrow's Straits as far as Melville Island ; and the Esquimaux, to the westward of Mackenzie's river, told Captain Franklin that they occasionally, though very rarely, visited that coast. The exact limit of their range to the westward is uncertain, but they are said not to be known on the islands in Behring's Straits, nor on the coast of Siberia to the eastward of Tchutskoinoss. They are not mentioned by Langsdorff and other visiters of the north-west coast of America ; nor did Captain Beechey meet with any in his late voyage to Icy Cape. None were seen on the coast between the Mackenzie and Copper Mine River ; and Pennant informs us, that they are unknown along the shores of the White Sea, which is an inlet of a similar character.

The Polar Bear being able to procure its food in the depth of even an Arctic winter, there is not the same necessity for its hibernating that exists in the case of the Black Bear, which feeds chiefly on vegetable matters ; and it is probable, that, although they may all retire occasionally to caverns in the snow, the pregnant females alone seclude themselves for the entire winter. It is mentioned in Le Roy's narrative of the residence of four Russian seamen for six winters in Spitzbergen, and also in the account of Barentz's winter in Nova Zembla, that the Bears disappeared with the sun, and returned again with that luminary, after an absence in the one case of four months, and in the other of three. Their retirement has been considered by some as a proof of their hibernation ; but, I think, the most probable explanation of it is that they went out to sea in search of food. Polar Bears were seen in the course of the two winters that Captain Parry remained on the coast of Melville Peninsula ; and the Esquimaux of that quarter derive a considerable portion of their subsistence not only from the flesh of the female Bears, which they dig together with their cubs from under the snow, but also from the males that they kill when roaming at large at all periods of the winter. Hearne states with more precision, and, I believe, from actual observation, that the males leave the land in the winter time and go out on the ice to the edge of the open water in search of seals, whilst the females burrow in deep snow-drifts from the end of December to the end of March, remaining without food, and bringing forth their young during that period ; that when they leave their dens in March, their young, which are generally two in number, are not larger than rabbits, and make a foot-mark in the snow no bigger than a crown piece. He also informs us that the males are found in company with the females in August, and then exhibit great attachment to them. Mr. Andrew Graham's observations, written before the publication of Hearne's Narrative, confirm the account given by that traveller. "In winter," says he, "the White Bear sleeps like other species of

F

the genus, but takes up its residence in a different situation, generally under the declivities of rocks, or at the foot of a bank, where the snow drifts over it to a great depth; a small hole, for the admission of fresh air, is constantly observed in the dome of its den. This, however, has regard solely to the she-Bear, which retires to her winter-quarters in November, where she lives without food, brings forth two young about Christmas, and leaves the den in the month of March, when the cubs are as large as a shepherd's dog. If perchance her offspring are tired, they ascend the back of the dam, where they ride secure either in water or ashore. Though they sometimes go nearly thirty miles from the sea in winter, they always come down to the shores in the spring with their cubs, where they subsist on seals and sea-weed. The he-Bear wanders about the marshes and adjacent parts until November, and then goes out to the sea upon the ice, and preys upon seals. They are very fat, and though very inoffensive if not meddled with, they are very fierce when provoked*.''

Captain Lyons records the Esquimaux account of the hibernation of the Polar Bear in the following words: "From Ooyarrakhioo, a most intelligent man, I obtained an account of the Bear, which is too interesting to be passed over in silence. ' At the commencement of winter, the pregnant Bears are very fat, and always solitary. When a heavy fall of snow sets in, the animal seeks some hollow place in which she can lie down, and remains quiet, while the snow covers her. Sometimes she will wait until a quantity of snow has fallen, and then digs herself a cave: at all events, it seems necessary that she should be covered by and lie amongst the snow. She now goes to sleep, and does not wake until the spring sun is pretty high, when she brings forth two cubs. The cave, by this time, has become much larger, by the effect of the animal's warmth and breath, so that the cubs have room enough to move, and they acquire considerable strength by continually sucking. The dam at length becomes so thin and weak, that it is with great difficulty she extricates herself, when the sun is powerful enough to throw a strong glare through the snow which roofs the den.' The Esquimaux affirm that during this long confinement the Bear has no evacuations, and is herself the means of preventing them by stopping all the natural passages with moss, grass, or earth. The natives find and kill the Bears during their confinement by means of dogs, which scent them through the snow, and begin scratching and howling very eagerly. As it would be unsafe to make a large opening, a long trench is cut of sufficient width to enable a man to look down, and see where the Bear's head lies,

* GRAHAM, MSS. p. 20.

and he then selects a mortal part into which he thrusts his spear. The old one being killed, the hole is broken open, and the young cubs may be taken out by the hand, as, having tasted no blood and never having been at liberty, they are then very harmless and quiet. Females which are not pregnant roam throughout the whole winter in the same manner as the males. The coupling time is May."

The flesh of the Polar Bear is, as stated by Captain Phipps (Lord Mulgrave), exceedingly coarse. The Russian sailors who wintered in Spitzbergen, found it, on the other hand, much more agreeable to the taste than the flesh of the rein-deer. I quote this fact here, not to show that there was any thing peculiarly gross in the taste of the Russians, but to have an opportunity of remarking, that when people have fed for a long time solely upon lean animal food, the desire for fat meat becomes so insatiable, that they can consume a large quantity of un-mixed, and even oily fat, without nausea. Our seamen relish the paws of the Bear, and the Esquimaux prefer its flesh at all times to that of the seal. Instances are recorded of the liver of the Polar Bear having poisoned people.

The reader who is desirous of fuller accounts of the manners and habits of this very curious animal will be gratified by turning to Marten's Spitzbergen, Fabricius' Fauna Grœnlandica, Pennant's Arctic Zoology, and Scoresby's Account of the Arctic Regions. I subjoin some well-authenticated measurements of Polar Bears.

	Captain Phipps.		Captain Ross.		Captain Lyon.	
	Feet.	Inches.	Feet.	Inches.	Feet.	Inches.
Length from nose to tail	7	1	7	10	8	7½
„ the shoulder-blade	2	3	2	0	0	0
Height at the shoulder	4	3	4	1	4	9
Girth near the fore-legs	7	0	6	0	7	11
„ of the neck	2	1	3	2	3	4½
Breadth of the fore-paw	0	7	0	10	0	10
„ hind-paw	0	0	0	8¼	0	0
Length from nostrils to hind head	0	0	1	6	1	6
„ of fore-claws	0	0	0	2½	0	2½
„ of hind-claws	0	0	0	1¾	0	1½
„ of tail	0	0	0	4	0	5
Weight	610 lbs.		1160 lbs.		1600 lbs *.	

* Captain Lyon states that his specimen was unusually large.

[11.]   1. PROCYON LOTOR. (Cuvier.) *The Raccoon.*

GENUS Procyon. STORR. CUVIER.
Ursus lotor. LIN. GMELIN. vol. i. p. 103.
Le Raton. BUFFON, vol. viii. pp. 337, t. xliii.
Raccoon Bear. PENNANT's *Arct. Zool.*, vol. i. p. 69.
Procyon lotor. CUVIER's *Règne An.*, vol. i. p. 143. SABINE, *Frankl. Jour.*, p. 649. HARLAN. *Faun.*, p. 53.
The Raccoon. GODMAN's *Nat. Hist.*, vol. i. p. 163.

This animal inhabits the southern parts of the fur districts, being found as far north as Red River, in latitude 50°, from which quarter about one hundred skins are procured annually by the Hudson's Bay Company. If there is no mistake as to the identity of the species, the Raccoon extends farther north on the shores of the Pacific than it does on the eastern side of the Rocky Mountains. Dixon and Portlock obtained cloaks of Raccoon skins from the natives of Cook's River, in latitude 60°, and skins, supposed to be of the Raccoon, were also seen at Nootka Sound by Captain Cook. Lewis and Clarke expressly state that the Raccoon, at the mouth of the Columbia, is the same with the animal so common in the United States. Desmarest says that the Raccoon extends as far south as Paraguay. It is an animal, with a fox-like countenance, but with much of the gait of a Bear, and being partially plantigrade, it was classed by Linnæus in the genus *Ursus.* In the wild state, it sleeps by day, comes from its retreat in the evening, and prowls in the night in search of roots, fruits, green corn, birds and insects. It is said to eat merely the brain, or suck the blood of such birds as it kills. At low water it frequents the sea-shore to feed on crabs and oysters. It is fond of dipping its food into water before it eats, which occasioned Linnæus to give it the specific name of *lotor.* It climbs trees with facility. The fur of the Raccoon is used in the manufacture of hats, and its flesh, when it has been fed on vegetables, is reported to be good. The live animal is often seen in English menageries.

DESCRIPTION.

The Raccoon has a round *head,* with a narrow, tapering nose, which projects considerably beyond the mouth. The end of the nose is naked and black, and it possesses much flexibility. The lips are also black. The eyes are round and moderately large; the pupils circular. The low, erect ears are elliptical, with their tips much rounded, and, together with their edges, are of a soiled white colour. The whiskers are strong. The muzzle is covered with short hairs, of a

MELES LABRADORIA.

Published by John Murray January 1829.

soiled white colour. This pale colour passes in the form of a band round the cheek and over the eyes. A dark mark includes the eye and cheek, on each side, and there is also a mark of a similar colour between the eyes, continued from the forehead. The dark colour is produced by a mixture of grey, dark-brown and black hairs. The back is grizzled, its fur consisting of dirty white hairs, ringed with black. The belly is considerably paler. The *tail* is bushy, like the brush of a fox, and has a dirty white colour, with about six dark rings round it. The *extremities* are short, and all the feet have five toes, armed with long, strong claws, fitted for burrowing. There is a fulness of the skin on the flanks, which adds to the apparent shortness of the limbs. The animal walks on the hind and fore toes, but when it sits, brings the whole hind sole to the ground, and it often assumes an erect posture like a Bear.

Carver quaintly describes the Raccoon as having the limbs of a beaver, the body of a badger, the head of a fox, the nose of a dog, the tail of a cat, and sharp claws, by which it climbs trees like a monkey.

DIMENSIONS.

	Feet.	Inches.		Feet.	Inches.
Length of head and body . . .	2	0	Length of tail (vertebræ) . . .	0	9½
„ head	0	6	Height of the back	1	1

[12.] 1. MELES LABRADORIA. (Sabine.) *American Badger*.

GENUS. Meles. BRISSON.
Carcajou. BUFFON, tom. vi. p. 117, pl. 23. (édit. de Paris en 36 vol. 1749–1789.) *Quadrupèdes enlum.* 295.
Common Badger. PENNANT's *Arctic Zoology*, vol. i. p. 71.
Badger var. β. American. PENNANT's *Hist. Quadr.*, vol. ii. p. 15.
Ursus Labradoricus. LIN. GMELIN, vol. i. p. 102.
Prarow. GASS's *Journal*, p. 34.
Blaireau. LEWIS AND CLARKE's *Voyage, &c.* vol. i. pp. 50, 137, 213.
Taxus Labradoricus. SAY, *Long's Exped.*, vol. i. p. 261.
Meles Labradoria. SABINE, *Franklin's First Journ.*, p. 649. HARLAN's *Fauna*, p. 57.
American Badger. GODMAN's *Nat. Hist.*, vol. i. p. 179.
Blaireau d'Amérique. F. CUVIER, *Hist. Nat. des Mamm.* cum figurâ.
Brairo et Siffleur. FRENCH CANADIANS.
Nannaspachæ-neeskæshew. Mistonusk, (also *awawteekæoo*, " the animal that digs,") CREE INDIANS.
Chocartoosh. PAWNEES.

PLATE 2.

Buffon, in the body of his great work, doubts whether the Badger be an inhabitant of the American continent, notwithstanding that M. Brisson had

described a small quadruped from New York, under the name of *Meles alba* *.
Brisson's animal, according to M. Desmarest, proved to be merely an albino
variety of the Raccoon †; but Buffon afterwards, in the first addition to his
article on the Glutton, described the skin of a true Badger, which he received,
it is said, from Labrador, under the misapplied name of *Carcajou* ‡. His
specimen was imperfect, having only four toes the fifth having been rubbed
off, as he supposes, in stuffing ; and Gmelin, who adopted the opinion of
Schreber in considering it to be a distinct species from the European Badger,
carelessly allowed "*palmis tetradactylis*" to form part of the specific character.
Shaw pointed out the differences between the two species more perfectly,
and his observations have been confirmed and extended by Mr. Sabine, who
described a specimen, obtained on the plains of the Saskatchewan by Captain
Franklin's party.   Kalm says that he saw the common Badger in Pennsylvania,
where it is known by the name of the Ground Hog §.   I suspect, however, that
there is some mistake in his observation, because recent American naturalists do
not mention it as an inhabitant of that state ; and the appellation of Ground
Hog is applied by the country people to the marmots and several other animals that
have the habit of burrowing in the earth ‖.   If there be, indeed, a true Badger on
the Atlantic coast, it must differ in habits, and be perhaps a distinct species from
the one described by Mr. Sabine, which inhabits a district of country very different
in character.   For the same reason, I have some doubts of Buffon's specimen
having come from Labrador ¶.   The Blaireaux, seen by Lewis and Clarke on the
plains of the Missouri, are doubtless specifically the same with those of the adjoining and similar Saskatchewan country ; and even the Brairo, which the same
travellers describe as an inhabitant of the open plains, and sometimes of the
woods, of Columbia, presents no character, in their account of it, which denote
it to be distinct from the Saskatchewan animal, except the curious and perhaps
accidental circumstance of a double nail, like the Beaver's, on one of the toes of
each hind foot.

The Meles Labradoria frequents the sandy plains or prairies which skirt the
Rocky Mountains as far north as the banks of the Peace River, and sources of

* BRISSON, *Règne An.*, p. 255.
† DESMAREST'S *Mamm.*, p. 168 and 174.
‡ The name of *Carcajou* belongs properly to the Wolverene.
§ KALM's *Travels*, vol. i. p. 189.
‖ GASS, indeed, in noticing the Badgers of the Missouri, says that they are about the size of a ground hog, and
nearly of the same colour.—GASS's *Journal*, p. 34.
¶ Buffon says " qu'il venoit du pays des Esquimaux," but in fact it may have been brought actually from the banks
of the Saskatchewan by some of the Canadian fur hunters.

the River of the Mountains, in latitude 58°. It abounds on the plains watered by the Missouri, but its exact southern range has not, as far as I know, been defined by any traveller. The sandy prairies, in the neighbourhood of Carlton-house, on the banks of the Saskatchewan, and also on the Red River, that flows into Lake Winipeg, are perforated by innumerable Badger-holes, which are a great annoyance to horsemen, particularly when the ground is covered with snow. These holes are partly dug by the Badgers for habitations; but the greater number of them are merely enlargements of the burrows of the *Arctomys Hoodii* and *Richardsonii*, which the Badgers dig up and prey upon.

Whilst the ground is covered with snow, the Badger rarely or never comes from its hole; and I suppose that in that climate it passes the winter from the beginning of November to April in a torpid state. Indeed, as it obtains the small animals on which it feeds by surprising them in their burrows, it has little chance of digging them out at a time when the ground is frozen into a solid rock. Like the Bears, the Badgers do not lose much flesh during their long hibernation, for, on coming abroad in the spring, they are observed to be very fat. As they pair, however, at that season, they soon become lean.

This Badger is a slow and timid animal, taking to the first earth it comes to when pursued; and as it makes its way through the sandy soil with the rapidity of a mole, it soon places itself out of the reach of danger. The strength of its fore-feet and claws is so great, that one which had insinuated only its head and shoulders into a hole, resisted the utmost efforts of two stout young men who endeavoured to drag it out by the hind legs and tail, until one of them fired the contents of his fowling-piece into its body. Early in the spring, however, when they first begin to stir abroad, they may be easily caught by pouring water into their holes; for the ground being frozen at that period, the water does not escape through the sand, but soon fills the hole, and its tenant is obliged to come out.

The American Badger appears to be a more carnivorous animal than the European one. A female which I killed had a small marmot, nearly entire, together with some field-mice, in its stomach. It had also been eating some vegetable matters.

DESCRIPTION

Of a female American Badger, killed at Carlton-house, in the latter end of April.

Its fur is very soft and fine, about three inches and a half long on the back, of a purplish-brown colour from the roots upwards, variegated with narrow black rings near its summits, and tipped with white, producing a pleasant and somewhat mottled or hoary gray colour, but exhibiting no brown tints when the fur lies smooth. The upper surface of the head is of a darker colour, bisected by a narrow white line, which runs from the nose to the nape of the neck. This white stripe is bounded by dark fur, which gradually fades into gray and white as it approaches the ears. A grayish-brown patch includes the eye, and extends to the tip of the nose. There is also a brownish patch on the cheek before the ears, but the rest of the cheek, the under-jaw, and the throat are white. Its belly is thinly coated with coarser whitish hairs, its legs are of a blackish-brown colour, and its tail of a dirty brown. It is low on its legs, has a broad, fleshy body, a flattish head, and very short, round ears. Its claws are long, strong, and of a pale horn colour. The molar teeth are remarkably smooth and flat on the crowns for an omnivorous animal. It measures two feet six inches from the nose to the root of the tail, and the length of its tail is six inches.

The European Badger differs totally from the American one, in its dark-coloured, much coarser, and shorter fur; in the well-defined white lines on the head, in its more conspicuous ears, which are black tipped with white, and in having a larger head. The differences betwixt these animals are detailed in the following quotation from Mr. Sabine :—

" The American Badger is generally less in size, and of a lighter make ; the head, though equally long, is not so sharp towards the nose, and the markings on the fur are remarkably different ; a narrow white line runs from between the eyes towards the back, the rest of the upper part of the head is brown, the throat and whole under jaw are white, the cheeks partly so ; a semicircular brown spot is placed between the light part of the cheeks and the ears ; the white marking extends in a triangular form a little above the eyes, and below the eyes in a line towards the fore part of the mouth, but the whole eye lies within the dark colour of the upper part of the head, which colour runs in a sharp angle at the corner of the eye into the white. The European Badger has three broad white marks; one on the top of the head and one on each side, and between them are two broad black lines, which include the eyes and ears ; and the whole under parts of the throat and jaw are black. The upper parts of the body and sides in the American animal are covered with rather long, fine, grayish hairs, which in the other are darker, coarser, and longer ; the under parts in the former are lighter than the upper, in the latter they are darker ; the legs in the first are dark brown, in the other quite black; and though the animal is of larger size generally, its nails, which are dark, are smaller than the light horn-coloured nails of the American species ; and, finally, the tail of the European Badger is longer than that of the American."

The specimen of the American Badger which Mr. Sabine comments upon, was two feet two inches long, excluding the tail, which was three inches long. Buffon's specimen was two feet four inches in length, exclusive of the tail. He remarks that it had one molar tooth of a side fewer than the *European* Badger, and he compares the colour of its fur to that of the Canada lynx. The specimen of *Meles Labradoria*, in the Zoological Museum, certainly very much resembles the lynx in the colours of the upper aspect of its body.   The *Tlacoyotl,* seu *Coyotlhumuli* of Fernandez, seems to be very like the *Meles Labradoria.*   " Inveni in agro Tetzcocano animal pilosum valde, vocatum Tlalcoyotl, duas longum spithamas, unguibus melis, aut Quauhpecotli similibus, brevibus cruribus et nigro vestitis pilo, brevissima cauda, corpore toto ex albo vergente in fulvum, sed dorso, ac superna parte capitis, et colli nigris, lineaque distinctis candenti; caput est parvum, rostrum tenue, et longiusculum, canini exerti, ac vita victusque eadem quæ Quauhpecotli, i. e. vorax est, nullisque parcit oblatis escis, nec tamen editur ejus caro*."   The Ytzcuintecuani, and the Quauhpecotli, or Meles montanus or Texon, of the same author, are probably of nearly allied genera.   The latter has a long tail.

[13.]   1. GULO LUSCUS.   (Sabine.)   *The Wolverene.*

GENUS. Gulo. STORR. CUVIER.
Carcajou.   LA HONTAN, *Voyage,* vol. i. p. 81.   *An.* 1703.
Quickhatch or Wolverene.   ELLIS's *Voy. Hudson's Bay,* p. 42, *t.* iv.   *An.* 1748.   EDWARDS, *t.* 103.
Ursus luscus.   LINN. *Syst.,* p. 71.   LINN. GMELIN, vol. i. p. 103.
Ursus Freti Hudsonis.   BRISSON, *Quadr.,* p. 188.   *An.* 1756.
Wolverene.   PENNANT's *Hist. Quadr.,* vol. ii. p. 8, *t.* 8.   *Arctic Zool.,* vol. i. p. 66.   HEARNE's *Journey,* p. 372.
Wolverin, Quiquihatch, or Carcajou.   GRAHAM's *MSS.,* p. 13.
Gulo arcticus, var. A. Glouton wolverene.   DESMAREST, *Mamm.,* p. 174.
Gulo luscus.   SABINE (CAPT.) *Suppl. Parry's First Voy.* clxxxiv.   SABINE (Mr.) *Franklin's First Journey,* p. 650.
       RICHARDSON's *Append., Parry's Second Voy.,* p. 292.
Kablee-arioo, ESQUIMAUX.   Naghai-eh. CHEPEWYANS.
Ommeethatsees, okeecoohagew, and okeecoohawgees.   CREE INDIANS.
Carcajou.   FRENCH CANADIANS.   Quickehatch.   ENGLISH RESIDENTS AT HUDSON's BAY.

The *glutton* of the northern parts of Europe, the *rossomak* of the Russians, has attracted the attention of naturalists from the many extravagant stories which have been told of its extraordinary voracity, and of its method of procuring relief when over-distended with food.   Olaus Magnus, who, according to Buffon, is the earliest writer that mentions this animal, has collected the popular notions of its habits, though without giving full credit to them himself; and his account has

* FRANCISCI FERNANDEZ *(Phillipi Secundi Prim. Medico) Hist. Anim. Novæ Hisp.,* cap. xxxvii. p. 12.

been copied by subsequent authors, almost without alteration. " The glutton," says he, " (*Latinice gulo*) is the most noted of all the animals which inhabit the north of Sweden, for its insatiable appetite, whence it has obtained the appellation of *jerf*, in the language of that country, of *wilfras*, in German, and *rosomaka*, in Sclavonian." " It is wont, when it has found the carcase of some large beast, to eat until its belly is distended like a drum, when it rids itself of its load by squeezing its body betwixt two trees growing near together, and again returning to its repast, soon requires to have recourse to the same means of relief*." This trait in its character, however, is treated as a fiction by the traveller Gmelin, who, in journeying through Siberia, had an opportunity of acquiring a knowledge of its habits from the hunters. Buffon, on the authority of the reports of preceding authors, describes it as a ferocious animal, which approaches man without fear, and attacks the larger quadrupeds without hesitation ; but he states that its pace is so slow that it can take its prey only by surprise, to accomplish which it employs an extraordinary degree of cunning. He terms it the " quadruped-vulture," and repeats the statement of Isbrand, that it is accustomed to ascend a tree and lie in wait for the elks and rein-deer, dropping on their backs as they pass, and adhering so firmly by its claws, that all their efforts to dislodge it are in vain, and they speedily fall a prey to its voracity. It is even said to entice the rein-deer to come beneath the tree in which it lies concealed, by throwing down the moss which that animal is fond of, and that the arctic fox is its jackal or provider †. This character seems to be entirely fictitious, and to have partly originated in the name of " glutton " having been given occasionally to the lynxes and sloths. I have, however, thought proper to recapitulate it here, previous to stating that it is very dissimilar to the habits of the American Wolverene, which is by many able naturalists considered to be the same species.

The *Wolverene* is first noticed by La Hontan, who says that it is very like a badger, but is larger and fiercer. He calls it the *carcajou*, which is the appellation by which it is still known to the French Canadian voyagers. Subsequent writers, however, have occasionally, through mistake, given the same name to the American Badger, and also to several species of *felis*, whence doubts have been excited as to the animal actually meant by La Hontan. The European labourers in the service of the Hudson's Bay Company term it Quickehatch, which is evidently derived from its Cree or Algonquin name of *okee-coo-haw-gees*, and

* OLAUS MAGNUS, *Gent. Septen.*, p. 138.

† A similar account has been told of the foxes in Canada driving the moose deer to a spot where the karkajou, described as having a long tail, is posted.—*Voyage d'Amérique*, vol. i. p. 272. An. 1723.

without being disposed to rely strongly on etymological inquiries, I am inclined to refer the Carcajou, or, as it is sometimes pronounced Carcayou, of the *Coureurs des bois*, to the same source (*okee-coo-haw-gew*). Many other Knisteneaux or Cree terms have been adopted into the vernacular language of the Canadian voyagers.

The Wolverene is a carnivorous animal, which feeds chiefly upon the carcases of beasts that have been killed by accident. It has great strength, and annoys the natives by destroying their hoards of provision, and demolishing their marten traps. It is so suspicious, that it will rarely enter a trap itself, but beginning behind, pulls it to pieces, scatters the logs of which it is built, and then carries off the bait. It feeds also on meadow mice, marmots, and other *rodentia*, and occasionally on disabled quadrupeds of a larger size. I have seen one chasing an American hare, which was at the same time harassed by a snowy owl. It resembles the bear in its gait, and is not fleet; but it is very industrious, and no doubt feeds well, as it is generally fat. It is much abroad in the winter, and the track of its journey in a single night may be often traced for many miles. From the shortness of its legs, it makes its way through loose snow with difficulty, but when it falls upon the beaten track of a marten trapper, it will pursue it for a long way. Mr. Graham observes that " the Wolverenes are extremely mischievous, and do more damage to the small-fur trade than all the other rapacious animals conjointly. They will follow the marten hunter's path round a line of traps extending forty, fifty, or sixty miles, and render the whole unserviceable, merely to come at the baits, which are generally the head of a partridge or a bit of dried venison. They are not fond of the martens themselves, but never fail of tearing them in pieces or of burying them in the snow by the side of the path, at a considerable distance from the trap. Drifts of snow often conceal the repositories thus made of the martens from the hunter, in which case they furnish a regale to the hungry fox, whose sagacious nostril guides him unerringly to the spot. Two or three foxes are often seen following the Wolverene for this purpose *."

The Wolverene is said to be a great destroyer of beavers, but it must be only in the summer, when those industrious animals are at work on land, that it can surprise them. An attempt to break open their house in the winter, even supposing it possible for the claws of a Wolverene to penetrate the thick mud walls when frozen as hard as stone, would only have the effect of driving the beavers into the water to seek for shelter in their vaults on the borders of the dam. The

* Graham's *MSS.*, p. 13.

G 2

Wolverene, although it is reported to defend itself with boldness and success against the attack of other quadrupeds, flies from the face of man, and makes but a poor fight with a hunter, who requires no other arms than a stick to kill it.

It brings forth from two to four young once a year. The cubs are covered with a downy fur, of a pale or cream colour. It is found throughout the whole northern parts of the American continent, from the coast of Labrador and Davis' Straits to the shores of the Pacific and the islands of Alaska. It even visits the islands of the Polar sea, its bones having been found in Melville island, nearly in latitude 75°. It is not rare in Canada, but the extent of its range to the southward is not mentioned by American writers.

DESCRIPTION.

This animal has a broad compact head, which is suddenly rounded off on every side to form the nose. In the shape of its jaws it resembles a dog. Its ears are low, rounded, and much hid by the surrounding fur. The back is arched; the tail low and bushy; the legs thick and short, and the whole aspect of the animal indicates strength, without much activity. The *fur* bears a great similarity to that of the black bear, but is not so long, nor of so much value. It is in general of a dark brown colour, passing in the height of winter almost into black. A pale reddish-brown band, more or less distinct in different individuals, and sometimes fading into soiled brownish-white, commences behind the shoulder, and running along the flanks, turns up on the hip and unites with its fellow on the rump. The short tail is thickly covered with long black hair. There are some white markings on the throat and between the fore-legs, which are not constant in size or number. The legs are brownish-black. This animal places its feet on the ground much in the manner of a bear, and imprints a track on the snow or sand, which is often mistaken for that of the bear by Europeans on their first arrival in the fur countries. The Indians distinguish the tracks at the first glance by the length of the steps. The claws are strong and sharp.

DIMENSIONS.

				Feet.	Inches.
Length of head and body	.	.	.	2	6
„ tail (vertebræ)	.	.	.	0	7
„ tail with fur		.	.	0	10

[14.]   1. MUSTELA (PUTORIUS) VULGARIS.   *The Common Weasel.*

GENUS. Mustela. LINN. *Sub-genus.* Putorius. CUVIER.
Mustela vulgaris. LIN. GMELIN, i., p. 99.
Mustela nivalis. LIN. *Fauna Suec.*, ii., p. 7.
Common Weasel. PENNANT, *Arctic Zool.* i., p. 75.
Putorius vulgaris. CUVIER, *Règne An.*
Mustela vulgaris. HARLAN. *Faun.*, i., p. 61.
No. 49. MUSEUM ZOOL. SOCIETY.

It is stated in Arctic Zoology, that this species inhabits the Hudson's Bay countries, Newfoundland, and the United States, as far south as Carolina, becoming in cold districts white in winter, like the Ermine.   It is omitted in Godman's account of the animals of the United States; and the Prince of Musignano is of opinion that what has been considered as the common weasel in the United States, is merely the ermine in its summer dress.   Both species, however, are indubitably inhabitants of the American continent, the ermine extending to the most remote arctic districts, and the Weasel as far to the north, at least, as the Saskatchewan river.   Captain Bayfield presented the Zoological Society with specimens of the Common Weasel, killed on the borders of Lake Superior, which agree in all respects with the European species, and I obtained similar specimens at Carlton House.

DESCRIPTION.

The Weasel very much resembles the ermine; but it is a much smaller animal, has a flatter forehead, a narrower and longer nose, and a much shorter tail. Its fur, short and of inferior quality, has, in summer, a dull yellowish-brown colour, deepening into chestnut brown on the upper part of the head and nose, and at the tip of the tail into blackish-brown.   The under parts are yellowish-white, as are also the whole of the feet, and the interior of the legs and thighs.   The entire of the under jaw is pure white, and the white extends half along the upper lip, terminating opposite the anterior part of the orbit, or at the posterior row of whiskers.   The upper part of the cheek, between the white at the angle of the mouth, and the orbit, is included in the brown colour of the head.   The tail is of the same colour above and below.   The brown and white colours join by a straight well-defined line on the sides of the neck and belly, the latter colour occupying nearly one-third less of the circumference of the body than the brown.   The claws are smaller and more curved than those of the ermine, and the extremities are more slender, but longer in proportion to its size.

DIMENSIONS

Of an old female killed at Carlton House.

		Inches.	Lines.		Inches.	Lines.
Length of the head and body	.	9	0	Height of the ear	0	2
„ head	. .. .	1	8	From tip of the nose to the anterior point		
„ tail (vertebræ) .	.	2	0	of the orbit	0	6
„ „ including hair	.	2	10			

The Weasels of the fur countries become white in winter like the Ermine, and are not distinguished from them by the traders.

[15.]   2. MUSTELA (PUTORIUS) ERMINEA. (Lin. Gmel.)
The Ermine, or Stoat.

Mustela erminea. LIN. GMELIN., i. p. 98.
Stoat-weasel. PENNANT, *Arctic Zool.*, i., p. 75.
Mustela erminea. PARRY's *First Voy.*, Suppl. clxxxv. FRANKLIN's *First Journey*, p. 652. PARRY's *Second Voy.*, App. p. 294. LYON's *Private Journal*, pp. 82—107.
Seegoos and Shacooshew. CREE INDIANS. Terreeya. ESQUIMAUX.

This well-known and very handsome little animal is a common inhabitant of America, from its most northern limits to the middle districts of the United States ; and many specimens, both in the brown winter and white summer fur, were brought home by the late expeditions of discovery.   It is a bold animal, and often domesticates itself in the habitations of the fur traders, where it may be heard the livelong night pursuing the white-footed mouse (*Mus leucopus*).   Captain Lyon mentions his having seen an ermine hunt the footsteps of mice, like a hound after a fox, and he also describes their mode of burrowing in the snow. " I now observed," says he, " a curious kind of burrow, made by the ermines, which was pushed up in the same manner as the tracks of moles through the earth in England.   These passages run in a serpentine direction, and near the hole or dwelling-place the circles are multiplied, as if to render the approach more intricate."   The same lively writer relates the manners of a captive ermine as follows :—" He was a fierce little fellow, and the instant he obtained day-light in his new dwelling, he flew at the bars, and shook them with the greatest fury, uttering a very shrill passionate cry, and emitting the strong musky smell which I formerly noticed. No threats or teasing could induce him to retire to the sleeping place, and whenever he did so of his own accord, the slightest rubbing on the bars was sufficient to bring him out to the attack of his tormentors. He soon took food from the hand, but not until he had first used every exertion to reach and bite the fingers which conveyed it. This boldness gave me great hopes of being able to keep my little captive alive through the winter, but he was killed by an accident."

According to Indian report, the ermine brings forth ten or twelve young at a time. In the time of Charlevoix, ermine-skins formed part of the *menues pelletries* exported from Canada; but their value at present is so little, that they do not repay the Hudson's Bay Company the expense of collecting; hence very few are brought to England from that quarter.

DESCRIPTION.

The ermine has a convex nose and forehead, a long slender body, and long cylindrical tail, with short and rather stout limbs. Its ears are low and rounded, and go more than half round the auditory opening. They are proportionably higher than the ears of the common weasel. In the winter time the fur in some specimens is of a pure white colour throughout, except on the end of the tail, which, together with a few of the anterior whiskers, are black. In other specimens there is a bright primrose-yellow tinge on the belly, the posterior part of the back, or the tail. The feet in the winter are clothed with hair on the soles, which projects so as to conceal the claws. In the summer the soles are nearly naked, and the fur on the upper parts resembles that of the common weasel in colour.

DIMENSIONS
Of a specimen killed at Fort Franklin, Great Bear Lake.

	Inches.	Lines.		Inches.	Lines.
Length of head and body	11	0	Length of head	2	3
,, tail (vertebræ)	4	0	Height of ear	0	3¼
,, ,, including fur	5	0	Distance between orbit and end of nose	0	6

In the neighbourhood of Carlton House there is a variety of a larger size, having a longer tail and longer fore-claws.

DIMENSIONS.

	Inches.	Lines.
Length of head and body	12	0
,, tail (vertebræ)	5	4
,, ,, including fur	6	6

[16.]   3. MUSTELA (PUTORIUS) VISON.   (Lin. Gmel.)
                   The Vison-Weasel.

> Otay. SAGARD THEODAT, *Hist. du Can.* p. 749. An. 1636.
> Foutereau. LA HONTAN, *Voyage*, i. p. 81. An. 1703.
> Mink. KALM, *Journ.*
> Le vison. BUFFON, xiii. p. 308, t. 43.
> Mustela vison. LIN. GMEL, i. p. 94.
> Minx. LAWSON, *Carol.*, p. 121.
> Mustela lutreola? FORSTER, *Phil. Trans.*, lxii. p. 371.
> Minx Otter. PENNANT, *Arctic Zool.*, i. p. 87.
> Vison-Weasel. IBID. i. p. 78.
> Jackash. HEARNE, *Journey*, p. 376. GRAHAM, *MSS.*, p. 6.
> Mustela vison. CUVIER, *Règne An.*, i. p. 150, t. 1. fig. 2.
> Mustela lutreola. SABINE, *Franklin's Journ.*, p. 652.
> Mustela vison et M. lutreocephala. HARLAN, *Fauna*, pp. 63, 65.
> The Mink. GODMAN, *Nat. Hist.*, i. p. 206.
> Vison Weasel. BRITISH MUSEUM.
> Shakwæshew or Atjackashew. CREE INDIANS.
> Mink. HUDSON'S BAY TRADERS. Foutereau. CANADIAN VOYAGERS.

This animal is very similar to the *mustela lutreola* of the north of Europe, in form; and the name of *mænk*, applied to the latter, is supposed by Pennant, with great probability, to have been transferred to the former by some Swedish colonists.   La Hontan mentions a sort of small amphibious Weasels, under the name of Foutereaux, which is the appellation of the *minks* to this day amongst the French Canadian voyagers.   Buffon described a specimen from Canada, preserved in the museum of M. Aubry, under the name of Vison, and gives a correct figure, except that the form of the tail of the specimen had been spoiled in mounting.   Pennant admits the Vison into his list of species, having had merely an imperfect view of M. Aubry's specimen through its glass-case, and not recognising it to be the same with his minx or lesser otter, which he considers as identical with the *Mustela lutreola*.   Forster, who received a Vison from Hudson's Bay, under the name of mink, expresses a doubt of its being the latter species; and Baron Cuvier has placed the European mink in his sub-genus Putorius, whilst he ranges the Vison amongst the true martens.   The Hudson's Bay Vison has the teeth of the polecats.   I have not been able to trace the origin of the term Vison; but a list of the furs exported to France, presented by a Montreal merchant to Kalm in 1749, informs us, that "the visons, or foutereaux, are a kind of martins that live in the water."   There is no animal of the genus *mustela* inhabiting the northern parts of America, which can be said to live in the water but the Vison ; the *fisher*, notwithstanding its name, being as much a land animal as the pine-martin.

The Vison passes much of its time in the water, and when pursued seeks shelter in that element in preference to endeavouring to escape by land, on which it travels slowly. It swims and dives well, and can remain a considerable time under water. Its short fur, forming a smooth glossy coat, its tail exactly like that of an otter, and the shortness of its legs, denote its aquatic habits. It preys upon small fish, fish-spawn, fresh-water mussels, &c., in the summer ; but in the winter, when its watery haunts are frozen over, it will hunt mice on land, or travel to a considerable distance through the snow in search of a rapid or fall, where there is still some open water. Under the article *Mustela Canadensis,* the mistakes which have arisen from the habits of this animal having been attributed to the Pekan or Fisher, are pointed out. The Vison, when irritated, exhales, next to the Skunk, the most fetid smell of any animal in the fur countries. The odour resides in a fluid secreted by two glands situated at the anus. It is not very timid when in the water, and will approach near to a canoe out of curiosity, diving however instantly on perceiving the flash of a gun, or any movement from whence it apprehends danger* It is easily tamed, and is capable of strong attachment. In a domestic state it is observed to sleep much in the day, and to be fond of warmth. One, which I saw in the possession of a Canadian woman, passed the day in her pocket, looking out occasionally when its attention was roused by any unusual noise. Like a cat, a tame Vison is easily offended, and will, on a sudden provocation, bite those who are most kind to it. It is fond of being caressed. The fur of the Vison is of little value, and at many of the remote parts their skins are taken by the traders from the Indians merely to accommodate the latter, but are afterwards burnt, as they will not repay the expense of carriage. The fur, however, is very fine, although short, and is likely, in the revolutions of fashion, to become valuable again.

We saw the Vison on the banks of Mackenzie's River as far north as latitude 66°, and there is every reason to believe that it ranges to the mouth of that river, in latitude 69°. It is a common animal throughout the whole breadth of the continent of America, and we are told by Pennant that it exists as far south as Carolina. It has from four to seven young at a time.

* It resembles a musk-rat in its mode of swimming, and is shot in the water in like manner, by the hunters, as La Hontan has remarked.

H

DESCRIPTION.

Dental formula, incisors $\frac{6}{6}$, canines $\frac{1-1}{1-1}$, grinders $\frac{4-4}{5-5}$ = 34.

The Vison has an anterior molar less in both jaws than the American pine-martin; but the teeth of the two species differ in shape merely in the antipenultimate or carnivorous tooth of the lower jaw having only a slightly salient angle, in place of the interior very minute point, which exists on the lower carnivorous tooth of the martin. *Size.*—Less than the pine-martin, but, from the greater length of its neck, it measures nearly as much from the nose to the tail. *Shape.*—The head is depressed and small; the nose short, flat, and thick; the eyes small, and far forward; the ears low, nearly semicircular, and covered with short fur. The neck is long, and the body is long and slender, and has much flexibility. The legs are short, and the toes are connected by short hairy webs that are completely concealed by the fur, which is as long on the feet both above and below, as on the legs. The claws are nearly straight, sharp, and white, and scarcely project beyond the fur. The tail is round and thick at the root, from whence it tapers gradually to the tip, exactly resembling the tail of an otter in form. In the prepared specimens, the part of the tail next the body is usually too slender, whilst towards the tip it is over-stuffed, causing the hairs to stand out, and giving it a bushy appearance contrary to nature.

The *fur* is short on the head, and is longest on the posterior part of the body and tail. It is of two sorts—a very dense down, and longer and stronger hairs. The tips of the latter form a smooth shining coat both on the body and tail, which completely conceals the down. The *colour* of the down is intermediate betwixt brown and gray, being nearly that which Werner denominates brocoli-brown. The colour of the surface of the fur is chocolate or umber-brown; a little paler on the head and belly, but deepening on the tail and posterior part of the back into blackish-brown. The lower jaw is white, with a narrow brown mark at the apex; and there are occasionally some white markings on the throat, but they are not constant either in number or size. The whiskers are of the same colour with the fur, and are shorter than the head, but remarkably strong. There are two brown-coloured glands situated in the hollow between the tuberosities of the ischium and the tail, which have each a small cavity capable of containing a garden pea, and lined by a white, wrinkled membrane. The fluid they secrete is very fetid.

DIMENSIONS.

	Inches.	Lines.		Inches.	Lines.
Length of head and body	17	0	Distance from centre of orbit to end of nose	1	0
„ tail, including fur	8	6	„ end of nose to auditory opening	2	5
„ head	3	3	Breadth between the ears	2	0

[17.]   4. Mustela Martes. (Linn.)   *The Pine-Martin.*

Genus. Mustela. Linn. *Sub-genus.* Mustela. Cuvier.
Mustela martes. Linn. Gmelin, vol. i. p. 95.
Pine-martin. Pennant's *Arctic Zool.*, vol. i. p. 77.
Mustela martes. Sabine, *Franklin's Journ.*, p. 651. Harlan's *Fauna*, p. 67.
Pine-Martin. Godman's *Nat. Hist.*, vol. i. p. 200.
Wawpeestan. Cree Indians. Wappanow. Monzonies.
Wawbeechins. Algonquins. Sable. American Fur-dealers.
Martin. Hudson's Bay Company's Lists.

The Pine-martin inhabits the woody districts in the northern parts of America, from the Atlantic to the Pacific, in great numbers, and has been observed to be particularly abundant where the trees have been killed by fire, but are still standing. It is very rare, as Hearne has remarked, in the district lying north of Churchill river, and east of Great Slave Lake, known by the name of Chepewyan or Barren Lands. A similar district, on the Asiatic side of Behring's Straits, twenty-five degrees of longitude in breadth, and inhabited by the Tchutski, is described by Pennant as equally unfrequented by the Martin, and for the same reason,—the want of trees. The limit of its northern range in America is like that of the woods, about the sixty-eighth degree of latitude, and it is said to be found as far south as New-England. Particular races of Martins, distinguished by the fineness and dark colours of their fur, appear to inhabit certain rocky districts. The rocky and mountainous but woody district of the Nipigon, on the north side of Lake Superior, has long been noted for its black and valuable Martin-skins.

The Martin preys on mice, hares, and partridges, and in summer, on small birds' eggs, &c. A partridge's head, with the feathers, is the best bait for the log traps in which this animal is taken. It does not reject carrion, and often destroys the hoards of meat and fish laid up by the natives, when they have accidentally left a crevice by which it can enter. The Martin, when its retreat is cut off, shows its teeth, sets up its hair, arches its back, and makes a hissing noise like a cat. It will seize a dog by the nose, and bite so hard, that unless the latter is accustomed to the combat, it suffers the little animal to escape. It may be easily tamed, and it soon acquires an attachment for its master, but it never becomes docile. Its flesh is occasionally eaten, though it is not prized by the Indians. The females are smaller than the males. They burrow in the ground, carry their young about six weeks, and bring forth from four to seven in a litter about the latter end of April. Mr. Graham says that this animal is sometimes troubled with epilepsy.

The fur of the Martin is fine, and it is used for trimmings, and also dyed so as to imitate sables and other expensive furs. Hence it has always been an important article of commerce. Upwards of one hundred thousand skins have long been collected annually in the fur countries.

DESCRIPTION.

The form of the Martin is well known. It has a pleasing aspect. Its fur is about an inch and a quarter long, of a pale, dull, grayish-brown, or hair-brown colour, from the roots upwards, dull yellowish-brown near the summit, and tipped with dark brown or black. The lustre of the surface of the fur is considerable. The hair of the tail is longer, coarser, and darker, than that of the body. At the tip of the tail its length is three inches, and it has a blackish colour. The yellowish-white markings on the throat vary in different individuals. The darkest skins are most prized. The fur is in the highest order in the winter time; in the beginning of summer, the dark tips of the hairs drop off, and the general colour of the fur is a pale orange-brown, with little lustre. The tips of the ears, at all times lighter than the rest of the fur, become very pale in the summer time. The natives remark that the fur of the Martin loses all its lustre, and consequently much of its value, upon the falling of the first shower of rain for the season. Length of the head and body from eighteen to twenty inches.

[18.]   5. MUSTELA CANADENSIS.   (Lin.)   *The Pekan*, or *Fisher*.

Le Pekan. BUFFON, vol. xiii. p. 304, t. xlii. Opt.
Mustela Canadensis. LINN. GMELIN, vol. i. p. 95.
Fisher. PENNANT's *Arct. Zool.*, vol i. p. 82. *Hist. Quadr.*, vol. ii. p. 238.
Mustela Pennanti. ERXLEBEIN, *Syst.*, p. 470.
Wejack. HEARNE's *Journ.* GRAHAM's *MSS.*
Fisher or Black-fox. LEWIS and CLARK, vol. iii. p. 25.
Fisher-weasel, or Pekan. WARDEN's *United States.*
Mustela Pennanti. SABINE, *Franklin's First Journ.*, p. 651.
Mustela Canadensis. HARLAN's *Fauna*, p. 65.
Pennant's Martin. GODMAN's *Nat. Hist.*, vol. i. p. 203.
Otchœk. CREE INDIANS. Woodshock. HUDSON's BAY COMPANY's SALE LISTS.
Wejack, or Fisher. FUR TRADERS. Pekan. CANADIAN VOYAGERS.

This animal has been described by authors under many appellations. A considerable number of its skins are annually imported into England by the Hudson's Bay Company, and exposed in their sales under the names of Woodshocks or

Fishers. The latter appellation, whatever its origin may have been, has led to much confusion in the history of the species, and has caused the habits of the *mustela vison* to be ascribed to it. Mr. Bartram, as quoted by Pennant, is the first written authority I can find for the name, and he distinctly says, " though they are not amphibious, and live on all kinds of lesser quadrupeds, they are called Fishers." Wejack, the appellation under which Hearne mentions it, is a corruption of its Cree or Knisteneaux name, *otchœk*, and the word Woodshock has a similar origin. It is universally termed *Pekan* by the Canadian fur-hunters, which may be considered as evidence of its being the animal described by Buffon. Pennant had only an imperfect view, through a glass case, of the Pekan, in a museum at Paris *, and does not appear to have recognised it in his *fisher*. Under the article *Pekan*, in Arctic Zoology, it is said that a skin of that species was sent from Hudson's Bay, by Mr. Graham, labelled with the name of *Jackash*. This name is given by the traders solely to the *mustela vison*, and I suspect that, through some accident, the label intended for a specimen of the latter animal had been affixed to a skin of the common Pine-martin. Hence the formation of a nominal species, by Pennant, and much of the confusion that has ensued. Large individuals of the common Pine-martin, in their summer dress, have a considerable resemblance to the Fisher, and might easily have been mistaken by Pennant for the animal he had imperfectly examined at Paris; and having once named it the Pekan, it followed that a true skin of the *Fisher*, also received from Mr. Graham, was described as a distinct species. Pennant actually says that his Pekan agrees in dimensions and white marks with the European martin.

The Pekan is a larger and stronger animal than any variety of the Pine-martin, but it has similar manners; climbing trees with facility, and preying principally on mice. It lives in the woods, preferring damp places in the vicinity of water, in which respect it differs from the martin, which is generally found in the driest spots of the pine forests. The Fisher is said to prey much on frogs in the summer season; but I have been informed that its favourite food is the Canada porcupine, which it kills by biting on the belly. It does not seek its food in the water, although, like the Pine-martin, it will feed on the hoards of frozen fish laid up by the residents.

It inhabits a wide extent of country, from Pennsylvania to Great Slave Lake, being thirty degrees of latitude, and I believe its range extends completely across the continent. It is found on the shores of the Pacific. It brings forth, once a year, from two to four young.

* *Arctic Zoology*, vol. i. p. 78.

DESCRIPTION.

The physiognomy of the Pekan is very different from that of the Martin. When the latter is threatened, its features resemble those of an enraged cat, but the expression of the Pekan's countenance approaches to that of a dog, although the apparent obliquity of its eyes give it a sinister look. The *head* has a strong, roundish, compact appearance, and contracts suddenly to form the nose, which terminates rather acutely. The ears, low and semicircular, are far apart, so as to leave a broad and slightly rounded forehead : they are smaller in proportion than the ears of the Pine-martin. The eyes, situated where the head curves in to form the nose, appear more oblique than they really are.

The *fur*, towards the roots, is fine and downy, and of a grayish or clove-brown colour ; yellowish-white upwards, and blackish-brown at the tips, with considerable lustre. It is short on the head, but on the body, particularly on the posterior parts, it is as long, though less fine than the fur of the Pine-martin. On the head, shoulders, and fore-part of the back, so much of the white is seen that they are quite hoary, but towards the tail the colour deepens into blackish-brown. The throat, belly, and legs are brownish-black ; the colour is lighter on the sides. There is a white spot very frequently between the fore-legs or on the throat, and another between the hind-legs, but these marks are not constant. The tail is clothed with long black fur. The chin and nose are tipped with brown. The ears, which are covered with short hairs, are pale anteriorly, dark brown behind, and have whitish margins. The *fore-legs* are short and strong. The toes on the fore and hind feet are connected at the base by a short web, which is covered on both sides with hair. The claws are strong, curved and sharp.

This animal is nearly twice the ordinary size of the Pine-martin, and has a longer tail ; and its fur is harsher, and much less valuable. Its body has the musky odour of the martin, but rather stronger. Some thousands of Pekans are annually killed in the Hudson's Bay countries, but they are less abundant than the pine-martins.

DIMENSIONS.

	Inches.		Inches.
Length of head and body . . .	23	Breadth from the tip of one ear to the tip of the other	7
,, head measured with a string from the		Height of ear	1
nose over the forehead to the nape of the neck	6½	Length of fore-leg and foot . . .	6
,, tail, including fur . . .	16	,, hind-leg, foot, and thigh . .	11

β. MUSTELA CANADENSIS, *varietas alba. White Pekan.*

This variety has the nose and feet brown ; the rest of the fur is white. Its dimensions are the same with those of the common variety.

There is a specimen in the Hudson's Bay Museum.

[19.] 1. MEPHITIS AMERICANA, *var.* Hudsonica.
The Hudson's Bay Skunk.

GENUS. Mephitis. CUVIER.
Skunk Weesel. PENNANT's *Arctic Zool.*, i., p. 85. HEARNE, *Journey*, p. 377.
Mephitis Americana. SABINE. *Franklin's Journ.*, p. 653. ZOOLOGICAL MUS. No. 68, 69.
Seecawk. CREE INDIANS.

This animal is prettily ornamented by a full bushy tail, and broad lateral white stripes, which contrast pleasingly with the white colours of the rest of the body. Its fur, although long, is coarse, and is but little valued in commerce. The Skunk is not an uncommon animal in the district it inhabits, which does not, I believe, extend to the north of latitude 56° or 57°. It exists in the rocky and woody parts of the country, but is still more frequent in the clumps of wood which skirt the sandy plains of the Saskatchewan. I have not been able to ascertain the southern range of this variety of Skunk; and, judging from Kalm's description, there appears to be a different one in Canada. The Skunk passes its winter in a hole, seldom stirring abroad, and then only for a short distance. It preys on mice, and in summer has been observed to feed much on frogs. It has a slow gait, and can be overtaken without difficulty, for it makes but a poor attempt to escape, putting its trust apparently in its power of discomfiting its pursuers by the discharge of a noisome fluid. This fluid, which is of a deep yellow colour, and is contained in a small bag placed at the root of the tail, emits one of the most powerful stenches in nature; and so durable, that the spot where a Skunk has been killed will retain the taint for many days. Mr. Graham says, that he knew several Indians who lost their eye-sight in consequence of inflammation, produced by this fluid having been thrown into them by the animal, which has the power of ejecting it to the distance of upwards of four feet. I have known a dead Skunk, thrown over the stockades of a trading post, produce instant nausea in several women in a house with closed doors upwards of a hundred yards distant. The odour has some resemblance to that of garlic, although much more disagreeable. One may, however, soon become familiarised with it: for, notwithstanding the disgust it produces at first, I have managed to skin a couple of recent specimens by recurring to the task at intervals. When care is taken not to soil the carcase with any of the strong-smelling fluid, the meat is considered by the natives to be excellent food.   It breeds once a year, and has from six to ten young at a time.   A considerable number of animals of the genus *Mephitis*, natives of America, resembling each other strongly in form and size, but differing in the number and variety of their stripes and markings, have been described by authors as so many distinct species.   Baron Cuvier thinks that the present state of our knowledge of these animals does not warrant us in considering them otherwise than as varieties of a single species, and of these varieties he enumerates fifteen*.   I have now seen a considerable number of specimens, killed to the north of the Great Lakes, none of which presented any important deviation in their markings from the one principally referred to in the description which follows; and M. Desmarest remarks, that " the varieties (if they are to be considered as such, and not as species) are, for the most part, sufficiently uniform in the same district of country in the disposition of their stripes."   The Hudson Bay variety, however, comes nearest to the description of the *Chinche* of Buffon ; the *Viverra Mephitis* of Gmelin, which is said to be an inhabitant of Chili.   The *Fiskatta*, or *Skunk*, of Kalm, which inhabits Canada, has a white dorsal line in addition to two lateral ones†.

DESCRIPTION.

The Skunk is low on its legs, with a broad fleshy body, wide forehead, and the general aspect rather of a wolverene than of a martin ;—*eyes* small ; *ears*, short and round.   A narrow white mœsial line runs from the tip of the nose to the occiput, where it dilates into a broad white mark.   It is again narrowed, and continues so until it passes the shoulders, when it forks, the branches running along the sides, and becoming much broader as they recede from each other.   They approach posteriorly, and unite on the rump, becoming at the same time narrower.   In some few specimens the white stripes do not unite behind, but disappear on the flanks.   The black dorsal space included by the stripes is egg-shaped, the narrow end of which is towards the shoulders.   The sides of the head and all the under parts are black.   The hair on the body is long.   The tail is covered with very long hair, and has generally two broad longitudinal white stripes above on a black ground.   Sometimes the black and white colours of the tail are irregularly mixed.   Its under surface is black.   The claws on the forefeet are very strong and long, being fitted for digging, and very unlike those of the martins.

---

* *Ossemens fossiles.*

† The earliest account of the Canada Skunk that I have met with, is by Sagard :—" Les enfans du diable," dit il, " que les Hurons appelle *Scangaresse*, et le commun de Montagnais, Babougi Manitou, ou Ouinesque, est une beste fort puante de la grandeur d'un chat ou d'une jeune renard, mais elle a la teste un peu moins aiguë, et la peau couverte d'un gros poil rude et enfumé, et sa grosse queuë retroussée de mesme, elle se cache en hyver sous la neige, et ne sort point qu'au commencement de la lune du mois de Mars laquelle les Montagnais nomment Ouiniscon pismi, qui signifie la Lune de la Ouinesque.   Cet animal, outre qu'il est de fort mauvaise odeur, est tres malicieux, et d'un laid regard."— F. G. Sagard Theodat, *Hist. du Canada*, p. 748.

[20.] 1. LUTRA CANADENSIS. (Sabine.) *The Canada Otter.*

GENUS. Lutra. RAY. CUVIER.
Loutre de Canada. BUFFON, vol. xiii. p. 326. t. 44.
Common Otter. PENNANT, *Arct. Zool.*, vol. i. p. 86.
Land Otter. WARDEN, *United States*, vol. i. p. 206.
Lutra Canadensis. SABINE, *Franklin's Journ.*, p. 653.
Lutra Brasiliensis. HARLAN, *Fauna*, p. 72.
The American Otter. GODMAN, *Nat. Hist.*, vol. i. p. 222.
Neekeek. CREE INDIANS. Capucca. INHABITANTS OF NOOTKA.

Buffon describes an Otter from Canada as differing from the European species merely in its greater size, and the colour of its fur. Ray had previously enumerated the *Saricovienne* of La Plata, or the *Carigueibeju* (Sarigoviou) of Brasil, as a species of his genus *lutra*, distinct from *Lutra vulgaris*. Pennant, in his History of Quadrupeds, following Linnæus and Brisson, refers the Brasilian Otter to the Sea Otter, of the following article; but, in his Arctic Zoology, he describes the Brasilian as a peculiar species confined to the warm parts of America; whilst he considers the Otter of the northern rivers as identified with the Common Otter of Europe. Baron Cuvier again unites the Canada and Brasil Otters under the name of *l'outre d'Amérique;* but the character ascribed by Margrave to the *lutra Brasiliensis,* of its tail and feet being of the same length, will not by any means apply to the Canada Otter, and I have therefore followed Mr. Sabine, in considering the subject of this article to be a species peculiar to the northern districts of America. M. Frederick Cuvier not only separates the Otter of Canada from that of South America, but also describes a distinct species inhabiting an intermediate district (*Lutra lataxina.*)*

The Canada Otter resembles the European species in its habits and food. In the winter season, it frequents rapids and falls, to have the advantage of open water; and when its usual haunts are frozen over, it will travel to a great distance through the snow, in search of a rapid that has resisted the severity of the weather. If seen, and pursued by hunters on these journies, it will throw itself forward on its belly, and slide through the snow for several yards, leaving a deep furrow behind it. This movement is repeated with so much rapidity, that even a swift

* *Dict. des Sciences Nat.*, xxvii.

I

runner on snow-shoes has much trouble in overtaking it. It also doubles on its track with much cunning, and dives under the snow to elude its pursuers. When closely pressed, it will turn and defend itself with great obstinacy. In the spring of 1826, at Great Bear Lake, the Otters frequently robbed our nets, which were set under the ice, at the distance of a few yards from a piece of open water. They generally carried off the heads of the fish, and left the bodies sticking in the net.

The Canada Otter has one litter annually about the middle of April, of from one to three young. It inhabits the Mackenzie and other rivers nearly to the Arctic sea; and there appears to be no difference betwixt the skins obtained on the shores of the Pacific, or in the neighbourhood of Hudson's Bay. Seven or eight thousand are imported annually into England.

DESCRIPTION.

The Canada Otter may be distinguished from the European species, by the fur of its belly being of the same shining brown colour with that of the back. It is a much larger animal, and has, in proportion, a shorter tail than the European one. Its fur very much resembles that of the beaver, having the same general colours, and, like it, consisting of a very fine waved and shining down, intermixed with longer and coarser hairs. Hearne remarks, that the colour and quality of its fur varies much with the season. In summer, when the hair is very short, it is almost black; but, as the winter advances, it turns to a beautiful reddish-brown, except a spot under the chin, which is gray. Otter-fur is nearly of the same fineness with beaver-wool, but being shorter, and not so well adapted for making felt, its price fluctuates more with the fashion.

The length of the Canada Otter is about five feet, including the tail, which measures eighteen inches.

[21.] 2. LUTRA (ENHYDRA) MARINA. (Erxlebein.) *The Sea Otter.*

GENUS. Lutra. RAY. *Sub-genus.* Enhydra. FLEMING.
Sea-Beaver. KRASCHENINIKOFF, *Hist. Kamsk.* (GRIEVE's *Trans.*), p. 131. *An.* 1764.
Lutra Marina. STELLER, *Nov. Com. Petrop.*, vol. xi. p. 367, t. xvi. ERXLEBEIN, *Syst. An.* 1777.
Sea-Otter. COOK's *Third Voy.*, vol. ii. p. 295. *An.* 1784. PENNANT's *Arctic Zool.*, vol. i. p. 88. *An.* 1784.
　　MEARES' *Voy.*, pp. 241–260. *An.* 1790. MENZIES, *Phil. Trans.*, p. 385. *An.* 1796.
Enhydra Marina. FLEMING's *Phil. Zool.*, vol. ii. p. 187. *An.* 1822.
Lutra Marina. HARLAN. *Fauna*, p. 73.
The Sea Otter. GODMAN's *Nat. Hist.*, vol. i. p. 228.
Kalan. KAMSKATDALES.

The Sea Otter inhabits the northern parts of the Pacific, from Kamskatcha to the Yellow Sea on the Asiatic side, and from Alaska to California on the American coast.　It seems to have more the manners of a seal than of the Land Otter.　It frequents rocks washed by the sea, and brings forth on land, but resides mostly in the water, and is occasionally seen very remote from the shore, sometimes, according to Pennant, more than a hundred leagues.　The fur of the Sea Otter being very handsome, was much esteemed by the Chinese, and, until the market at Canton was overstocked, prime skins brought extraordinary high prices.　The trade for a considerable period was in the hands of the Russians, who soon after the discovery of the north-west coast of America, by Beering and Tschirikow, sent mercantile expeditions thither.　Captain Cook's third voyage drew the attention of English speculators to that quarter, and vessels were freighted both by private adventurers and by the India Company, for the purpose of collecting furs on the American coast and conveying them to Canton.　Pennant, alluding to this traffic, says, " what a profitable trade (with China) might not a colony carry on, were it possible to penetrate to that part of America by means of rivers and lakes."　The event that Pennant wished for soon took place.　Sir Alexander Mackenzie having traversed the continent of America, and gained the coast of the Pacific, his partners in trade followed up his success, by establishing fur posts in New Caledonia, and a direct commerce with China; but the influx of furs into that market soon reduced their price.

DESCRIPTION.
[Extracted from MEARES' *Voyage.*]
The Sea Otter is furnished with a formidable set of teeth; its fore-paws are like those of the River Otter, but of much larger size, and greater strength; its hind-feet are skirted with a membrane, on which, as on the fore-feet, there grows a thick and coarse hair.　The fur varies

I 2

in beauty according to the age of the animal. The young cubs, of a few months old, are covered with a long, coarse, white hair, which protects the fine down that lies beneath it. The natives often pluck off this coarse hair, when the lower fur appears like velvet, of a beautiful brown colour. As they increase in size, the long hair falls off, and the fur becomes blackish, but still remains short. When the animal is full grown, it becomes of a jet black, and increases in beauty; the fur then thickens, and is thinly sprinkled with white hairs. When they are past the age of perfection, and verge towards old age, their skin changes into a dark brown, dingy colour, and of course diminishes in value. The skins of those killed in the winter are of a more beautiful black, and in every respect more perfect than those which are taken in the summer and autumn. The male Otter is beyond all comparison more beautiful than the female, and is distinguished by the superior jetty colour, as well as velvety appearance of his skin; whereas the head, throat and belly of the female, are not only covered with fur that is white, but which is also of a very coarse texture. The skins in the highest estimation are those which have the belly and throat plentifully interspersed with a kind of brilliant silver hairs, while the body is covered with a thick black fur, of extreme fineness, and a silky gloss*.

[22.]   1. CANIS LUPUS, OCCIDENTALIS.   *The American Wolf.*

GENUS. Canis. LINN.
Missouri Wolf. LEWIS and CLARK, vol. i. p. 283.
Canis Lupus. SABINE, *Franklin's Journ.*, p. 654. SABINE (CAPT.), *Parry's Voy., Suppl.*, clxxxv.
        RICHARDSON, *Parry's Second Voy., App.*, p. 295.
Wolf. LYON'S *Private Journal*, pp. 151, 339, &c.

The Common Wolves of the Old and New World have been generally supposed to be the same species—the *Canis lupus* of Linnæus. The American naturalists have, indeed, described some of the northern kinds of Wolf as distinct; but it never seems to have been doubted that a Wolf, possessing all the characters of the European Wolf, exists within the limits of the United States.

* Not having been on the coasts where the Sea Otter is produced, I can add nothing to its history from my own observation, and I have preferred taking the description of the fur from one who was engaged in the trade, to extracting a scientific account of the animal from systematic works, which are in the hands of every naturalist. In the narrative of Captain Cook's voyage, it is mentioned that a young Sea Otter brought on board had six lower incisors. Steller, and all succeeding systematic writers, describe it as having six incisors above and four below. Probably two of the lower ones drop out before the animal becomes adult.

The Wolf to which these characters have been ascribed, seems to be the "large brown Wolf" of Lewis and Clark, and, according to them, inhabits not only the Atlantic countries, but also the borders of the Pacific and the mountains which approach the Columbia river, between the Great Falls and rapids, but is not found on the Missouri to the westward of the Platte. I have seen none of these *Brown Wolves ;* but if their resemblance is so close to the European Wolf as Major Smith * states it to be, I have no hesitation in saying that they differ decidedly from the Wolf which inhabits the countries north of Canada. While attached to the late expeditions, I passed through thirty degrees of latitude and upwards of fifty of longitude on the American continent, and in the course of seven years travelled upwards of twenty thousand miles, during the whole of which time I had almost daily opportunities of observing the form and manners of the Wolves, but I saw none which had the gaunt appearance, the comparatively long jaw and tapering nose, the high ears, long legs, slender loins, and narrow feet of the Pyrenean Wolf.

In some of the districts which we traversed, the Wolves were very numerous, and varied greatly in the colour of their fur, some being white, others totally black, but the greater number were mixed gray and white, more or less tinged in parts with brown. These variations of colour, however, not being attended with any differences of form, nor peculiarity of habits, I deem them to be no more characteristic of proper species or even permanent varieties than colour would be in the domestic dog. All the northern Wolves, whatever their colours are, have certain characters in common wherein they differ from the European race ; and the Indian report of the extreme variations of colour being occasionally observed in Wolves of the same litter, strengthens my opinions.

DESCRIPTION.

The American Wolf of the northern districts is covered with long and comparatively fine fur, mixed with a large quantity of shorter woolly hair, and it has a more robust form than the European Wolf. Its muzzle is thicker and more obtuse, its head larger and rounder, and there is a sensible depression at the union of the nose and forehead. Its more arched forehead is comparatively broad, the space between the ears being greater than their height. The ears are shorter, wider at the base, and more acute, and have, consequently, a more conical form, whilst the greater length of the hair on the side of the neck of this Wolf makes them appear even shorter than they are. Its neck, covered with a bushy fur, appears short and thick. Its legs are rather short, its feet broad, with thick toes, and its tail is bushy, like the brush of a fox.

The European Wolf, on the contrary, has a coarser fur, with less of the soft wool inter-mixed with it. Its head is narrower, and tapers gradually to form the nose, which is pro-

* GRIFFITH's *Animal Kingdom*, vol. ii. p. 348.

duced on the same plane with the forehead.   Its ears are higher and somewhat nearer to each other ; their length exceeds the distance between the auditory opening and the eye.   Its loins are more slender, its legs longer, feet narrower, and its tail is more thinly clothed with fur.

The shorter ears, broader forehead, and thicker muzzle of the American Wolf, with the bushiness of the hair behind the cheek, give it a physiognomy more like the social visage of an Esquimaux dog than the sneaking aspect of an European Wolf.   Buffon enumerates black, tawny-gray and white, as the colours exhibited by the fur of the European Wolves.   In the American northern Wolves the gray colour predominates, and there is very little of the tawny hue.   The general arrangement of the patches of colour is, however, nearly the same in both races.

Notwithstanding the above enumeration of the peculiarities of the American Wolf, I do not mean to assert that the differences existing between it and its European congener are sufficiently permanent to constitute them, in the eye of the naturalist, distinct species.   The same kind of differences may be traced between the foxes and native races of the domestic dog of the new world and those of the old; the former possessing finer, denser, and longer fur, and broader feet, well calculated for running on the snow.   These remarks have been elicited by a comparison of live specimens of American and Pyrenean Wolves; but I have not had an opportunity of ascertaining whether the Lapland and Siberian Wolves, inhabiting a similar climate with the American ones, have similar peculiarities of form, or whether they differ in physiognomy from the Wolf of the south of Europe.   I have, therefore, in the present state of our knowledge, considered it unadvisable to designate the northern Wolf of America by a distinct specific appellation, lest I should unnecessarily add to the list of synonyms, which have already overburthened the science of Zoology.   The word *occidentalis,* which I have affixed to the Linnean name of *Canis lupus,* is to be considered as merely marking the geographical position of the peculiar race of Wolf which forms the subject of this article.   I have avoided adopting, as a specific name, any of the appellations founded on colour, because they could not with propriety be used to denote more than casual varieties of a species, in which the individuals shew such a variety in their markings.

Wolves are found in greater or less abundance in different districts, but they may be said to be very common throughout the northern regions ; their footmarks may be seen by the side of every stream, and a traveller can rarely pass a night in these wilds without hearing them howling around him.   They are very numerous on the sandy plains which, lying to the eastward of the Rocky Mountains, extend from the sources of the Peace and Saskatchewan rivers towards the

Missouri. There bands of them hang on the skirts of the buffalo herds, and prey upon the sick and straggling calves. They do not, under ordinary circumstances, venture to attack the full-grown animal: for the hunters informed me that they often see wolves walking through a herd of bulls without exciting the least alarm; and the marksmen, when they crawl towards a buffalo for the purpose of shooting it, occasionally wear a cap with two ears in imitation of the head of a wolf, knowing from experience that they will be suffered to approach nearer in that guise. On the Barren-grounds through which the Coppermine River flows, I had more than once an opportunity of seeing a single wolf in close pursuit of a rein-deer; and I witnessed a chace on Point Lake when covered with ice, which terminated in a fine buck rein-deer being overtaken by a large white wolf, and disabled by a bite in the flank. An Indian, who was concealed on the borders of the lake, ran in and cut the deer's throat with his knife, the wolf at once relinquishing his prey, and sneaking off. In the chase the poor deer urged its flight by great bounds, which for a time exceeded the speed of the wolf; but it stopped so frequently to gaze on its relentless enemy, that the latter, toiling on at a "long gallop,"* with its tongue lolling out of its mouth, gradually came up. After each hasty look, the poor deer redoubled its efforts to escape; but either exhausted by fatigue, or enervated by fear, it became, just before it was overtaken, scarcely able to keep its feet. The Wolves destroy many foxes, which they easily run down if they perceive them on a plain at any distance from their hiding places. In January, 1827, a wolf was seen to catch an Arctic fox within sight of Fort Franklin, and although immediately pursued by hunters on snow-shoes, it bore off its prey in its mouth without any apparent diminution of its speed †. The buffalo-hunters would be unable to preserve the

---

* Lord Byron's description of a chase by Wolves is so characteristic, that no apology is requisite for the insertion of the passage from whence this expression is borrowed:—

> " We rustled through the leaves like wind,
>   Left shrubs and trees and wolves behind;
>   By night I heard them on the track,
>   Their troop came hard upon our back,
>   With their long gallop which can tire
>   The hound's deep note and hunter's fire:
>   Where'er we flew they followed on,
>   Nor left us with the morning sun;
>   Behind I saw them scarce a rood,
>   At day-break winding through the wood;
>   And through the night had heard their feet,
>   Their stealing, rustling step repeat."
>
>   MAZEPPA.

† The same wolf continued for some days to prowl in the vicinity of the Fort, and even stole fish from a sledge, which two dogs were accustomed to draw home from the nets without a driver. As this kind of depredation could not be permitted to go on, the wolf was waylaid and killed. It proved to be a female, which accounted for the sledge-dogs not having been molested.

game they kill from the wolves, if the latter were not as fearful as they are rapacious. The simple precaution of tying a handkerchief to a branch, or of blowing up a bladder, and hanging it so as to wave in the wind, is sufficient to keep herds of Wolves at a distance *. At times, however, they are impelled by hunger to be more venturous, and they have been known to steal provisions from under a man's head in the night, and to come into a traveller's bivouac, and carry off some of his dogs. During our residence at Cumberland House in 1820, a wolf, which had been prowling round the Fort, and was wounded by a musket-ball and driven off, returned after it became dark, whilst the blood was still flowing from its wound, and carried off a dog from amongst fifty others, that howled piteously, but had not courage to unite in an attack on their enemy †. I was told of a poor Indian woman who was strangled by a Wolf, while her husband, who saw the attack, was hastening to her assistance; but this was the only instance of their destroying human life that came to my knowledge. As the winter advances and the snow becomes deep, the wolves being no longer able to hunt with success, suffer from hunger, and in severe seasons many die. In the spring of 1826 a large gray Wolf was driven by hunger to prowl amongst the Indian huts which were erected in the immediate vicinity of Fort Franklin, but not being successful in picking up aught to eat, it was found a few days afterwards lying dead on the snow near the Fort. Its extreme emaciation and the emptiness of its intestines shewed clearly that it died from inanition. The skin and cranium were brought to England, and presented to the Museum of the Edinburgh University; and a drawing from it is to be engraved for Mr. Wilson's beautiful Illustrations of Zoology.

The American Wolf burrows, and brings forth its young in earths with several outlets like those of a fox. I saw some of their burrows on the plains of the Saskatchewan, and also on the banks of the Coppermine River. The number of young in a litter varies from four or five to eight or nine. In Captain Parry's and Captain Franklin's narratives, instances are recorded of the female Wolves associating with the domestic dog; and we were informed that the Indians endeavour to improve their sledge-dogs by crossing the breed with wolves. The resemblance between the northern wolves and the domestic dog of the Indians is so great, that the size and strength of the Wolf seems to be the only difference. I

---

* The Wolves in the north of Europe are equally cautious. " To prevent the Wolves from destroying the reindeer, the Laplanders tie them to some tree, and it seldom happens that they are attacked in that situation: for the Wolf, being a suspicious animal, is afraid that there should be some snare laid for him, and that this is employed for a bait to draw him thither."—*Regnard.*

† The track in the snow shewed that it was the wounded Wolf which had returned.

have more than once mistaken a band of wolves for the dogs of a party of Indians; and the howl of the animals of both species is prolonged so exactly in the same key, that even the practised ear of an Indian fails at times to discriminate them.

The following notices, by Captain Lyons, of the wolves of Melville Peninsula, are good illustrations of the strength and habits of the northern wolves in general :—" A fine dog was lost in the afternoon. It had strayed to the hummocks ahead without its master, and Mr. Elder, who was near to the spot, saw five wolves rush at, attack, and devour it in an incredibly short space of time : before he could reach the place the carcase was torn in pieces, and he found only the lower part of one leg. The boldness of the wolves was altogether astonishing, as they were almost constantly seen amongst the hummocks, or lying quietly at no great distance in wait for dogs. From all we observed, I have no reason to suppose that they would attack a single unarmed man, both English and Esquimaux frequently passing them without a stick in their hands; the animals, however, exhibited no symptoms of fear, but rather a kind of tacit agreement not to be the beginners of a quarrel, even though they might have been certain of proving victorious."—" The wolves had now grown so bold as to come alongside, and on this night they broke into a snow-hut, in which a couple of newly purchased Esquimaux dogs were confined, and carried them off, but not without some difficulty, for in the day-light we found even the ceiling of the hut sprinkled with blood and hair. When the alarm was given, and the wolves were fired at, one of them was observed carrying a dead dog in his mouth, clear of the ground, at a canter, notwithstanding the animal was of his own weight. Before morning they tore a quantity of canvass off the observatory, and devoured it."—" The Esquimaux wolf-trap is made of strong slabs of ice, long and narrow; so that a fox can with difficulty turn himself in it, but a wolf must actually remain in the position in which he is taken. The door is a heavy portcullis of ice, sliding in two well-secured groves of the same substance, and is kept up by a line, which, passing over the top of the trap, is carried through a hole at the furthest extremity : to the end of the line is fastened a small hoop of whalebone, and to this any kind of flesh-bait is attached. From the slab which terminates the trap, a projection of ice, or a peg of wood or bone, points inwards near the bottom, and under this the hoop is lightly hooked; the slightest pull at the bait liberates it, the door falls in an instant, and the wolf is speared where he lies."

K

Var. A. Lupus griseus. *Common Gray Wolf.*

Grey Wolf. Cook's *Third Voyage*, vol. ii. p. 293. Lewis and Clarke, vol. i. p. 206, 283.
Common Gray Wolf. Schoolcraft's *Travels*, p. 285.
Canis lupus—griseus. Sabine, *Franklin's Voy.*, p. 654.
Canis lupus. Parry's *First, Second, and Third Voyages.*
Mahaygan. Cree Indians. Yes. Chepewyans.
Amarok. Esquimaux.

Pennant, in his Arctic Zoology, remarks, that "the wolves towards Hudson's Bay are of different colours—grey and white, and some black and white; the black hairs being mixed with the white chiefly along the back. In Canada they have been found entirely white." Lewis and Clark also say, "the large wolves of the Missouri are lower, shorter in the legs, and thicker than the Atlantic Wolf; their colour, which is not affected by the seasons, is of every variety of shade, from a gray or blackish-brown to a cream-coloured white." The gray, or rather the gray and white variety, is the Common Wolf from Lake Superior to the northern extremity of the Continent, and in the islands beyond it. It has been seen on the Atlantic coast from Nootka northwards.

The following description, by Mr. Sabine, of a specimen procured at Cumberland House, in latitude 54°, and deposited by Captain Franklin in the British Museum, will make the reader acquainted with its appearance :—

DESCRIPTION.

" It is very dissimilar in colour to the usual state of the (European) wolf, and is of a much greater size. The teeth are remarkably strong and large; the ears sharp and erect, thickly clothed with dark-brown hair, tipped with gray; above and below on the neck the hair is thick and bushy; the whole of the body is covered with a mixture of long gray and black hairs, having some few white ones intermixed on the back; the sides and belly are dark gray; the tail is bushy, gray tipped with brown; the legs are strong, covered with dark-brown hair; claws strong, short, and arched."

A specimen procured at Carlton House on the same river, and now in the Museum of the Zoological Society*, has the face, cheeks, throat, belly, hips, and tail, white, except a small part of the latter, adjoining the rump, where it is blackish. On the back and sides there is an intermixture of long black and white hairs, which, with the grayish wool that partially

---

* No. 33. *Catalogue of the Museum.*

appears, gives a general grayish hue to these parts, deepening along the dorsal line into black. The hair on the back part of the cheeks is very bushy. In other individuals which I have seen, the mixture of dark-gray with black and brown forms distinct patches on an almost white ground. There is generally a darker line along the spine.

The Gray Wolf differs in size in different districts, and even in the same district individuals differing much in height and strength, but (as far as one can judge from their teeth) of nearly equal age, are to be found. The wolves of the desert country lying to the north of Great Slave Lake, and much frequented by reindeer, are of great dimensions. Farther north again, in the islands of the Polar sea, visited by Captain Parry, they were generally smaller, their average height in that quarter being only about twenty-seven inches. Captain Sabine states those of Melville Island to be as big as a full-sized setter-dog.

DIMENSIONS

Of the prepared skin of a Gray Wolf killed at Cumberland House.

	Feet.	Inches.		Feet.	Inches.
Length of the head and body . .	4	0	Height to the top of the shoulder .	2	0
„ tail . . .	1	2			

DIMENSIONS

Of a Gray Wolf starved to death at Fort Franklin, April 1821, (measured before it was skinned.)

	Feet.	Inches.		Feet.	Inches.
Length of head and body . . .	4	2	haunches, the feet being flat on the ground, and the fur of the back pressed down .	2	8
„ tail (vertebræ) . .	1	7			
„ tail, including fur .	2	2	Height with the fur of the back in its natural		
Height at the fore-shoulder, and also at the			rough state	3	0

DIMENSIONS

Of a specimen as mounted in the Zoological Museum, which was procured at Carlton House.

	Feet.	Inches.		Feet.	Inches.
Length of head and body	4	4	Height of the back,—fur rough . .	2	5
„ tail (vertebræ) . . .	1	3½	Distance from top of the nose to the orbit .	0	4¼
„ tail, including the hair . .	1	6½	Height of the ear, measured on the inner		
Height of the back,—fur pressed down . .	2	2	side	0	3½

K 2

Var. B.   Lupus albus.   *White Wolf.*

White Wolf. Lewis and Clark, vol. i. p. 107 ; vol. iii. p. 263.
Canis lupus—albus. Sabine. *Franklin's Journ.*, p. 655.
White Wolf. Icones. Franklin's *Journ.*, p. 312. Lyon's *Private Journ.*, p. 297.

Wolves totally white are not uncommon in the most northern parts of America, particularly in districts nearly destitute of wood *.   They are occasionally seen also on the plains of the Missouri.   A Yellow Wolf, mentioned by Lewis and Clark, (vol. i. p. 40) may be perhaps classed with the white variety.

DESCRIPTION.

The White Wolf figured in Captain Franklin's narrative above referred to, was killed near Fort Enterprise, in February, 1821.   Its ears were short and erect.   Its fur was long and of a yellowish-white colour over the whole body, the nose alone having a slight tinge of gray.

Its Dimensions were as follows :—

	Feet.	Inches.		Feet.	Inches.
Length of head and body . . .	4	4	Girth before the hind legs . . .	2	0
„ tail	1	7	Length of fore-leg and foot with toe-nails	1	8
Height at both fore and hind quarters .	2	10	„ hind ditto ditto ditto	1	10
Girth behind the fore-legs . . .	2	6			

Var. C.   Lupus sticte.   *Pied Wolf.*

Wolves having black colours instead of gray, distributed in large patches on the sides, are sometimes seen in the fur countries, associated with the Common Gray Wolves.   On the banks of the Mackenzie, I saw five young wolves leaping and tumbling over each other, with all the playfulness of the puppies of the domestic dog, and it is not improbable that they were all of one litter.   One of them was pied, another entirely black, and the rest shewed the common gray colours.   I was unable to procure a specimen of the Pied Wolf.

* Muller informs us that white wolves are found on the Jenisei ; and Regnard says that the Lapland wolves " are almost all of a whitish-gray colour ; there are some of them white."   It is desirable that these Siberian or Lapland wolves should be compared with the Pyrenean races.

CANIS LUPUS VAR. NUBILUS.

Published by John Murray. January 1829.

Var. D. Lupus nubilus. *The Dusky Wolf.*

Canis nubilus. Say. *Long's Exped.*, vol. i. p. 333.
Dusky Wolf. Godman's *Nat. Hist.*, vol. i. p. 265.

Plate iii.

The Dusky Wolf differs little from the pure black variety which follows. It was considered by Mr. Say to be a distinct species, on account of its colour, a different physiognomy from that of the Common Red Wolf, its more robust form, and its shorter ears; but, with the exception of colour, these characters are common to all the varieties of the northern American Wolf, and are in fact some of the peculiarities that distinguish them from Pyrenean wolves.

DESCRIPTION.

Mr. Say's specimen had " a dusky colour; the hair cinereous at the base, then brownish-black, then gray, then black; the gray of the hairs combining with the black tips to produce a mottled appearance; the gray predominating on the sides. Ears short, deep brownish-black, with a patch of gray hair within. The under parts dusky ferruginous, grayish with long hairs between the thighs, and with a large white spot on the breast; the ferruginous colour very much narrowed on the neck, but dilated on the lower part of the cheeks; legs brownish-black, with a slight admixture of gray hairs, excepting on the anterior edge of the hind thighs, and the lower edgings of the toes, where the gray predominated; the *tail* was short, fusiform, a little tinged with ferruginous, black above, near the base, and at the tip; the top of the trunk hardly attaining to the *os calcis;* the longer hairs of the back, particularly over the shoulders, resembled a short, sparse mane."

DIMENSIONS.

	Feet.	Inches.		Feet.	Inches.
Length of head and body	4	3¾	Length from anterior canthus of the eye to		
„ the trunk of the tail . .	1	1	the middle of the tip of the nose .	0	5½
„ ear from anterior angle to the tip	0	3¾	Distance between the anterior angles of the		
„ from the anterior angle of the ear to the posterior canthus of the eye .	0	4¾	ears, rather more than . . .	0	3

The Dusky Wolf, figured in this work, was killed at Fort Resolution on Great Slave Lake, in latitude 61°, and is now preserved in the Museum of the Zoological Society.

DESCRIPTION.

It is covered with long hairs, intermixed with a thick woolly down, more or less conspicuous on different parts of the body. The long hairs are mostly black, but there are a few white ones interspersed. The wool or down is of a dull yellowish-gray colour; least of it is seen on the back, which has consequently nearly a black colour; the fur covering the spine is long, blackish towards the roots, and shining black at the tips; the sides are dusky-gray, many of the long hairs having apparently fallen off; the belly is blacker. The anterior parts of the legs are hoary from an intimate intermixture of black and white hairs, in which the former predominate. The posterior surfaces of the legs are covered with long, white and gray fur. The tail is black, with a few white hairs, and has a black tip. The wool on the tail is of the same colour with that on the body, but it is not visible until the long hairs are turned aside. The chin and extremity of the upper lip are white, and there are many long white hairs on the cheek. The feet are very hairy, the hair on the soles projecting beyond the claws.

DIMENSIONS

Of the specimen after it was mounted.

	Feet.	Inches.		Feet.	Inches.
Length of head and body . .	4	2	Distance between the ears . .	0	3½
„ tail (vertebræ) . .	1	2	„ from the top of the nose to the		
„ tail, with the fur . .	1	6	anterior part of the orbit .	0	5¼
Height, pressing down the fur .	2	2	„ between the eyes . .	0	2¾
„ of the ear on the inner side .	0	3½	Length of the head . . .	1	1

Var. E.   Lupus ater.   *Black American Wolf.*

Loup noir de Canada. Buffon, vol. ix. p. 364. t. 41. (malè.)
Black Wolf. Say, *Long's Exped.*, vol. i. p. 95. Franklin's *Journ.*, vol. i. p. 172.
          Griffith's *Anim. King.* cum Icone, vol. ii. p. 348. (opt.)
          Godman's *Nat. Hist.*, vol. i. p. 267. Icone ex Griffithii icone mutuato.
Canis Lycaon. Harlan's *Faun. Amer.*, p. 82.

We saw some Black Wolves on the banks of the Mackenzie, but they are more common on the river Saskatchewan, and in districts further south. Mr. Say informs us that they abound on the Missouri. The Indians do not consider them to be a distinct race, but report that one or more black whelps are occasionally found in a litter of a Gray Wolf. In conceding to their opinion, I do not mean to assert that the offspring of Black Wolves are not most frequently black. Five Black Wolves are mentioned by Say, as having been taken from one den; and Mr.

Hood, in Captain Franklin's Narrative, records an instance in which a Black Wolf was shot, and three black whelps taken from her den.

The Black Wolves differ in external appearance from the gray ones only in colour, and their haunts and habits are precisely the same. Buffon, in his description of a young Black Wolf from Canada, remarks that the ears were wider, further apart, and more pointed, than those of the European Wolf; the eyes smaller, and also further apart. The comparatively broad forehead, indicated by the greater distance between the ears and eyes, is, as I have already stated, common to all the varieties of the Wolf of the northern parts of America. The French naturalist's description of the behaviour of this Wolf, when turned out against a bull, is so characteristic, not only of the American Wolf, but also of the Indian dogs, when under the influence of fear, that I cannot resist quoting it at length :—" Cet animal," dit il, " avoit été pris fort jeune en Canada, et apporté en France par un Officier de Marine, qui le garda dans sa maison pendant quelque tems ; mais l'animal étant devenu féroce en grandissant, il fut mis au combat de taureau à Paris, où il ne montra pas beaucoup de courage lorsqu'on le fit entrer en lice mais dès que l'on approchoit de la loge où on le gardoit, il entroit en fureur, se jetoit brusquement en avant de toute la longueur de sa chaine, montroit les dents et aboyoit, non pas comme les chiens, mais seulement par des cris successifs et interrompus qu'il ne repetoit qu'après d'assez longs intervales."

I have frequently observed an Indian dog, after being worsted in combat, retreat into a corner, and howl at intervals for an hour together. They also howl piteously when apprehensive of punishment, and throw themselves into attitudes strongly resembling those exhibited by a wolf when caught in a trap. The plate given by Buffon is but an indifferent representation of the Black Wolf. The individual was not only young, but its fur, as is customary with animals in captivity, seems to have been in a bad state*. A most excellent etching, by Landseer, of a Black Wolf, kept in the Tower, has appeared in Griffith's translation of Cuvier's *Règne Animal*. Though it may be remarked, that even in this the fur is not represented in the fine condition which the animal exhibits during the winter in its native climate.

Linnæus has described the Black Wolf of Europe under the appellation of *Canis Lycaon ;* and Baron Cuvier and other naturalists have followed his example in speaking of it as a distinct species ; but authors have not clearly pointed out any

* The small size of this specimen, by the way, may have given rise to Gmelin's mistake in confounding it with the black fox under the name of *Canis Lycaon.*

difference between the *Canis Lycaon* of Europe, and the *Canis Lupus ;* though it has happened that the peculiar characters of the American Wolves have sometimes been ascribed to the *Canis Lycaon,* from the descriptions having been taken from American specimens. One can easily understand how Black Wolves accidentally congregating may produce an offspring of the same hue with themselves, until, by a concurrence of circumstances, the black variety is the predominating one in a particular district ; but the breed must be frequently contaminated by wolves of other colours. Pallas, in a letter to Pennant, says, " I have seen at Moscow about twenty spurious animals from dogs and Black Wolves. They are for the most part like wolves, except that some carry their tails higher, and have a kind of coarse barking. They multiply among themselves, and some of the whelps are grayish, rusty, or even of the whitish hue of the Arctic wolves *."

Black Wolves are more frequent in the southern parts of Europe than in the northern ; and to the south of the Pyrenees they are said to be more common than the ordinary species or variety†. In like manner the American Wolf is more common on the Missouri than farther north ; and it is reported to be plentiful in Florida, where, according to Bartram, the females are distinguished by a white spot on the breast ‡.

* *Arctic Zoology*, vol. i. p. 42.   † GRIFFITH, *Anim. Kingd.*, vol. ii. p. 348.
‡ WARDEN, *United States*, vol. i. p. 207.   DESMAREST, *Mammalogie*, p. 198.

PLATE 4.

CANIS LATRANS.

Published by John Murray. January 1829.

[23.]   2. CANIS LATRANS.   (Say.)   *The Prairie Wolf.*

Small Wolves. DU PRATZ, *Louisiana*, vol. ii. p. 54.
Prairie Wolf. GASS's *Journal, &c.*, p. 56.
Prairie Wolf and Burrowing Dog. LEWIS and CLARK, vol. i. pp. 102, 134, 283 ; vol. iii. pp. 28, 238.
        SCHOOLCRAFT's *Travels*, p. 285.
Canis latrans. SAY. *Long's Exped.*, p. 27, *note*, p. 332.
Cased Wolves. HUDSON BAY COMPANY's LISTS.
Meesteh-chaggoneesh. CREE INDIANS.

PLATE IV.

This animal has long been known to voyagers on the Missouri and Saskatchewan as distinct from the Common Wolf. It is mentioned in Mr. Graham's MSS., and its skins have always formed part of the Hudson Bay Company's importations, under the name of Cased Wolves *. Lewis and Clark give a good descrip‧tion of it (vol. i. p. 283), and Mr. Say has added the specific name. The Prairie Wolf has much resemblance to the common Gray Wolf in colour, but differs from it so much in size, voice and manners, that it is fully entitled to rank as a distinct species. It inhabits the plains of the Missouri and Saskatchewan, and also, though in smaller numbers, those of the Columbia. On the banks of the Saskatchewan, these animals start from the earth in great numbers on hearing the report of a gun, and gather around the hunter in expectation of getting the offal of the animal he has slaughtered. They hunt in packs, and are much more fleet than the Common Wolf. I was informed by a gentleman who has resided forty years on the Saskatchewan, and is an experienced hunter, that the only animal on the plains which he could not overtake, when mounted on a good horse, was the Prong-horned Antelope, and that the meesteh-chaggoneesh or Prairie Wolf was the next in speed. The Canadian stag is less fleet ; and as to the red fox, it is soon run down.

The northern range of the Prairie Wolf is about the fifty-fifth degree of latitude, and it probably extends southwards to Mexico. It associates in greater numbers than the Gray Wolf of the same districts ; hunts in packs, and brings forth its young in burrows on the open plain remote from the woods.

The Prairie Wolf, described by Lewis and Clark and Mr. Say, has a narrower

---

* The skins are not split open like the large Wolf skins, but stripped off and inverted or cased, like the skin of a fox or rabbit.

L

muzzle and longer ears than the one found on the Saskatchewan, but the difference is not so great as to enable us to speak of them as distinct varieties.

DESCRIPTION.

The fur of the Prairie Wolf is of the same quality with that of the Gray Wolf, and consists of long hairs, with a thick wool at their base. The wool has a smoky or dull lead colour; the long hairs on the back are white either for their whole length, or they are merely tipped with black. The prevailing colour along the spine is dark blackish-gray, sprinkled with white hairs. Its cheeks, upper lip, chin, throat, belly, and insides of the thighs are white. There is a light brown tint upon the upper surface of the nose, on the forehead, and between the ears, on the shoulders, on the sides, where it is mixed with gray, and on the outsides of the thighs and legs. The tail is gray and brown, with a black tip. Some individuals have a broad black mark on the shins of the fore-legs, like the European wolf. The ears are short, erect, and roundish, white anteriorly and brown behind. The tail is bushy, and is clothed like the body with wool and long hair. Some specimens want the brown tints, and have more of the gray colour.

DIMENSIONS
Of No. 34, Zoological Museum, killed on the Saskatchewan.

	Feet.	Inches.		Feet.	Inches.
Length of head and body . . .	3	0	Distance between the end of the nose and		
„ tail (vertebræ) . . .	1	0	the anterior angle of the eye	0	$3\frac{3}{4}$
„ tail, with the fur . . .	1	$2\frac{1}{2}$	„ between the ears . . .	0	$3\frac{1}{4}$
„ the ear, measured behind	0	$2\frac{1}{2}$	„ between the eyes . . .	0	$2\frac{1}{4}$

DIMENSIONS
Of a scull of the Saskatchewan sort.

	Inches.	Lines.		Inches.	Lines.
From the incisors to the junction of the occipital and sagittal crests by callipers	8	0	The total length of the nose in a recent specimen ought to be about . .	4	0
From ditto to ditto, following the curvature of forehead	8	11	Greatest breadth of the scull, i. e. from the outside of one zygomatic arch to the out-		
Length of nasal bones . . .	2	7	side of the other . . .	5	0
The os incisivum projects beyond the nasal bones	1	3	From incisors along the base of the scull to the inferior margin of the occipital foramen	7	0

There is a second specimen in the Zoological Museum, brought from the plains of the Columbia, which has the narrow muzzle and long ears of Say's specimens.

[24.]   3. Canis familiaris. (Linn.) *Domestic Dog.*

Canis f. var. A. borealis. (Desmarest.) *Esquimaux Dog.*

Canis familiaris var. N. borealis. Desmarest. *Mamm.*, p. 194.
Eskimaux Dog. Captain Lyon's *Private Journal*, p. 332. 244.
Icones*, Parry's *Second Voyage*, p. 290 and 358.

The great resemblance which the Domestic Dogs of the aboriginal tribes of America bear to the wolves of the same country, was remarked by the earliest settlers from Europe†, and has induced some naturalists of much observation to consider them to be merely half-tamed wolves‡. Without entering at all into the question of the origin of the Domestic Dog, I may state that the resemblance between the wolves and the dogs of those Indian nations, who still preserve their ancient mode of life, continues to be very remarkable, and it is nowhere more so, than at the very northern extremity of the Continent, the Esquimaux dogs being not only extremely like the gray wolves of the Arctic circle, in form and colour, but also nearly equalling them in size. The dog has generally a shorter tail than the wolf, and carries it more frequently curled over the hip, but the latter practice is not totally unknown to the wolf, although that animal, when under the observation of man, being generally apprehensive of danger or on the watch, seldom displays this mark of satisfaction. I have, however, seen a family of wolves playing together, occasionally carry their tails curled upwards.

In the Museum of the Zoological Society there is a specimen of an Esquimaux dog, which was brought originally from Baffin's Bay by Captain Ross's expedition, and which was afterwards the faithful attendant of Captain Parry's crews during the memorable winter of 1819, which they passed on Melville Island. The great likeness of this specimen to a gray wolf from Carlton House, preserved in the same case, must be obvious to every one who has seen them, although their birth-places lie upwards of twenty degrees of latitude apart, and they are, therefore, not so favourable examples to shew the resemblance as if they had been natives of the same district.

* In L'Hist. Nat. des Mammifères, there is a plate and description of a supposed Esquimaux Dog, a present to the Jardin du Roi ; but Capt. Sabine informs us that it is a cross between a real Esquimaux bitch and a Newfoundland dog.—*Appendix, Parry's First Voyage*, p. clxxxvi.   † Smith, *Virginia.*   ‡ Kalm.

DESCRIPTION.

The Dog has short conical ears like the American wolf, but its nose is still shorter than that of the latter animal. Its nose, cheeks, belly, and legs, are white. The fore-legs are destitute of the black mark above the wrist, which characterises the European wolf, and which is visible in some American wolves, but not in all. The top of the head and the back are almost black; but there is a narrow white line down the spine of the back, which I have not noticed in any coloured wolf. Its sides are thinly covered with long, black, and some white hairs, and there is a shorter dense coat of yellowish-gray wool, like that of the wolf, which is partly visible. The tail, like the back, is clothed with black and white hairs, the latter predominating at its tip. There is a thick wool on the tail concealed by the longer hairs.

DIMENSIONS
Of the specimen in the Zoological Museum.

	Feet.	Inches.				Feet.	Inches.
Length from the end of the nose to the tail	4	3	Height of the ears (inside)	.	.	0	3
„ of the tail (vertebræ) .	1	2	Breadth between the eyes	.		0	$2\frac{1}{2}$
„ „ including the fur at tip	1	5	„ „ ears	.	.	0	$4\frac{1}{4}$
„ from end of the nose to the orbit	0	4					

Captain Lyon had so many opportunities of studying the habits of the Esquimaux dog, and his account of them is so much to the purpose, that I think it advantageous to the reader to have it repeated here in his own words :—

" Having myself possessed, during our second winter, a team of eleven very fine animals, I was enabled to become better acquainted with their good qualities than could possibly have been the case by the casual visits of the Esquimaux to the ships. The form of the Esquimaux Dog is very similar to that of our shepherd's dogs in England, but he is more muscular and broad-chested, owing to the constant and severe work to which he is brought up. His ears are pointed, and the aspect of the head is somewhat savage. In size, a fine dog is about the height of the Newfoundland breed, but broad like a mastiff in every part except the nose. The hair of the coat is in summer, as well as in winter, very long, but during the cold season a soft downy under covering is found, which does not appear in warm weather. Young dogs are put into harness as soon as they can walk, and being tied up, soon acquire a habit of pulling, in their attempts to recover their liberty, or to roam in quest of their mother. When about two months old, they are put into the sledge with the grown dogs, and sometimes eight or ten little ones are under the charge of some steady old animal, where, with frequent and sometimes cruel beatings, they soon receive a competent education. Every dog is distinguished by a particular name, and the angry repetition of it has an effect as instantaneous as an application of the whip, which instrument is of an immense length, having a lash of from eighteen to twenty-four feet, while the handle is of one foot only. With this, by throwing it on one side or the other of the leader, and repeating certain words, the animals are guided or stopped. Wăh-aya, āyă, whooă, to the right. A-wha, a-wha, a-whut, to the left. A-look, turn, and whooă, stop. When the sledge is stopped, they are all taught to lie down, by throwing the whip gently over their backs, and they will remain in this position even for hours, until their master returns to them. A walrus is frequently drawn along by three or four of these dogs, and seals are sometimes carried home in the same manner; though I have in some instances seen a dog bring home the greater part of a seal in panniers placed across his back. The latter mode of conveyance is often used in summer, and the dogs also carry skins or furniture overland to the sledges when their masters are going on any expedition. It might be supposed that in so cold a climate these animals had peculiar periods of gestation, like the wild creatures; but, on the contrary, they bear young at every season of the year, and seldom exceed five at a litter. Cold has very little effect on them; for, although the dogs at the huts slept within the snow passages, mine at the ships had no shelter, but lay alongside, with the thermometer at 42° and 44°, and with as little concern as if the weather had been mild. I found, by several experiments, that three of my dogs could draw me on a sledge, weighing 100lbs., at the rate of one mile in six minutes; and as a proof of the strength of a well-grown dog, my leader drew 196lbs. singly, and to the same distance, in eight minutes. At another time, seven of my dogs ran a mile in four minutes, drawing a heavy sledge full of men. Afterwards, in carrying stores to the Fury, one mile distant, nine dogs drew 1611lbs., in the space of nine minutes. My sledge was on runners neither shod nor iced; but had the runners been iced, at least 40lbs. might have been added for each dog."

In another passage Captain Lyon says, "Our eleven dogs were large, and even majestic looking animals; and an old one, of peculiar sagacity, was placed at their head by having a longer trace, so as to lead them through the safest and driest places; these animals having such a dread of water, as to receive a severe beating before they would swim a foot. The leader was instant in obeying the voice of the driver, who never beat, but repeatedly called to him by name. When the dogs slackened their pace, the sight of a seal or bird was sufficient to put them instantly to their full speed; and even though none of these might be seen on the ice, the cry of ' a seal!'—' a bear!'—' a bird!' &c., was enough to give play to the legs and voices of the whole pack. It was a beautiful sight to observe the two sledges racing at full speed to the same object, the dogs and men in full cry, and the vehicles splashing through the holes of water with the velocity and spirit of rival stage-coaches.   There is something of the spirit of professed whips in these wild races ; for the young men delight in passing each other's sledge, and jockeying the hinder one by crossing the path.   In passing on different routes the right hand is yielded, and should an unexperienced driver endeavour to take the left, he would have some difficulty in persuading his team to do so.   The only unpleasant circumstance attending these races is, that a poor dog is sometimes entangled and thrown down, when the sledge, with perhaps a heavy load, is unavoidably drawn over his body.   The driver sits on the fore-part of the vehicle, from whence he jumps when requisite to pull it clear of any impediments which may lie in the way, and he also guides it by pressing either foot on the ice.   The voice and long whip answer all the purposes of reins, and the dogs can be made to turn a corner as dexterously as horses, though not in such an orderly manner, since they are constantly fighting ; and I do not recollect to have seen one receive a flogging without instantly wreaking his passion on the ears of his neighbours.   The cries of the men are not more melodious than those of the animals ; and their wild looks and gestures, when animated, give them an appearance of devils driving wolves before them.   Our dogs had eaten nothing for forty-eight hours, and could not have gone over less than seventy miles of ground ; yet they returned, to all appearance, as fresh and active as when they first set out."

The Esquimaux dogs are likewise useful to their masters in discovering, by the scent, the winter retreats which the bears make under the snow.

Canis f. var. B. lagopus. *Hare Indian Dog.*

Plate v.

This variety of Dog is cultivated at present, so far as I know, only by the Hare Indians, and other tribes that frequent the borders of Great Bear Lake and the banks of the Mackenzie.   It is used by them solely in the chase, being too small to be useful as a beast of burthen or draught.

CANIS FAMILIARIS VAR. LAGOPUS.

Published by John Murray. January 1829

DESCRIPTION.

The Hare Indian Dog has a mild countenance, with, at times, an expression of demureness. It has a small head, slender muzzle; erect, thickish ears; somewhat oblique eyes; rather slender legs, and a broad hairy foot, with a bushy tail, which it usually carries curled over its right hip. It is covered with long hair, particularly about the shoulders, and at the roots of the hair, both on the body and tail, there is a thick wool. The hair on the top of the head is long, and on the posterior part of the cheek it is not only long, but being also directed backwards, it gives the animal, when the fur is in prime order, the appearance of having a ruff round the neck. Its face, muzzle, belly, and legs, are of a pure white colour, and there is a white central line passing over the crown of the head and the occiput. The anterior surface of the ear is white, the posterior yellowish-gray or fawn-colour. The end of the nose, the eyelashes, the roof of the mouth, and part of the gums, are black. There is a dark patch over the eye. On the back and sides there are larger patches of dark blackish-gray or lead-colour mixed with fawn-colour and white, not definite in form, but running into each other. The tail is bushy, white beneath and at the tip. The feet are covered with hair which almost conceals the claws. Some long hairs between the toes project over the soles, but there are naked callous protuberances at the root of the toes and on the soles, even in the winter time, as in all the wolves described in the preceding pages. The American foxes, on the contrary, have the whole of their soles densely covered with hair in the winter. Its ears are proportionably nearer each other than those of the Esquimaux dog.

The size of the Hare Indian Dog is inferior to that of the prairie wolf, but rather exceeds that of the red American fox. Its resemblance, however, to the former is so great, that, on comparing live specimens, I could detect no marked difference in form, (except the smallness of its cranium,) nor in the fineness of the fur, and arrangement of its spots of colour. The length of the fur on the neck, back part of the cheeks, and top of the head, was the same in both species. It, in fact, bears the same relation to the prairie wolf that the Esquimaux Dog does to the great gray wolf. It is not, however, a breed that is cultivated in the districts frequented by the prairie wolf, being now confined to the northern tribes, who have been taught the use of fire-arms within a very few years. Before that weapon was introduced by the fur-traders, a dog, so well calculated by the lightness of its body and the breadth of its paws, for passing over the snow, must have been invaluable for running down game, and it is reasonable to conclude that it was then generally spread amongst the Indian tribes north of the Great Lakes.

The Hare Indian Dog is very playful, has an affectionate disposition, and is soon gained by kindness. It is not, however, very docile, and dislikes confinement of every kind. It is very fond of being caressed, rubs its back against the hand like a cat, and soon makes an acquaintance with a stranger. Like a wild animal, it is very mindful of an injury, nor does it, like a spaniel, crouch under the lash; but if it is conscious of having deserved punishment, it will hover round the tent of its master the whole day, without coming within his reach, even when he calls it. Its howl, when hurt or afraid, is that of the wolf; but when it sees any unusual object, it makes a singular attempt at barking, commencing by a kind of growl, which is not, however, unpleasant, and ending in a prolonged howl. Its voice is very much like that of the prairie wolf. The larger dogs, which we had for draught at Fort Franklin, and which were of the mongrel breed in common use at the fur-posts, used to pursue the Hare Indian Dogs for the purpose of devouring them; but the latter far outstripped them in speed, and easily made their escape. A young puppy, which I purchased from the Hare Indians, became greatly attached to me, and when about seven months old ran on the snow by the side of my sledge for nine hundred miles, without suffering from fatigue. During this march it frequently, of its own accord, carried a small twig or one of my mittens for a mile or two; but, although very gentle in its manners, it shewed little aptitude in learning any of the arts which the Newfoundland dogs so speedily acquire, of fetching and carrying when ordered. This Dog was killed and eaten by an Indian, on the Saskatchewan, who pretended that he mistook it for a fox.

CANIS F. var. C. CANADENSIS. *North American Dog.*

Attim. CREE INDIANS.	Animous. ALGONQUINS.
Watts. SLOUACCOUSS INDIANS.	Shong. STONE INDIANS.
Hudther. FALL INDIANS.	Ametoo. BLACK-FEET INDIANS.
Hey. SARSEES, or CIRCEES.	Thling. CHEPEWYANS.

By the above title I wish to designate the kind of Dog which is most generally cultivated by the native tribes of Canada, and the Hudson's Bay countries. It is intermediate in size and form between the two preceding varieties; and by those who consider the domestic races of dog to be derived from wild animals, this might be termed the offspring of a cross between the prairie and gray wolves. This breed wants the strength of the Esquimaux dog, and does not possess the affectionate and playful disposition of the Hare Indian variety. It is used at certain seasons in the chase, and by some tribes as a beast of burthen or draught; but it has all the sneaking habits of the wolf, with less courage, and without the intelligence of that animal. It unites with its companions to assail a stranger on his approach to the hut of its master; retreats on the least show of resistance, or endeavours to get behind him, and silently snaps at his legs. When opposed to another dog, it curls the upper lip very much, shews the whole of its teeth, and snarls for a long time, before it ventures to bite. A little Scotch terrier, that accompanied us on the last expedition, disconcerted the largest of them by the smartness of his attack, and used to send an animal, more than four times his own size, howling away, although the density of its woolly covering prevented his short teeth from wounding the skin. When they fight among themselves, the dog that is vanquished, is not unfrequently torn in pieces by the rest of the pack. They hunt the larger domestic animals in packs, snapping at their heels and harassing them until worn out, but scarcely ever venture to seize them by the throat. All the dogs of a camp assemble at night to howl in unison, particularly when the moon shines bright.

The fur of the North American Dog is similar to that of the Esquimaux breed, and of the wolves. The prevailing colours are black and gray, mixed with white. Some of them are entirely black. Their thick woolly coat forms an admirable protection against the cold, and when they are fat they can lie all night on the snow without inconvenience during the most intense cold. In the summer time they are fond of making deep holes in sandy ground ; and this habit is retained by the mongrel breed which the Canadian voyagers rear for draught. These often burrow completely underneath the out-buildings of a fort, and will in a single night make their way beneath the door of a store-house, if the precaution has not been taken of flooring it with wood. The flesh of these Dogs is esteemed before that of almost any other animal by the Canadian voyagers, and is eaten by some of the Indian tribes on the Saskatchewan and shores of the Great Lakes; but the Chepewyan tribes hold the practice in abhorrence, because they consider themselves to be descendants of a dog. I quote Theodat's account of this variety of dog, written in the year 1630, because it applies pretty correctly to the North American dogs of the present day, and shews that at that early period, and perhaps even before the arrival of Europeans, they formed an esteemed article of the food of the natives :—

" Les chiens du Canada sont un peu differens des nostres, sinon au naturel et au sentiment, qui ne leur est point mauvais. Ils hurlent plustost qu'ils n'abayent, et ont tous les oreilles droictes comme renards, mais au reste tout semblables au matins de mediocre grandeur de nos villageois ; ils arrestent l'eslan et descouvurent la

M

giste de la beste, et sont de fort petite despence a leur maistre, mais au reste plus propre à la cuisine qu'à tout autre service.   La chair en est assez bonne et sent aucunement le porc ; j'en mangeois assez peu souvent, car un telle viande est fort estimée dans le pays, c'est pourquoy je n'en avois pas si souvent que j'eusse bien desiré.   Ils sont fort importuns dans les cabanes, marchent sur vous, et s'ils recontrent le pot au descouvert ils ont incontinent leur museau aigu dans *la sagamité,* qui n'en est pas estimée moins nette*.''

Canis f.   var. D.   Novæ Caledoniæ.   *Carrier Indian Dog.*

Scacah.   Carrier Indians.

The Attnah or Carrier Indians of New Caledonia possess a variety of dog which differs from the other northern races.   Mr. M'Vicar made me a present of an excellent example of this breed, which I intended to have brought to England ; but it was stolen, and fell a sacrifice to the desire which a party of Canadian voyagers had to partake of a meal of dog's flesh.   I regretted its loss the more, as, intending to have a portrait taken of it, I had neglected to draw up a detailed account of its characteristic features.

DESCRIPTION.

It had erect ears, and a head large in proportion, even when compared with the Esquimaux dog.   Its body was long, and its legs short.   Its fur was rather shorter and sleeker than that of the other native dogs, and its body was studded with small spots of various colours.   There was a good deal of intelligence in its countenance, mixed with wildness.   It was extremely active, and could leap to a great height.   The Carrier Indians use it in the chase.   It was of the size of a large turnspit-dog, and had somewhat of the same form of body ; but it had straight legs, and its erect ears gave it a different physiognomy.

* Sagard Theodat.   *Canada,* p. 757.

[25.] 4. Canis (Vulpes) lagopus. (Linn.) *The Arctic Fox.*

Genus, Canis, Linn. *Sub-genus,* Vulpes. Desmarest.
Pied foxes. James, *Voy.,* p. 50. An. 1633.
Canis lagopus. Linn. *Syst.,* vol. i. p. 59. Forster, *Phil. Trans.,* 62, p. 370.
Arctic fox. Pennant, *Arctic Zool.,* vol. i., p. 42. Hearne, *Journey,* p. 363.
Greenland dog. Pennant, *Hist. Quadr.,* vol. i. p. 257 ? a young individual.
Canis lagopus. Captain Sabine, *Parry's First Voy. Suppl.* clxxxvii. Mr. Sabine, *Franklin's Jour.,*
p. 658. Richardson, *Parry's Second Voy.,* App. p. 299. Harlan, *Fauna,* p. 92.
Isatis and Arctic fox. Godman, *Nat. Hist.,* vol. i. p. 268.
Stone fox. Auctorum.
Terreeanee-arioo. Esquimaux of Melville-Peninsula. Terienniak. Greenlanders.
Wappeeskeeshew-makkeeshew. Cree Indians. Peszi. Russians.

DESCRIPTION.

The Arctic Fox in its full *winter dress* is entirely of a pure white colour, or white with a slight tinge of yellow, except at the tip of the tail, where there are a few black hairs intermixed. Before the eyes, and on the lower jaw, the hair is short and sleek; on the forehead and posterior part of the cheeks, it becomes considerably longer, and on the occiput and neck it equals the ears in height, and is intermixed with a soft wool or down. There is so much of the wool on the body, that it gives the fur the character of that of the Polar Hare. The ears are of a rounded form, and are covered with shorter hairs than the neighbouring parts: the shortest hair is on their edges, and it terminates so evenly with that on the back and front of the ear, as to seem as if it had been trimmed with a pair of scissars, and to render the ear thicker in appearance than it really is *. The long fur on the posterior part of the cheeks is directed backwards, and contributes to give a peculiar cast to the physiognomy, and an apparent great thickness to the neck, which features are common to the foxes, wolves, and native races of the domestic dog in the northern parts of America, and distinguish them from their congeners of the Old World. The *vibrissæ* about the mouth are very strong, and are in some specimens nearly white, in others of a dusky-brown colour. The hair on the body is long, particularly on the sides. It is rather longer on the belly than on the back, but not so close and woolly. It is denser, and coarser on the tail than elsewhere. The shoulders and thighs are protected by long fur, but the fore-parts of the legs are covered with short hair, the hind-legs having the shortest and smoothest coat. On the posterior surface of the legs the hair is longer. The soles of the feet are covered with very dense woolly hair, of a dirty white colour, giving them that resemblance to the feet of a hare which is the origin of the Linnean name for the species.

* This fox has shorter and rounder ears than any variety of the red fox.

M 2

In most specimens, the fur has a bluish-gray colour at its roots on the back, the shoulders, and outside of the thighs, but particularly on the neck and tail. The proportion of the length of the fur so coloured, varies with the individual and the season of the year. In some it is confined to so small a space at the roots as to be scarcely perceptible, and in others it is so great as to tarnish the whiteness of its surface. At almost all times the short hair clothing the posterior surface and margins of the ears is of a dark brownish-gray colour for half its length, so as to give them a bluish or blackish tinge, whenever it is ruffled. No naked callous places exist on the soles of this fox in the winter time. The claws are long, compressed, slightly arched, and have a light horn colour.

Summer dress.—In the months of April or May, when the snow begins to disappear, the long white fur falls off, and is replaced by shorter hair, which is more or less coloured. A specimen, which was killed at York Factory on Hudson's Bay, in August, is described by Mr. Sabine as follows. " The head and chin are brown, having some fine white hairs scattered through the fur; the ears, externally, are coloured like the head ; within they are white: a similar brown colour extends along the back to the tail, and from the back is continued down the outside of all the legs, but on the latter a few white hairs are inter-mixed; the whole under parts, and the insides of the legs are dingy white ; the tail is brownish above, becoming whiter at the end, and is entirely white beneath."—On the approach of winter the fur lengthens, the white hairs increase in number, all the hairs become white at the tips, but retain more or less of the bluish or brownish-gray colour at the roots, until the fur is in *prime* winter order, when it is of its full length, and almost every where of a pure white colour from the roots to the tips.—The fur on the soles of the feet becomes thinner and shorter in the summer time, and several naked callous places then appear, but they are not so large as those which exist on the soles of the other North American Foxes at the same season of the year.

It is necessary to observe that the majority only of the Arctic Foxes acquire the pure white dress even in winter; many have a little duskiness on the nose, and others, probably young individuals, remain more or less coloured on the body all the year. On the other hand, a pure white Arctic Fox is occasionally met with in the middle of summer, and forms the variety named *kakkortak* by the Greenlanders. Hearne states that the Arctic Foxes, " when young, are almost all over of a sooty black ; but as the fall advances, the belly, sides, and tail turn to a light ash-colour ; the back, legs, some part of the face, and the tip of the tail, change to a lead colour, and when the winter sets in, they become perfectly white. There are few of them which have not a few dark hairs at the tip of the tail, all the winter*."

* Although I am not aware that a comparison between recent specimens of the Arctic Foxes of the New and Old Worlds has been made so as to prove their identity of form, yet their perfect similarity of habits, and in the series of variations in their fur, may lead us to conclude that the species is the same on both continents. The Siberian hunters informed Gmelin " that they often found gray and white individuals in the same litter, and that the first have at birth a very deep gray colour, the latter a yellowish tint, the hairs being in both very short. Towards the end of the summer, when the hair begins to increase in length, foxes are often met with, having a brown streak along the back, crossed by a similar one at the shoulders. These individuals, sometimes termed *cross foxes*, become at length entirely white." All the different species of fox seem liable to produce a crucigerous variety.

DIMENSIONS

Of a full-grown specimen.

	Feet.	Inches.		Feet.	Inches.
Length of the head and body . .	2	1	Distance from the wrist to the end of the middle fore-claw . . .	0	3½
„ tail (vertebræ) . . .	1	0	„ from the tip of the nose to the anterior angle of the eye . .	0	2¼
„ tail with the fur . . .	1	2	„ between the anterior angles of the eyes	0	1⅔
„ head . , . .	0	5¾	„ between the ears . .	0	2⅓
Height of ears anteriorly . . .	0	2			
„ ears posteriorly . . ,	0	1¾			
Breadth of the ears near their base . .	0	2			
Distance from the heel to the tip of the middle hind-claw . , . .	0	5½			

Captain Lyon, during two winters passed on the Melville peninsula, studied with attention the manners of several of these animals, which were taken and kept as pets; and his account contains so many interesting facts respecting their natural history, which are recorded nowhere else, that I shall make no apology for copying it into this work. " In form, the Arctic Fox bears a great resemblance to our European species, although considerably smaller; and, owing to the great quantity of white woolly hair with which it is clothed, is somewhat like a little shock dog. The brush is full and large, affording an admirable covering for the nose and feet, to which it acts as a muff when the animal sleeps. Although the head is not so pointed as in our English Reynard, yet it has completely the air of cunning which is so observable in all species of foxes. The eyes are bright, piercing, and of a clear hazel. The face of the female was always remarked to be shorter than that of the male, and it has less of cunning and more of mildness in its general expression. The ears are short, and thickly covered with hair, having the appearance of being doubled at the edges, or rather of having been cropped. The cheeks are ornamented by a projecting ruff, which extends from behind the ears quite round the lower part of the face, to which it gives a very pleasing appearance.

" The legs are rather long than otherwise, and shew great strength of muscle. The feet, which are large, are armed with strong claws. When the animal is standing still, the hind-legs are so placed as to give the idea of weakness in the loins, which is certainly not the case, as few creatures can make more powerful leaps. The general weight was about eight pounds, although some were as low as seven, and a few as high as nine pounds and a half when in good case.

" The Arctic Fox is an extremely cleanly animal, being very careful not to dirt those places in which he eats or sleeps. No unpleasant smell is to be perceived, even in a male, which is a remarkable circumstance. To come unawares on one of these creatures is, in my opinion, impossible; for even when in an apparently sound sleep, they open their eyes at the slightest noise which is made near them, although they pay no attention to sounds when at a short distance. The general time of rest is during the daylight, in which they appear listless and inactive ; but the night no sooner sets in than all their faculties are awakened ; they commence their gambols, and continue in unceasing and rapid motion until the morning. While hunting for food, they are mute, but when in captivity or irritated, they utter a short growl like that of a young puppy. It is a singular fact, that their bark is so undulated as to give an idea that the animal is at a distance, although at the very moment he lies at your feet. Although the rage of a newly-caught fox is quite ungovernable, yet it very rarely happened that on two being put together they quarrelled. A confinement of a few hours often sufficed to quiet these creatures ; and some instances occurred of their being perfectly tame, although timid, from the first moment of their captivity. On the other hand, there were some which, after months of coaxing, never became more tractable. These we supposed were old ones.

" Their first impulse on receiving food is to hide it as soon as possible, even though suffering from hunger, and having no fellow-prisoners of whose honesty they are doubtful. In this case snow is of great assistance, as being easily piled over their stores, and then forcibly pressed down by the nose. I frequently observed my dog-fox, when no snow was attainable, gather his chain into his mouth, and in that manner carefully coil it so as to hide the meat. On moving away, satisfied with his operations, he of course had drawn it after him again, and some-times with great patience repeated his labours five or six times, until in a passion, he has been constrained to eat his food without its having been rendered luscious by previous concealment. Snow is the substitute for water to these creatures, and on a large lump being given to them, they break it in pieces with their feet, and roll on it with great delight. When the snow was slightly scattered on the decks, they did not lick it up as dogs are accustomed to do, but by repeatedly pressing with their nose, collected small lumps at its extremity, and then drew it into the mouth with the assistance of the tongue." In another passage, Captain Lyon, alluding to the above mentioned dog-fox, says, " He was small and not perfectly white ; but his tameness was so remarkable, that I could not afford to kill him, but confined him on deck in a small hutch, with a scope of chain. The little animal astonished us very much by his extraordinary sagacity : for, during the first day, finding himself much tormented by being drawn out repeatedly by his chain, he at length, whenever he retreated to his hut, took this carefully up in his mouth, and drew it so completely after him, that no one who valued his fingers would endeavour to take hold of the end attached to the staple."

Hearne says, that, when taken young, the Arctic fox may be domesticated in some degree, but he never saw one that was fond of being caressed; and they are always impatient of confinement. Notwithstanding the degree of intelligence which the anecdotes related by Captain Lyon shew them to possess, they are unlike the red fox, in being extremely unsuspicious; and instances are related of their standing by, while the hunter is preparing the trap, and running headlong into it the moment he retires a few paces. Captain Lyon received fifteen from a single trap in four hours. The voice of the Arctic fox is a kind of yelp, and when a man approaches their breeding places, they put their heads out of their burrows, and bark at him, allowing him to come so near that they may be easily shot. They appear to have the power of decoying other animals within their reach, by imitating their voices. "While tenting, we observed a fox prowling on a hill side, and heard him for several hours afterwards in different places, imitating the cry of a brent goose *." They feed on eggs, young birds, blubber, and carrion of any kind; but their principal food seems to be lemmings of different species.

The Arctic fox is an inhabitant of the most northern lands hitherto discovered, and in North America their southern limit appears to be about latitude 50°. They are numerous on the shores of Hudson's Bay, north of Churchill, and exist also in Behring's Straits; but the brown variety, mentioned in the following pages, is the more common one in the latter quarter. They breed on the sea coast, and chiefly within the Arctic circle, forming burrows in sandy spots,—not solitary, like the red fox, but in little villages, twenty or thirty burrows being constructed adjoining to each other. We saw one of these villages on Point Turnagain, in latitude 68½°. Towards the middle of winter they retire to the southward, evidently in search of food, keeping as much as possible on the coast, and going much further to the southward in districts where the coast-line is in the direction of their march. Captain Parry relates that the Arctic foxes, which were previously numerous, began to retire from Melville Peninsula in November, and that by January few remained. Towards the centre of the Continent, in latitude 65°, they are seen only in the winter, and then not in numbers; they are very scarce in latitude 61°, and at Carlton House, in latitude 53°, only two were seen in forty years. On the coast of Hudson's Bay, however, according to Hearne, they arrive at Churchill, in latitude 59°, about the middle of October, and afterwards receive reinforcements from the northward, until

* Lyon's *Private Journal,* p. 424.

their numbers almost exceed credibility.   Many are captured there by the hunters, and the greater part of the survivors cross the Churchill River as soon as it is frozen over, and continue their journey along the coast to Nelson and Severn Rivers.   In like manner they extend their migrations along the whole Labrador coast to the Gulf of St. Lawrence.   Most of those which travel far to the southward are destroyed by rapacious animals ; and the few which survive to the spring, breed in their new quarters, instead of returning to the north.   The colonies they found are, however, soon extirpated by their numerous enemies.   A few breed at Churchill, and some young ones are occasionally seen in the vicinity of York factory. There are from three to five young ones in a litter.

The Esquimaux take the Arctic foxes in traps, which are described by Captain Parry as being " extremely simple and ingenious.   They consist of a small circular arched hut built of stones, having a square aperture at the top, but quite close and secure in every other part.   This aperture is closed by some blades of whale-bone, which, though in reality only fixed to the stones at one end, appear to form a secure footing, especially when the deception is assisted by a little snow laid on them.   The bait is so placed that the animal must come upon this platform to get at it ; when the latter, unable to bear the weight, bends downwards, and after pre-cipitating the fox into the trap, which is made too deep to allow of his escape, returns by its elasticity to its former position, so that several may then be caught successively."   They are also taken in the wolf-traps of ice, described in page 65, and all the rocky islands lying off the mouth of the Coppermine River are studded with square traps, built of stone by the Esquimaux, wherein the fox is killed by a flat stone falling upon him when he pulls at the bait.

The fur of the Arctic fox is of small value in commerce when compared with that of any variety of the red fox.   Its flesh, on the other hand, particularly when young, is edible ; whilst that of the red fox is rank and disagreeable.   Captain Franklin's party agreed with Hearne, in comparing the flavour of a young Arctic fox to that of the American hare.   Captain Lyon considered it to resemble the flesh of a kid.

Canis lag. var. β. fuliginosus. *Sooty Fox.*

Canis Lagopi varietas. Pallas, *glires*, p. 12. An. 1778.
Sooty Dog. Pennant, *Hist. Quad.*, vol. i. p. 257.
Kernektak. Fabricius, *Fauna Grœnl.*, p. 20. An. 1790.
Canis fuliginosus. Shaw, *Zool.*, vol. i. p. 331.
Blue Fox (and Canis fuliginosus.) Mackenzie's *Travels in Iceland*, p. 337.
Le chien brun. Desmarest, *Mamm. in notis*, p. 205.
Tree-innœuck-kannortoot (black foxes.) Esquimaux.

This is evidently a mere variety of the Arctic fox, similar to the black variety of the red fox, although more common. It has the form and stature of the Arctic fox, and may be easily distinguished from the black or silver fox of commerce, by its round ears, and its very inferior fur. It differs from the ordinary summer or winter states of the Arctic fox, in being almost entirely of an uniform blackish-brown colour.

DESCRIPTION.

One killed on Winter Island, in lat. 66°, on the 16th December, had the longest and darkest fur on the belly, the colour there being black with a slight tinge of brown. The face, from a sprinkling of short white hairs, was hoary, and there were a very few white hairs on the back, not sufficient, however, to vary the colour, unless on close inspection. The whole fur on the body was long, had a considerable lustre, and when blown inside, exhibited a bright ash-grey colour towards its roots. The fur on the soles of the feet was of a grayish-white colour, and as bushy as on the feet of the white variety in the winter time. The claws were of the same size and colour as those of the pure white variety, and differed in form from those of the red fox. This individual had attained the full size of the Arctic fox.

On Captain Franklin's last Expedition, similar specimens were seen in the summer near the mouth of the Mackenzie. A specimen procured by Captain Beechey in Behring's Straits, differed merely in having longer and finer fur, of a pure chestnut colour, without any admixture of white hairs. The face was yellowish-brown, with a white margin to the upper lip. The rounded ears were covered with silky fur on each side, and with shorter and paler fur on the margins. The fur on the tail was coarser, woolly, curled, and somewhat matted, and of a dull yellowish-brown, altogether very unlike the silky brush of the black variety of *Canis fulvus.* The tip of the tail in this specimen was of the same colour with the rest of it. The anterior surfaces of the legs were covered with yellowish-brown fur, forming a smooth shining coat, and the soles with dense yellowish-white, woolly hair.

N

Otho Fabricius gives a clear account of the sooty variety of the Arctic fox in the following passage:—" There are two kinds of Arctic fox," says he : " one bluish-black, with white wool on the soles of the feet, and occasionally white whiskers, is named by the Greenlanders *keknektak ;* the other entirely white with the exception of the naked tip of the nose, which is called by the same people *kakkortak.* They are by no means different species, for they couple together ; and one variety produces young having the colours of the other ; nay, I can even bear witness that the bluish individual will become white, and a white one bluish, according to its age."

Pennant considers the Sooty Fox to be a distinct species which is numerous in Iceland ; and Sir George Mackenzie describes it as varying considerably in the shades of its fur, from a light brownish, or bluish-gray, to a colour nearly approaching to black. He quotes the authority of Horrebow for their being brought from Greenland to Iceland occasionally on fields of ice. The Greenland fox, No. 164 of Pennant's History of Quadrupeds, described as of a brown colour above, white beneath, with feet furred beneath, rounded ears, and as being of very small stature, seems to be nothing more than a young Arctic fox in its common autumn dress.

On one occasion during our late coasting voyage round the northern extremity of America, after cooking our supper on a sandy beach, we had retired to repose in the boats, anchored near the shore, when two Sooty Foxes came to the spot where the fire had been made, and carrying off all the scraps of meat that were left there, buried them in the sand above high-water mark. We observed that they hid every piece in a separate place, and that they carried the largest pieces farthest off. A little Scotch terrier dog that accompanied us had precisely the same habit. It attended us closely at our meals ; and receiving much more from the men than it could eat, it carried the surplus always to the distance of two or three hundred yards, and hid it carefully, never putting two pieces in the same spot. I have quoted in page 86, Captain Lyon's observation of the Arctic fox seldom eating its food until it had been adapted to its taste by concealment.

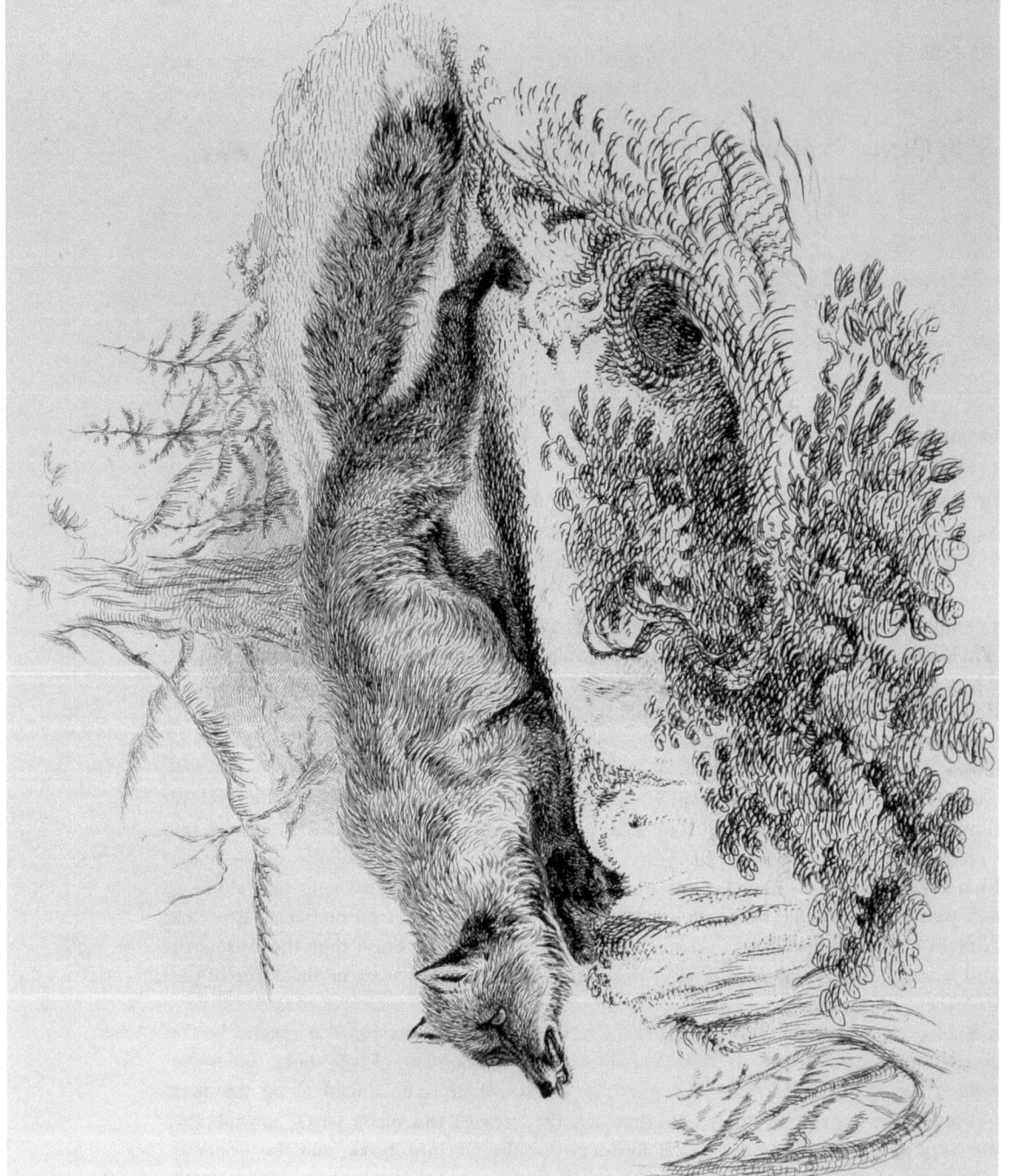

CANIS (VULPES) FULVUS.

Published by John Murray. January 1829.

[26.] 5. CANIS (VULPES) FULVUS. (Desmarest.) *The American Fox.*

European Fox. PENNANT, *Arct. Zool.*, vol. i. p. 45.
Red or large Fox. HUTCHINS, *MSS.*
Large Red Fox of the plains. LEWIS and CLARK, vol. iii. p. 29.
Renard de Virginie. PALISOT de BEAUVOIS, *Bul. Soc. Phil.*
Canis fulvus. DESMAREST, *Mamm.*, p. 203.
Red Fox. SABINE, *Franklin's Journ.*, p. 656.
Makkeeshew. CREE INDIANS.

PLATE VI.

This animal is very plentiful in the wooded districts of the fur countries, about eight thousand being annually imported into England from thence. It bears a strong resemblance to the Common European Fox, and until De Beauvois pointed out its peculiarities, it was considered to belong to that species.

DESCRIPTION.

On comparing a fine specimen of the English Fox with an American Red Fox, each were observed to have dark markings on the sides of the muzzle, posterior parts of the ears, and fore-part of the legs ; the tails of both have an intermixture of black hairs, and are tipped with white. The Red Fox, however, differs in its long and very fine fur, and in the brilliancy of its colours. Its cheeks are rounder, its nose thicker, shorter, and more truncated. Its eyes are nearer to each other. Its ears shorter, the hair on its legs is a great deal longer, and its feet are much more woolly beneath, the hair extending beyond the claws, which are shorter than those of the European Fox. In short the Red Fox differs from the European one in nearly the same characters that distinguish the gray American wolf from the Pyrenean one—in the breadth and capacity of its feet for running on the snow, the quantity of long hair clothing the back part of the cheeks, which in conjunction with the shorter ears and nose give the head a more compact appearance. The Red Fox has a much finer brush than the European one, and it is altogether a larger animal. Desmarest mentions differences in the form of the sculls of the two species.

Mr. Sabine describes a skin of the Red Fox in prime order as having " a general bright ferruginous colour on the head, back, and sides, less brilliant towards the tail; under the chin white ; the throat and neck a dark gray ; and this colour is continued along the first part of the belly in a stripe of less width than on the breast ; the under parts, towards the tail, are very pale red ; the fronts of the fore-legs and the feet are black, and the fronts of the lower parts of the hind-legs are also black ; the tail is very bushy, but less ferruginous than the body, the hairs mostly terminated with black, and more so towards the extremity

N 2

than near the root, giving the whole a dark appearance; a few of the hairs at the end are lighter, but it is not tipped with white." The colour of the tip of the tail differs in different specimens.

An individual, killed at Great Slave Lake, had its head and shoulders of a bright reddish-orange colour, which towards the rump acquired a gray tint by an intermixture of black and white hairs. On the tail the red hairs were mixed with gray and black ones, the tip was white. The soles of its feet were completely covered with fur. In the summer time the fur of the soles is worn off, and naked callous places appear, but they are not so large as in the English Fox. This specimen is now in the Museum of the Zoological Society, and from it the accompanying etching was made.

Its Dimensions are,

		Feet.	Inches.			Feet.	Inches.
Length of head and body	. .	2	9	Height at the shoulders	. .	1	1
,, head alone .	. .	0	8	,, of ears		0	2¾
,, tail (vertebræ)	. .	1	2	Distance from the end of the nose to the			
,, tail with the fur	. .	1	6	anterior angle of the eyes	. .	0	3

The Red Fox burrows in the summer, and in the winter takes shelter under a fallen tree. It brings forth four young about the beginning of May. They are covered at birth with a soft downy fur, of a yellowish-gray colour, the orange coloured hair not beginning to appear until they are five or six weeks old. Even the Indian hunters do not know the cubs at an early age from those of the Cross or Silver Foxes, and I therefore cannot now place the reliance I was once induced to do on their report of young cross foxes being found in the burrows of the Red Fox. I procured four cubs, a fortnight old, which several hunters said were cross foxes, but which proved eventually to be the red variety. These little creatures began very early to make burrows in the sandy floor of the house in which I kept them, and used to hide themselves during the day. They were, however, very tame, came when I called them, and would take food from my hand and carry it to their different places of concealment, never eating it when overlooked. I entertained hopes of bringing them to England, but they made their escape on the journey to the coast.

The Red Foxes prey much on the smaller animals of the rat family, but they are fond of fish, and reject no kind of animal food that comes in their way. They are taken in steel traps, and also in fall-traps, made of logs, but much nicety is required in setting them, as the animal is very suspicious. Some of the best fox-hunters in the fur countries ascribe their success to the use of *assafœtida, castoreum,* and other strong smelling substances with which they rub their traps and the small twigs set up in the neighbourhood, alleging that foxes are fond of such perfumes. The Red Fox hunts for its food chiefly in the night, but it is also frequently seen iu the day-time. In the winter time their tracts are most frequent on the borders of lakes, which they quarter much like a pointer dog. They turn aside to almost every stump or twig sticking up through the snow, and void their urine on it like a dog.

The Red Fox does not possess the wind of its English congener. It runs for about a hundred yards with great swiftness, but its strength is exhausted in the first burst, and it is soon overtaken by a wolf or a mounted huntsman. Its flesh is ill tasted, and is eaten only through necessity.

CANIS FULVUS, var. β. DECUSSATUS. *American Cross Fox.*

Renard barrè ou Tsinantontonque. SAGARD THEODAT, *Canada*, p. 745.
European Fox, var. β. Cross Fox. PENNANT, *Arctic Zool.*, vol. i. p. 46.
Canis decussatus. GEOFFROY, *Collect. du Mus.* SABINE, *Franklin's Journ.*, p. 656. HARLAN, *Fauna*, p. 88.
Cross Fox. HUDSON BAY COMPANY'S LISTS.
Beloduschki. RUSSIANS.

I am inclined to adhere to the opinion of the Indians in considering the Cross Fox of the fur traders to be a mere variety of the Red Fox, as I found on inquiry that the gradations of colour between characteristic specimens of the Cross and Red Fox are so small, that the hunters are often in doubt with respect to the proper denomination of a skin, and I was frequently told "This is not a cross fox *yet,* but it is becoming so." The *Canis crucigera* of Gesner, which is considered by Baron Cuvier to be a mere variety of the European Fox, differs from the latter animal in the same way that the American Cross Fox does from the red one, and there is also a crucigerous variety of the Isatis or Arctic Fox. Mr. Hutchins, however, remarks that the Cross Fox does not exceed the size of the European one, and is smaller than the Red Fox. If there really be a difference of size between the red and cross races inhabiting the same districts, they ought, perhaps, to be considered distinct species.

The fur of the Cross Fox is valuable, and some years ago it was worth four or five guineas a skin, whilst that of the Red Fox did not bring more than fifteen shillings. The difference of value seems to depend principally on the colour, for some of the red foxes appear to have as long and as fine fur.

Of a very characteristic Specimen of the Cross Fox, quoted from Mr. Sabine.

" The front of the head is gray, composed of black and white hairs, the latter predominating on the forehead ; the ears are covered with soft black fur behind ; and with long yellowish hairs within ; the back of the neck and shoulders are pale ferruginous, crossed with dark stripes, one extending from the head to the back, the other passing the first at right angles over the shoulders ; the rest of the back is gray, composed of black fur, tipped with white : the sides are pale ferruginous, running into the gray of the back ; the chin and all the under parts, as well as the legs, are black, a few of the hairs being tipped with white ; the under part of the tail and the parts of the body adjacent are pale yellow, the gray colour of the back extends to the upper part of the tail, at the commencement—the rest of the tail is dark above and lighter beneath, being tipped with white. The character of the fur is thick and long." The quantity of red fur, and the brightness of its colours, vary in different specimens; and the cruciform markings are scarcely apparent in some specimens, which from the fineness of their fur are acknowledged to be Cross Foxes.

CANIS FULVUS, var. γ. ARGENTATUS.   *Black or Silver Fox.*

Renard noir ou Hahyuha. SAGARD THEODAT, *Canada*, p. 744.
European Fox var. a. Black, PENNANT, *Arctic Zool.*, vol. i. p. 46.
Renard noir ou Argentè. GEOFFROY, *Collect. du Museum.*
Grizzle Fox. HUTCHINS, *MSS.*
Renard Argentè. F. CUVIER, *Mamm. lith.* 5 livr.
Canis argentatus. DESMAREST, *Mammal.*, p. 203. SABINE, *Franklin's Journey*, p. 657. HARLAN, *Fauna*, p. 88.
The Black or Silver Fox. GODMAN, *Nat. Hist.*, vol. i. p. 274.
Tschernoburi. RUSSIANS.

This variety is more rare than the Cross Fox, a greater number than four or five being seldom taken in a season at any one post in the fur countries *, though the hunters no sooner find out the haunts of one than they use every art to catch it, because its fur fetches six times the price of any other fur produced in North America. La Hontan speaks of a black fox skin as being in his time worth its

* Foxes of a corresponding colour seem to be equally rare in Europe. " The Silver or Black Fox is so rare, that seldom more than three or four are taken in the course of a year on the Lofoden islands, and I never heard of its having been met with in the other parts of Norway."—A. DE CAPELL BROOKE, *Travels in Norway*, p. 285.

weight in gold. Although, from what I observed, I do not think that the Black Fox displays more cunning in avoiding the snare than the red one, yet its rarity, and the eagerness of the hunters to take it, cause them to think it peculiarly shy. " It is to be remarked," says Pennant, " that the more desirable the fur is, the more cunning and difficult to be taken is the fox that owns it." Mr. Hutchins also informs us " the blacker the fur the lesser the fox," but neither is this latter remark consonant with my own observation.

Mr. F. Cuvier mentions that the smell of the American Black Fox is very disagreeable, but differs a little from that of the Common Fox of Europe. He thinks the identity of the American species with the Black Fox of the north of Europe doubtful. The Black Fox of America inhabits the same districts with the Red Fox, and is never seen far within the barren grounds. In some instances, however, the Sooty Fox described in page 89 may have been mistaken for it.

DESCRIPTION.

The *Canis argentatus* is sometimes found entirely of a shining black colour, with the exception of the end of the tail, which is white. It is more common to observe it with parts of its fur hoary from an intermixture of hairs tipped with white. A very fine specimen preserved in the Hudson's Bay Museum has the head and back hoary, most of the long hairs on those parts being white from the tip for a considerable way down. The downy fur at the root of the longer hairs has a dark blackish-brown colour. The nose, legs, sides of the neck, and all the under parts, are dusky, approaching to black. The tail is black. Its ears are erect, triangular, but not very acute, and are covered with a soft fur of a brownish-black colour. In some individuals the fur, which in most parts is hoary, has a shining black colour unmixed with white, from the crown of the head to the middle of the back and down the outside of the shoulders, being an approach to the cruciform arrangement. Like the two preceding varieties, the Black and Silver foxes have the soles of their feet thickly covered with wool in the winter, no callous spots being then visible.

[27.]   3. CANIS (VULPES) VIRGINIANUS. (Gmelin.)   *The Gray Fox.*

Gray Fox.   CATESBY, *Carolina*, vol. ii. p. 78. t. 78.   KALM, *Travels*, (*Pinkerton's Coll.*) vol. xiii. p. 467.
     PENNANT, *Arctic Zool.*, vol. i. p. 48.
Canis virginianus. GMELIN, *Syst.*, vol. i. p. 74.   SABINE, *Franklin's Journey*, p. 654.   HARLAN,
     Fauna, p. 89.
Virginian fox.   SHAW, *Zool.*, vol. i. p. 325.
Canis cinereo-argenteus.   SAY, *Long's Exped.*, vol. ii. p. 340.
The Gray Fox (Canis cinereo-argentatus).   GODMAN's *Nat. Hist.*, vol. i. p. 280.

This animal, which is said to be the most common species of fox in the southern parts of the United States, did not come under our notice on the late Expeditions, but it is here introduced to mark its most northern limit.   Its skins are sometimes included amongst the Hudson Bay Company's importations from their most southern Canadian posts.   Kalm says that the Gray Foxes are very common in Pennsylvania, and in the southern provinces ; but scarce in the northern ones, on which account the French call them Virginian foxes.   He also says that they are smaller, less destructive, less active, and have a less rank smell than the European foxes.   The Gray Fox has been confounded by some writers with the Cross Fox, which it much resembles, though it is smaller in size ; by others with the Kit Fox, which has also gray colours.   Dr. Godman informs us that the chase of this animal affords more pleasure to the American sportsmen than that of the Red Fox *, " because it does not immediately forsake its haunts and run for miles in one direction, but after various doublings, is generally killed near the place where it first started !"   Catesby, on the contrary, says that " they give no diversion to the sportsmen, for after a mile's chase they run up a tree."   The same author informs us that they breed in hollow trees.   Langsdorff relates that in California he saw a great number of foxes following the cows, and living upon the most friendly terms with the young calves.

* He alludes here, not to the *Canis fulvus*, but to the *C. vulpes vulgaris* of this work.

[28.]   4. Canis (vulpes vulgaris) vulpes? (Linn. ?)   *The Fox?*

Canis vulpes. Harlan, *Fauna*, p. 86.
Canis fulvus. Godman, *Nat. Hist.*, vol. i. p. 276.

M. Frederick Cuvier and M. Desmarest, who admit and describe the American Red Fox (*C. fulvus*) as a distinct species, state the *Common Fox* to be also an inhabitant of North America.   It does not exist in the countries north of Canada lying to the eastward of the Rocky Mountains, and consequently did not come under our notice on the late Expeditions ; but it is admitted into this work, as being most probably an inhabitant of New Caledonia.   Several of the voyagers who have visited the Atlantic coast of North America mention two kinds of red fox skins, in possession of the natives ; the one having a fine, long, silky fur, of a reddish-yellow colour (*C. fulvus?*) ; the other of a smaller size, having shorter and coarser fur, and less lively tints of colour (*C. vulpes?*).   I think it very probable that an investigation into the characters of the American foxes will shew that the reddish Fox of the Atlantic states is a variety of the *Canis cinereus*, which has been mistaken for the European Fox.   Dr. Godman states that these reddish foxes "are numerous in the middle and southern states of the Union, and are every where notorious depredators 'on the poultry yards."   Kalm says, " the red foxes are very scarce here (New York) ; they are entirely the same with the European sort.   Mr. Bartram, and several others, assured me, that, according to the unanimous testimony of the Indians, this kind of fox never was seen in the country before the Europeans settled in it.   But of the manner of their coming over I have two different accounts : Mr. Bartram, and several other people, were told by the Indians, that these foxes came into America soon after the arrival of the Europeans, after an extraordinary cold winter, when all the sea to the northward was frozen.   But Mr. Evans and some others assured me that the following account was still known by the people.   A gentleman of fortune in New England, who had much inclination for hunting, brought over a great number of foxes from Europe, and let them loose in his territories, that he might be able to indulge his passion for hunting.   This, it is said, happened at the very beginning of New England's being peopled with European inhabitants.   These foxes were believed to have so multiplied, that all the red foxes in the country were their

o

offspring." Kalm* considers neither of these accounts as satisfactory ; indeed, as to the former, none of the Indian tribes inhabiting New England could possibly have any knowledge of the state of the sea to the north, and to this day the tribes dwelling even twenty degrees of latitude nearer its shores are equally ignorant of it. The Esquimaux alone inhabit the coast, and it is unlikely that any accounts from them could be transmitted through ten or twelve intermediate nations, most of whom have been from time immemorial at war with their neighbours.

[29.]   6. CANIS (VULPES) CINEREO-ARGENTATUS. *The Kit Fox.*

Archithinew Fox. HUTCHINS, *MSS.* PENNANT, *Arct. Zool. Suppl.,* p. 52.
Kit Fox, or small burrowing Fox of the plain. LEWIS & CLARK, vol. i. p. 400 ; vol. iii. pp. 29, 282.
Canis velox. SAY, *Long's Exped.,* vol. ii. p. 339. HARLAN, *Fauna,* p. 91.
Canis cinereo-argentatus. SABINE, *Franklin's Journ.,* p. 658.
Le Renard tricolor. F. CUVIER, *Hist. Nat. des Mamm.*
The Swift Fox. GODMAN, *Nat. Hist.,* vol. i. p. 282.
Kit Fox. FUR TRADERS. Chien de Prairie†. CANADIAN VOYAGERS.

The species which forms the subject of this article burrows in the open plains extending from the Saskatchewan to the Missouri, and, according to Lewis and Clark, also in the plains of the Columbia. Mr. Sabine has referred it to the *Canis cinereo-argentatus* of Schreber, or the *Fulvous-necked Fox* of Shaw, and most probably correctly ; but many points with regard to its synonyms require to be cleared up, as authors in their descriptions appear to have confounded it with the gray or Virginian Fox. Schreber himself may have partly produced the error, by terming the animal *C. griseus* in his text, and *C. cinereo-argentatus* on the plate. It has long been known to the Hudson's Bay fur-traders, its skins forming a portion of their annual exports, under the name of *kit foxes.* It is, as Mr. Sabine justly remarks, the smallest of the American foxes ; but the measurement that he gives of two feet for the length of the head and body, being taken from a hunter's skin, which is always much stretched, is too great. I was unsuccessful in my endeavours to procure a recent specimen of this interesting little quadruped,

* KALM'S *Travels,* (*Pinkerton's Tr.*) pages 13 and 467.
† The name of " Prairie Dog" is bestowed also on the Louisiana marmot, and on other animals.

although I saw many hunters' skins. The Saskatchewan river is the northern limit of its range. Its burrows are formed in the open part of the plains, at a distance from the woody country. According to Mr. Say, it excels even the antelope in fleetness; and Lewis and Clark inform us that it is extremely vigilant, and betakes itself, on the slightest alarm, to its burrows, which are very deep. It seems to be the American representative of the *corsac,* inhabiting similar districts; and possibly like the corsac its fur changes its colours with the seasons.

DESCRIPTION
OF A HUNTER'S SKIN :—

The nose is considerably shorter, and the face broader than in other foxes. The upper part of the nose is covered by very short hairs of a pale yellowish, or wood, brown colour, on each side of which there is an oval patch of brownish fur, rendered hoary by many of its hairs being tipped with white. The whiskers are strong, and of a black colour, fading into brown at their tips. The portion of the lip anterior to them is brownish white; and the whole upper lip is margined by a stripe of white hairs about half an inch wide. There is, however, a narrow blackish-brown line between the white and the posterior angle of the mouth, which is prolonged round the margin of the lower lip. The upper part of the head, including the cheeks and orbits, the superior surface of the neck, the back and hips, are covered with fur of a pleasant grizzled colour, produced by an intermixture of hairs tipped with brown, black, and white. On the crown of the head, the yellowish-brown predominates, the white is equally diffused through it, there is no dark central line, and the grizzled colour unites gradually before the eyes with the unmixed fawn colour of the nose. The white hairs prevail immediately round the orbits, and there is much white on the cheeks. On the neck, where the fur lies smooth, the white, with a slight intermixture of black, is the colour of the surface, the yellowish-brown being seen only through the interstices of the longer hairs. Towards the rump less of the brown is seen, and more of the black hairs, but the white tips still predominate. The fur on the parts just enumerated appears, when blown aside, of a deep clove-brown, or brownish-gray colour from the roots for three-fourths of its length upwards; it is then yellowish-brown, followed by a very narrow ring of black, a larger ring of pure white, and generally a minute black tip. There are also, particularly towards the posterior part of the back, many interspersed hairs considerably longer than the others, which are black from the root to the tip. The breadth of the grizzled colours on the neck does not exceed the distance between the ears, but it gradually widens from the shoulders backwards. The sides of the neck, the shoulders and flanks, are of a dull reddish-orange, or very pale tile-red colour. The fur on these parts is longer but not so dense as that on the back, and is bluish-gray one half of its length, and reddish the remainder. On the flanks there are a few intermixed black hairs. The lower jaw is white, with a tinge of blackish-brown on its margins and towards its extremity. The chest exhibits the same reddish-orange colour with the sides; the throat and belly and inner surfaces of the extremities

O 2

are white. The fur on all the under parts is white towards the roots. The outside of the fore-legs and posterior parts of the hind-legs are brownish-orange, the upper surfaces of the feet are white; and the soles are covered, with the exception of some callous spots, with brownish wool. The *tail* is full of woolly hair, but tapers at the end. The prevailing colour of its upper surface is yellowish-gray, with a considerable intermixture of black, and a few white hairs; the under surface is brownish-orange, and the whole of the tip is black. The fur of the tail is ash-gray towards its base.

DIMENSIONS of a Hunter's Skin :—

	Inches.	Lines.		Inches.	Lines.
Length of head and body .	22	0	Distance from the tip of the nose to the anterior angle of the eye .	2	0
,,　　tail . .	10	0	,,　between the anterior angles of the eyes	1	2
,,　　tail including the fur at its tip	12	6	Length of whiskers . .	3	0

The dimensions of a specimen described by M. F. Cuvier, are—

	Inches.	Lines.		Inches.	Lines.
Length of head and body (English measure)	21	6	Length of tail . .	14	0
,,　head . .	5	0	Height of back . .	13	0

Mr. Say gives the dimensions of the cranium of his specimen, taken by calipers, as follows :—

	Inches.		Inches.
The entire length from the insertion of the superior incisors to the tip of the occipital ridge is rather more than	$4\frac{3}{10}$	Between the insertion of the lateral muscles at the junction of the frontal and parietal bones .	$\frac{5}{10}$
The least distance between the orbital cavities .	$\frac{9}{10}$	Greatest breadth of this space on the parietal bones	$1\frac{4}{10}$

[30.] 1. FELIS CANADENSIS. (Geoffroy.) *Canada Lynx.*

GENUS. Felis. LINN.
Loup-cervier. (Anarisqua.) SAGARD THEODAT, *Canada*, p. 747. An. 1636.
Loup-cervier or Lynx. DOBBS, *Hudson's Bay*, p. 41. An. 1744.
Cat-Lynx. PENNANT, *Arctic Zool.*, vol. i. p. 50.
Cat or Pishu. HUTCHINS, *MSS.*
Lynx or Wild Cat. HEARNE, *Journey*, p. 366. MACKENZIE, *Journey*, p. 106, &c.
Felis Canadensis. GEOFFROY, *Ann. du Mus.* SABINE, *Franklin's Journ.*, p. 659.
ZOOLOGICAL MUSEUM. No. 72.
Peeshoo. CREE INDIANS and CANADIAN VOYAGERS.

This is the only species of the genus which exists north of the Great Lakes, and eastward of the Rocky Mountains. It is rare on the sea-coast, and does not frequent the Barren Grounds, but it is not uncommon in the woody districts of the interior, since from seven to nine thousand are annually procured by the Hudson's Bay Company. It is found on the Mackenzie River, as far north as latitude 66°. It is a timid creature, incapable of attacking any of the larger quadrupeds; but well armed for the capture of the American hare, on which it chiefly preys. Its large paws, slender loins, and long, but thick, hind legs, with large buttocks, scarcely relieved by a short thick tail, give it an awkward, clumsy appearance. It makes a poor fight when it is surprised by a hunter in a tree; for, though it spits like a cat, and sets its hair up, it is easily destroyed by a blow on the back with a slender stick; and it never attacks a man. Its gait is by bounds, straightforward, with the back a little arched, and lighting on all the feet at once. It swims well, and will cross the arm of a lake two miles wide; but it is not swift on land. It breeds once a year, and has two young at a time. The natives eat its flesh, which is white and tender, but rather flavourless, much resembling that of the American hare.

The early French writers on Canada, who ascribed to it the habit of dropping from trees on the backs of deer, and destroying them by tearing their throats and drinking their blood, gave it the name of *Loup cervier.* The French Canadians now term it indifferently *Le Chat,* or *Le Peeshoo.* The mistake of Charlevoix in applying to it the appellation of *Carcajou,* which is proper to the wolverene, has produced some confusion of synonyms amongst subsequent writers. Pennant considered it as identical with the Lynx of the Old World; Geoffroy St. Hilaire named it as a distinct species; and Temminck has again, under the name of *Felis borealis,* described the species as the same in both hemispheres.

DESCRIPTION.

The *head* is round, the nose obtuse, and the face has much of the form of that of the domestic cat; but the facial line is more convex between the eyes. The *ears* are erect, triangular, and tipped by an upright, slender tuft of coarse, black hairs; they are placed about their own breadth apart, and on their posterior surface they have a dark mark beneath the tip, which is continued near both margins downwards towards their bases. On the *body* and *extremities* the fur is hoary, most of the hairs being tipped with white; on the crown of the head, and for a broad space down the middle of the back, there is a considerable intermixture of blackish-brown, and on the sides and legs, of pale wood-brown. In some specimens these colours produce an indistinct mottling, but in general there are no defined markings. A rufous tinge is also occasionally present about the nape of the neck, and on the posterior part of the thigh. The *tail* is coloured like the back, except the tip, which is black. The *fur* is close and fine on the back; longer and paler on the belly. When blown aside, it shews, on the middle of the back, a dark liver-brown colour from the roots to near the tips; but on the sides, it is, for the greatest part of its length, of a pale yellowish-brown, being merely a little darker near the roots. The *legs* are thick; the toes very thick and furry, and are armed with very sharp, curved, awl-shaped, white claws, shorter than the fur. There are four toes on each foot, those on the hind feet being rather the largest, but both feet have much spread.

DIMENSIONS
Of a prepared specimen.

	Feet.	Inches.		Feet.	Inches.
Length of the head and body . .	3	1	Height of the ear without the tuft measured behind	0	2
Height of the back	1	4½			
Length of the tail (vertebræ) . . .	0	4	Distance from the tip of the nose to the fore-part of the ear . . .	0	4¾
„ tail, with the fur . .	0	4½			
„ black tuft on the ear . .	0	1½			

[31.]   2. FELIS RUFA. (Guldensted ?) *Bay Lynx?*

Bay Lynx. PENNANT, *Hist. Quadr.*, No. 171 ? *Arctic Zool.*, vol. i., p. 51 ?
Felis rufa. GULDENSTED, *Act. Petrop.*, 20. p. 449 ?

Mr. Douglas brought a specimen of a Lynx from the Columbia River, that is reported to have the same habits with the Canada Lynx, which it much resembles in size and form.   No variety, however, of the latter inhabiting the fur countries to the eastward of the Rocky Mountains, presents the dark colour of the back, and the bright wood-brown on the sides, with the black spots on the belly, and the transverse black marks on the legs, exhibited by this one.   The hunters consider it to be quite distinct from the Canada Lynx.   Neither does it correspond entirely with the descriptions given by authors of the Bay Lynx, although it much resembles that animal in the markings about the face.   Mr. Douglas thinks that there are more than one non-descript animal of this genus, which inhabit the countries bordering on the Columbia.   The skins procured in that quarter are generally carried to the China market, without passing through the hands of European furriers ; hence, they are not likely to have come under the inspection of M. Temminck, who has so well described the Atlantic species.

DESCRIPTION.

Size and general aspect, exclusive of colour, that of a small Canada Lynx.   The *colour* of the hind head and of a broad dorsal stripe is blackish-brown, a little grizzled by a considerable number of the tips of the hairs being of a pale wood-brown.   On these parts the fur is hair-brown at its roots, and blackish-brown for the greater part of its length.   On the dorsal aspect of the neck the fur is reddish-brown from the base to near the tips, where the longer hairs are ringed with wood-brown and black ; the colour of the surface is produced by an intimate mixture of the two latter, but, as on the back, there are neither spots nor streaks. The *forehead* has a hoary brown colour, with dark brownish markings.   The eyelashes are black, the upper and under eyelids, most of the whiskers, and part of the upper lip, are white. The fur on the cheeks is yellowish-brown, with white tips, and there is a dark stripe under the eye, another on the back part of the cheek, and a third at the angle of the mouth.   The ears are lined interiorly with pale hairs, and are covered posteriorly with blackish-brown fur. The tufts on their tips, if any existed, have fallen off.   The *sides* of the neck and the *flanks* are pale chestnut, brown, rendered hoary by white tips equally but sparingly diffused.   The same colour prevails on the shoulders, and outer aspects of the fore and hind extremities, prettily varied by a number of short transverse stripes of blackish-brown ; the insides of the extremities are paler, but exhibit the same transverse dark marks. The *under jaw* is white, except the tip of the chin, which is brownish. The throat is white. The chest is coloured and spotted nearly like the shoulders. The belly is white, tinged anteriorly with brown, and marked throughout with rather large blackish-brown spots. The *tail* is reddish-brown, with some blackish markings above; and is white underneath. It is shaped like the tail of the Canada Lynx.

DIMENSIONS.

	Inches.	Lines.		Inches.	Lines.
Length of the head and body .	33	0	Length of the head, including the curvature		
„ tail (vertebræ) . .	4	0	of the forehead	6	6
			Height of the ear	2	0

[32.]   3. FELIS FASCIATA.   (Rafinesque.)   *Banded Lynx.*

The Tiger cat. LEWIS and CLARK, vol. iii. p. 28.
Lynx fasciatus. " RAFINESQUE, *Am. Month. Mag.* 1817, p. 46." DESMAREST, *Mamm.*
HARLAN, *Fauna*, p. 100.

I possess no other information respecting this animal than what is contained in the following description of it by Lewis and Clark.   It seems to bear considerable resemblance to the Canada Lynx, but differs from it, and from the preceding species, in the transverse dorsal stripes.

DESCRIPTION.

" The tiger cat inhabits the borders of the plains and the woody country in the neighbourhood of the Pacific. It is of a size larger than the wild cat of the United States, and much the same in form, agility, and ferocity. The colour of the back, neck, and sides is of a reddish-brown, irregularly variegated with small spots of dark-brown: the tail is about two inches long, and nearly white, except the extremity, which is black. It terminates abruptly, as if it had been amputated. The belly is white, beautifully variegated with small black spots ; the legs are of the same colour with the sides, and the back is marked transversely with black stripes : the ears are black on the outer side, covered with fine, short hair, except at the upper point, which is furnished with a pencil of hair, fine, straight, and black, three-fourths of an inch in length. The hair of this animal is long and fine, far exceeding that of the wild cat of the United States."

[33.]   1. Castor fiber, Americanus.   *The American Beaver.*

Genus. Castor. Linn.
Castor. Sagard Theodat, *Canada*, p. 767.
Castor fiber, Linn. *Syst.*
Beaver Castor. Pennant, *Arct. Zool.*, vol. i. p. 98.
Castor ordinaire. Desmarest, *Mamm.*
Castor Americanus. F. Cuvier.
Castor fiber. Harlan, *Fauna*, vol. i. p. 122.
The Beaver. Godman, *Nat. Hist.*, vol. ii. p. 21.
Ammisk. Cree Indians. Ttsoutayè. Hurons.

DESCRIPTION.

The Beaver has the form of an oval sack.   The greatest girth of its body is just before the hind legs, and it tapers gradually on every side from thence to the obtuse muzzle.   The hind legs are situated far forward, and the part of the body that projects behind them tapers pretty suddenly to the setting on of the flat scaly tail.   The *incisors* are smooth and orange-coloured anteriorly, and posteriorly they are narrower and white.   The *nose* is very obtuse both vertically and horizontally.   The eye is small, and is situated rather nearer to the ear than to the end of the nose; the pupil is almost closed in a strong light.   The *ears* are short, thick, rounded, and well clothed with short fur; the animal closes the auditory openings by folding them vertically.   The *fur* consists of a dense coat of somewhat waved, shining, smoke-gray down, concealed by a long coarse hair, which lies smooth, and, when in season, has a shining chestnut-brown colour.   In summer, the fur, previous to falling off, changes its colour to a pale yellowish-brown, and some of the winter specimens have a very dark hue, approaching to blackish-brown.   The *tail* is tongue-shaped, and is covered with oval, angular scales, which are not tiled, and are smallest along the margin of the tail.   They are not hard; some scattered hairs spring from their interstices, and the root of the tail is covered for a short space with finer but shorter hair than that of the back.   The tail is flat horizontally.   The *fore-extremities* are small and very short.   The toes are well separated, and, with the palms, are very flexible.   They are used like hands in conveying food to the mouth, but are so short that the animal is obliged to incline its head towards them.   The fore-claws are somewhat compressed, strong, and fitted for digging.   The middle one is the largest, the one on each side of it somewhat shorter, and the outermost and innermost are the two shortest.   The three exterior ones wear down, whilst the other two remain sharp.   The hind-feet have long, hard, and callous soles, and their long toes are connected by a web, which extends even beyond the roots of the nails.   The second toe has two nails, the under one of which is rounded with a cutting edge, and lies nearly at right angles to the upper one;—there is a less perfect double nail on the inner toe.   The other toes have simple

P

nails, which bear a striking resemblance to those of the human hand, but are rather more compressed. When the animal sits erect, it rests on its hind-legs and tail ; but when it walks, it puts its fore-feet to the ground, and, then arching its body, brings forward its hind-legs, and trails the tail behind it. The whole hind sole touches the ground as it walks, but only the toes of the fore-feet. When the fore-feet are lifted, the toes are curled together, or the fist is closed, which gives them a peculiar appearance of weakness, like a paralytic hand ; they are spread out when they touch the ground. The motions of the Beaver are slow when it is not pursued.

DIMENSIONS
Of a full-grown Beaver (Spec. 420, Zool. Mus.), killed at Great Slave Lake.

	Inches.	Lines.		Inches.	Lines.
Length of head and body . . .	40	0	Distance from tip of nose to anterior part of the eye	2	10
„ head alone . .	7	3			
„ tail, scaly part . . .	11	6	„ the posterior part of the orbit to anterior part of the ear . . .	2	5

DIMENSIONS
Of a recent specimen of what is termed, by the Fur Traders, a three-quarter Beaver.

	Inches.		Inches.
Length of head and body	30	Length of fore-feet	3
„ tail, scaly part . . .	10	„ of the sole of the hind foot . .	7
Circumference before hind legs . . .	30	Greatest breadth of the tail	5½
„ immediately behind fore-legs .	18		

The weight of a full-grown Beaver is about twenty-four pounds.

I have not had an opportunity of dissecting a Beaver; but I was informed by the hunters, that both males and females are furnished with one pair of little bags, containing *castoreum*, and also with a second pair of smaller ones betwixt the former and the anus, which are filled with a white fatty matter, of the consistence of butter, and exhaling a strong odour. This latter substance is not an article of trade; but the Indians occasionally eat it, and also mingle a little with their tobacco when they smoke. I did not learn the purpose that this secretion is destined to serve in the economy of the animal; but from the circumstance of small ponds when inhabited by Beavers being tainted with its peculiar odour, it seems probable that it affords a dressing to the fur of these aquatic animals. The *castoreum*, in its recent state, has an orange colour, which deepens, as it dries, into bright reddish-brown. During the drying, which is allowed to go on in the shade, a gummy matter exudes through the sack, which the Indians delight in eating. The male and female castoreum is of the same value, ten pairs of bags of either kind being reckoned to an Indian as equal to one beaver-skin. The castoreum is never adulterated in the fur countries. The flesh of the Beaver is much prized by the Indians and Canadian Voyagers, especially when it is roasted in the skin, after the hair has been singed off. In some districts it requires all the influence of the Fur Trader to restrain the hunters from sacrificing a considerable quantity of beaver fur every year, to secure the enjoyment of this luxury; and Indians of note have generally one or two feasts in a season, wherein a roasted Beaver is the prime dish. Hearne terms it delicious food. It resembles pork in its flavour, but the lean is dark-coloured, the fat oily, and it requires a strong stomach to sustain a full meal of it. The tail, which is considered a great luxury, consists of a grisly kind of fat, as rich, but not so nauseating, as the fat of the body.

The Beaver attains its full size in about three years; but breeds before that time. According to Indian report, it pairs in February, and after carrying its young about ten weeks, brings forth from four to eight or nine cubs, towards the middle or end of May. Hearne states the usual number of young, produced by the Beaver at a time, to be from two to five, and that he saw six only in two instances, although he had witnessed the capture of some hundreds in a gravid state *. The female has eight teats. In the pairing season the call of the Beaver is a kind of groan; but the voice of the cubs, which are very playful, resembles the cry of an infant. When the Beaver cuts down a tree it gnaws it all round, cutting it however somewhat higher on the one side than the other, by which the direction of its fall is determined. The stump is conical, and of such a height as a Beaver, sitting on his hind quarters, could make. The largest tree I observed cut down by them was about the thickness of a man's thigh (that is, six or seven inches in diameter;) but Mr. Graham says, that he has seen them cut a tree which was ten inches in diameter.

Pennant fixes the southern range of the American Beaver in latitude 30°, in Louisiana, not far from the Gulf of Mexico; whilst Say mentions the confluence of the Ohio and Mississippi as their limit, which is about seven degrees further to the northward. In high latitudes they are confined to the wooded districts, there not being even willows enough for their subsistence on the banks of the small lakes and rivulets of the Barren Grounds. Their most northern range is perhaps on the banks of the Mackenzie, which is the largest American river that discharges itself into the Polar Sea, and is also the best wooded, owing to the quantity of alluvial soil deposited on its banks. Beavers occur in that quarter as high as $67\frac{1}{2}$° or 68° of latitude, and their range from east to west extends from one side of the continent to the other, with the exception of the Barren districts. They are pretty numerous in the country lying immediately to

* I was informed by a hunter, that the Indians are accustomed, on breaking up a lodge of Beavers, to open the old female as soon as they kill her, for the purpose of ascertaining, by counting the dilatations of the tubular uterus, what number of young may be expected to be found in the washes Hearne also says, " On examining the womb of a Beaver, when not with young, there is always found a hardish round knob, for every young one of the last litter."

the northward of Fort Franklin ; and from the swampy and impracticable nature of the country, they are not likely to be soon eradicated from thence.   The Iroquois are the greatest Beaver takers in Canada, and their hunters now allot the beaver districts amongst themselves, and endeavour to preserve these animals from extinction, by trenching the beaver-dams of any one quarter only once in four or five years, and taking care to leave always a pair at least in a dam to breed. Further north the Indians, when they break up a beaver lodge, destroy, as far as they are able, both young and old, and the numbers of Beaver are consequently now very much reduced.   Gangs of Iroquois were also introduced into the fur countries to the north some years ago ; and by setting traps, which destroyed indiscriminately Beaver of all sizes, they almost extirpated the species from their hunting grounds.   The Hudson's Bay Company are, however, endeavouring to remedy this evil, by laying plans to insure an adequate supply of the very useful beaver-fur, although it is not likely that it can ever be so plentiful as it was formerly.   In the year 1743, the imports of beaver-skins into the ports of London and Rochelle, amounted to upwards of 150,000 ; and there is reason to suppose that a considerable additional quantity was at that period introduced, illicitly, into Great Britain.   In 1827, the importation of beaver-skins into London, from more than four times the extent of fur country than that which was occupied in 1743, did not much exceed 50,000.

In some seasons a great mortality occurs amongst the Beavers from some unknown cause, many being found dead in their lodges.   Towards the north, the fur of the Beaver is better, and continues in prime order through a greater portion of the year.   At Great Slave Lake, in latitude 61°, July, August, and September, are the only months in which the beaver-fur of inferior quality is procured.   In commerce, beaver-skins, cut open, stretched to a hoop, and dried in the ordinary manner, are named *beaver-parchment*, and form by far the greatest part of the importation. · When the beaver-skins have been made into dresses, and worn by the Indians, it is termed *beaver-coat ;* and, though it may have been in use a whole season, it still brings a good price.   Inferior sized skins are named *beaver-cub.* An incisor tooth of a beaver is fixed in a wooden handle by the northern Indians, and used with great dexterity to cut bone.   This was the instrument with which that people fashioned the horns of the rein-deer into spear-heads and fish-gigs ; but these bone weapons are now generally replaced by iron, and the beaver tooth has been supplanted by an English file.

The best account of the manners of the Beaver, and the most free of extravagancies, is that given by Hearne ; and it agrees so exactly with the information I received from the Indian hunters, that were I to record the latter it would appear to be borrowed almost entirely from that traveller. I therefore prefer giving it in Hearne's own words.

" The beaver being so plentiful, the attention of my companions was chiefly engaged on them, as they not only furnished delicious food, but their skins proved a valuable acquisition, being a principal article of trade as well as a serviceable one for clothing. The situation of the beaver-houses is various. Where the beavers are numerous, they are found to inhabit lakes, ponds, and rivers, as well as those narrow creeks which connect the numerous lakes with which this country abounds ; but the two latter are generally chosen by them when the depth of water and other circumstances are suitable, as they have then the advantage of a current to convey wood and other necessaries to their habitations, and because in general they are more difficult to be taken than those that are built in standing water. They always choose those parts that have such a depth of water as will resist the frost in winter, and prevent it from freezing to the bottom. The beavers that build their houses in small rivers or creeks, in which water is liable to be drained off when the back supplies are dried up by the frost, are wonderfully taught by instinct to provide against that evil, by making a dam quite across the river, at a convenient distance from their houses. The beaver dams differ in shape according to the nature of the place in which they are built. If the water in the river or creek have but little motion, the dam is almost straight ; but when the current is more rapid, it is always made with a considerable curve, convex toward the stream. The materials made use of are drift-wood, green willows, birch, and poplars, if they can be got ; also, mud and stones intermixed in such a manner as must evidently contribute to the strength of the dam ; but there is no other order or method observed in the dams, except that of the work being carried on with a regular sweep, and all the parts being made of equal strength. In places which have been long frequented by beavers undisturbed, their dams, by frequent repairing, become a solid bank, capable of resisting a great force both of water and ice ; and as the willow, poplar, and birch generally take root and shoot up, they by degrees form a kind of regular planted hedge, which I have seen in some places so tall, that birds have built their nests among the branches.

" The beaver-houses are built of the same materials as their dams, and are always proportioned in size to the number of inhabitants, which seldom exceeds four old, and six or eight young ones ; though, by chance, I have seen above double that number. Instead of order or regulation being observed in rearing their houses, they are of much ruder structure than their dams ; for, notwithstanding the sagacity of these animals, it has never been observed that they aim at any other convenience in their houses, than to have a dry place to lie on ; and there they usually eat their victuals, which they occasionally take out of the water. It frequently happens, that some of the large houses are found to have one or more partitions, if they deserve that appellation ; but it is no more than a part of the main building, left by the sagacity of the beaver to support the roof. On such occasions, it is common for those different apartments, as some are pleased to call them, to have no communication with each other but by water ; so that, in fact, they may be called double or treble houses, rather than different apartments of the same house. I have seen a large beaver-house built in a small island, that had near a dozen apartments under one roof ; and, two or three of these only excepted, none of them had any communication with each other but by water. As there were beavers enough to inhabit each apartment, it is more than probable that each family knew their own, and always entered at their own doors, without any further connection with their neighbours than a friendly intercourse, and to join their united labours in erecting their separate habitations, and building their dams where required. Travellers who assert that the beavers have two doors to their houses, one on the land side and the other next the water, seem to be less acquainted with these animals than others who assign them an elegant suite of apartments. Such a construction would render their houses of no use, either to protect them from their enemies, or guard them against the extreme cold of winter.

" So far are the beavers from driving stakes into the ground when building their houses, that they lay most of the wood crosswise, and nearly horizontal, and without any other order than that of leaving a hollow or cavity in the middle ; when any unnecessary branches project inward, they cut them off with their teeth, and throw them in among the rest, to prevent the mud from falling through the roof. It is a mistaken notion, that the wood work is first completed and then plastered ; for the whole of their houses as well as their dams are, from the foundation, one mass of mud and wood, mixed with stones, if they can be procured. The mud is always taken from the edge of the bank, or the bottom of the creek or pond, near the door of the house ; and though their forepaws are so small, yet it is held close up between them under their throat, that they carry both mud and stones, while they always drag the wood with their teeth. All their work is executed in the night ; and they are so expeditious, that in the course of one night I have known them to have collected as much mud as amounted to some thousands of their little handfuls. It is a great piece of policy in those animals to cover the outside of their houses every fall with fresh mud, and as late as possible in the autumn, even when the frost becomes pretty severe, as by this means it soon freezes as hard as a stone, and prevents their common enemy, the wolverene, from disturbing them during the winter. And as they are frequently seen to walk over their work, and sometimes to give a flap with their tail, particularly when plunging into the water, this has, without doubt, given rise to the vulgar opinion that they used their tails as a trowel, with which they plaster their houses; whereas that flapping of the tail is no more than a custom which they always preserve, even when they become tame and domestic, and more particularly so when they are startled.

" Their food consists of a large root *, something resembling a cabbage-stalk, which grows at the bottom of the lakes and rivers. They also eat the bark of trees, particularly those of the poplar, birch, and willow; but the ice preventing them from getting to the land in the winter, they have not any barks to feed on in that season, except that of such sticks as they cut down in summer, and throw into the water opposite the doors of their houses; and as they generally eat a great deal, the roots above-mentioned constitute a principal part of their food during the winter. In summer, they vary their diet, by eating various kinds of herbage, and such berries as grow near their haunts during that season. When the ice breaks up in the spring, the beavers always leave their houses, and rove about until a little before the fall of the leaf, when they return again to their old habitations, and lay in their winter-stock of wood. They seldom begin to repair the houses till the frost commences, and never finish the outer coat till the cold is pretty severe, as hath been already mentioned. When they erect a new habitation, they begin felling the wood early in summer, but seldom begin to build until the middle or latter end of August, and never complete it till the cold weather be set in.

" Persons who attempt to take beaver in winter should be thoroughly acquainted with their manner of life, otherwise they will have endless trouble to effect their purpose, because they have always a number of holes in the banks, which serve them as places of retreat when any injury is offered to their houses; and in general it is in those holes that they are taken. When the beaver which are situated in a small river or creek are to be taken, the Indians sometimes find

* Root of *Nuphar luteum.*—J. R.

it necessary to stake the river across, to prevent them from passing ; after which they endeavour to find out all their holes or places of retreat in the banks. This requires much practice and experience to accomplish, and is performed in the following manner. Every man being furnished with an ice-chisel, lashes it to the end of a small staff, about four or five feet long ; he then walks along the edge of the banks, and keeps knocking his chisel against the ice. Those who are acquainted with that kind of work, well know by the sound of the ice when they are opposite to any of the beavers' holes or vaults. As soon as they suspect any, they cut a hole through the ice big enough to admit an old beaver, and in this manner proceed till they have found out all their places of retreat, or at least as many of them as possible. While the principal men are thus employed, some of the understrappers, and the women, are busy in breaking open the house, which at times is no easy task, for I have frequently known these houses to be five or six feet thick ; and one, in particular, was more than eight feet thick in the crown. When the beavers find that their habitations are invaded, they fly to their holes in the banks for shelter ; and on being perceived by the Indians, which is easily done by attending to the motion of the water, they block up the entrance with stakes of wood, and then haul the beaver out of its hole, either by hand if they can reach it, or with a large hook made for that purpose, which is fastened to the end of a long stick. In this kind of hunting, every man has the sole right to all the beaver caught by him in the holes or vaults ; and as this is a constant rule, each person takes care to mark such as he discovers by sticking up a branch of a tree by which he may know them. All that are caught in the house are the property of the person who finds it. The beaver is an animal which cannot keep under water long at a time, so that when their houses are broke open, and all their places of retreat discovered, they have but one choice left, as it may be called, either to be taken in their house or their vaults : in general they prefer the latter ; for where there is one beaver caught in the house, many thousands are taken in the vaults in the banks. Sometimes they are caught in nets, and in summer, very frequently, in traps.

" In respect to the beaver dunging in their houses, as some persons assert, it is quite wrong, as they always plunge into water to do it. I am the better enabled to make this assertion, from having kept several of them till they became so domesticated as to answer to their name, and follow those to whom they were accustomed, in the same manner as a dog would do ; and they were as much pleased at being fondled as any animal I ever saw. In cold weather they were kept in my own sitting room, where they were the constant companions of the

Indian women and children, and were so fond of their company, that when the Indians were absent for any considerable time, the Beaver discovered great signs of uneasiness; and on their return, shewed equal marks of pleasure by fondling on them, crawling into their laps, lying on their backs, sitting erect like a squirrel, and behaving like children who see their parents but seldom. In general, during the winter, they lived on the same food as the women did, and were remarkably fond of rice and plum-pudding: they would eat partridges and fresh venison very freely, but I never tried them with fish, though I have heard they will at times prey on them. In fact there are few graminivorous animals that may not be brought to be carnivorous."

Castor fiber, var. B, nigra. *Black Beaver.*

Castor fort noir. SAGARD THEODAT, *Canada*, p. 767.
Castor fiber. var. B. Castor noir. DESMAREST, *Mamm.* p. 278.

Beaver, entirely black, but not differing in any other respect from those of the ordinary dark brown colour, are of occasional occurrence. I saw one or two which were kept as curiosities. Hearne, in speaking of this variety, says, " Black Beaver, and that of a beautiful gloss, are not uncommon; perhaps they are more plentiful at Churchill than at any other Factory in the Bay; but it is rare to get more than twelve or fifteen of their skins in the course of one year's trade."

Q

Castor fiber, var. C, varia. *Spotted Beaver.*

Castor fiber, var. D. Castor varié. Desmarest, *Mamm.*

This variety is more rare than the preceding, and never came under my notice. Mr. Say mentions that an Indian had in the course of his life caught three specimens of Beaver with a large white spot on their breasts.

Castor fiber, var. D, alba. *White Beaver.*

White beaver. Dobbs, *Hudson's Bay*, p. 40.
Castor fiber, var. C. Castor blanc. Desmarest, *Mamm.*
Castor albus. Brisson, *Reg. An.*

An albino variety of the Beaver is of very rare occurrence. Hearne saw but one in the course of twenty years, and it had many reddish and brown hairs along the ridge of the back,—its sides and belly were of a glossy silvery white. When the Indians find an individual of this kind, they convert the skin into a medicine bag, and are very unwilling to dispose of it.

[34.] 1. FIBER ZIBETHICUS. (Cuvier.) *The Musquash.*

GENUS Arvicola, *Sub-genus* Fiber. CUVIER.
Rat-musqué. SAGARD THEODAT, *Canada*, p. 771.
Castor Zibethicus. LINN. *Syst.*, xii. 1. p. 79.
L'Ondatra. BUFFON, tom. x. p. 1.
Musk-rat. LAWSON, *Carolina*, p. 120.
Musk-beaver. PENNANT, *Arctic Zool.*, vol. i. p. 106.
Musquash. JOSSELYN, *New England.* HEARNE, *Journey*, p. 379.
Mus Zibethicus. LIN. GMELIN, vol. i. p. 125.
Fiber Zibethicus. SABINE, *Franklin's Journey*, p. 659. HARLAN, *Fauna*, p. 132.
Musk-rat. GODMAN, *Nat. Hist.*, vol. ii. p. 58.
Ondathra. HURONS.
Musquash, watsuss, or wachusk, also peesquaw-tupeyew (the animal that sits on the ice in a round form). CREE INDIANS.

DESCRIPTION.

The Musquash, or, as it is often named, the Musk-rat, has a thick flattish body, with a short head, indistinct neck, thighs hid in the body, very short legs, and large hind-feet. Its tail is compressed laterally, and has a length nearly equal to that of the body, excluding the head. It is furnished with large yellowish *incisors*, of which the upper ones are flatly rounded anteriorly, without grooves, and obliquely truncated on the cutting edge. The lower ones are chiselled away posteriorly, so as to come nearly to a point at the extremity, and are somewhat longer than the upper ones. The *lips*, covered with coarse hair, turn inwards. The *nose* is short, thick, and obtuse, and is covered with short hair. The *eyes*, small and lateral, are much hidden by the fur. The *ears*, low and oblong, are covered with hair like that on the adjoining parts of the head, and are not conspicuous.

The *fur* much resembles that of the beaver, but is shorter; the down is coarser and of much less value, and the long hairs are less strong and shining, and do not form so close a coat. Although the fur of the Musquash resists the water when the animal is alive, it is easily wetted immediately after death. The fur on the upper parts is somewhat longer than that beneath. Its colour externally is a dark umber-brown on the whole upper part of the head, including the ears, between the shoulders and on the back. The sides, anterior part of the belly, middle of the breast, lateral parts of the neck, and the cheeks, are of a shining yellowish-brown hue, the tint being deepest on the sides, but fading on the belly and cheeks into light wood-brown. The chin, throat, sides of the chest, and posterior part of the abdomen, are ash-gray, owing to the down being intermixed with few long hairs, and, consequently, more visible on those parts. The down on the upper parts exhibits, when blown aside, a dark lead-gray or blackish-gray colour from the roots upwards, the tips alone being tinged with brown; on the under surface of the animal the down has a lighter bluish-gray colour, and its tips are brownish-gray.

Q 2

The *fore-extremities* are short, only the wrist and fingers being visible beyond the body. They are covered exteriorly by a short, smooth, shining coat of hair, to the roots of the nails. Interiorly some long hairs curve over the wrist, and the palms and inner surfaces of the toes are naked. The nails are short, conical, very slightly curved, and much compressed. The second toe is the longest, the third is very nearly equal to it, the first is a little shorter, and rises higher up, and the thumb is by much the shortest and furthest back of all. The thumb has, however, a conspicuous phalanx, and its claw is as long and is of the same shape with those of the fingers, so that it ought not to be termed merely rudimentary.

Nearly as little of the *hind-legs* appears as of the fore ones, but the feet are very much longer. The metatarsal bones are considerably longer than the toes, and the latter are separated their whole length. The hind-feet are turned obliquely inwards. Exteriorly, they have a shining, smooth, hairy covering, similar to that on the fore-feet, and, like it, of a grayish-brown colour; but the margins of the soles and toes are furnished with an even row of long, shining, pale grayish-white hairs, curving inwards. The under surface of the feet is naked from the heel to the claws. The inner and outer toes arise nearly opposite to each other, and are about the same length. The remaining three arise from longer metatarsal bones, and their phalanges are nearly equal in length to each other. The two middle ones are united by a web, for about half the length of their first phalanges, and there is also a short web between the third and fourth toes. The *claws* of all the hind-toes are rather large, conical, slightly arched, thin, whitish, and excavated underneath. The hairs of the hind-feet do not extend much beyond the roots of the nails. From the shortness of its extremities, the Musquash runs badly, and is easily overtaken on land; but it swims and dives well, though it cannot continue long under water.

Its *tail* is compressed, convex on the sides, with its acute edges in a vertical plane. It is covered with a thin, sleek coat of short hairs, which allow a number of small, roundish scales, well separated from each other, to appear through them. Both hairs and scales are of a dusky-brown colour. The acute margins of the tail are covered with a close line of longer hairs, those on the upper edge being of a dark-brown colour, and those on the under one of a soiled-white. The tail is rather thicker in the middle than at the root, and it tapers gradually from its middle to its extremity, which is not acute.

DIMENSIONS.

	Inches.	Lines.		Inches.	Lines.
Length of head and body	14	0	Length of fore-feet to the end of middle claw	1	3
„ tail	8	6	„ hind feet, from heel to the end of		
„ head	3	4	the middle claw	3	2
„ whiskers	2	0	„ hind claws	0	6
„ lower incisors	0	9	„ from end of the nose to the eye	1	3

There is a considerable variation in the size of individuals, as is common in all the species of the Linnean genus *mus*.

The Musquashes have a strong musky smell, particularly the male ones in the spring time; yet their flesh is eaten by the Indians, and when it is fat they prize it for a time, but are said to tire of it soon—it somewhat resembles flabby pork.

In latitude 55°, the Musquash has three litters in the course of the summer, and from three to seven young at a litter. They begin to breed before they attain their full growth. The districts in which they are most abundant are subject to inundations, which, covering all the low grounds, leave no resting places for these animals, and destroy great numbers; in severe winters, also, they are sometimes almost extirpated from certain parts of the country by the freezing up of the swamps, which they inhabit. In such cases, being deprived of their usual food, they are driven by famine to destroy each other. They are likewise subject at uncertain intervals to a great mortality from some unknown cause. Their great fecundity, however, enables them to recover these losses in a very few years, although the deaths are at times so numerous, that a fur-post, where the Musquash is the principal return, is not unfrequently abandoned until they have recruited.

The southern limit of the range of the Musquash may be stated to be some-where about latitude 30°. Bartram informs us, that they exist in the northern parts of Georgia and Florida; and we have ascertained that they extend northwards nearly to the mouth of the Mackenzie, in latitude 69°. Their favourite abodes are small grassy lakes or swamps, or the grassy borders of slow-flowing streams where there is a muddy bottom. They feed chiefly on vegetable matters; and in northern districts principally on the roots and tender shoots of the bulrush and reed-mace, and on the leaves of various carices and aquatic grasses. The sweet-flag (*acorus calamus*), of whose roots, according to Pennant, they are very fond, does not grow to the northward of Lake Winipeg. In the summer, they frequent rivers, for the purpose it is said of feeding upon the fresh-water mussels (*Unio*). We often saw small collections of mussel-shells on the banks of the larger rivers, which we were told had been left by them.

In the autumn, before the shallow lakes and swamps freeze over, the Musquash builds its house of mud, giving it a conical form, and a sufficient base to raise the chamber above the level of the water. The chosen spot is generally amongst long grass, which is incorporated with the walls of the house, from the mud being deposited amongst it, but the animal does not appear to make any kind of composition or mortar by tempering the mud and grass together. There is, however, a dry bed of grass deposited in the chamber. The entrance is under water. When ice forms over the surface of the swamp, the Musquash makes breathing holes through it, and protects them from the frost by a covering of mud. In severe winters, however, these holes freeze up, in spite of their coverings, and many of the animals die. It is to be remarked that the small grassy lakes selected by the Musquash for its residence, are never so firmly frozen nor covered with such thick ice as deeper and clearer water. The Indians kill these animals by spearing them through the walls of their houses, making their approach with great caution, for the Musquashes take to the water when alarmed by a noise on the ice. An experienced hunter is so well acquainted with the direction of the chamber, and the position in which its inmates lie, that he can transfix four or five at a time. As soon as, from the motion of the spear, it is evident that an animal is struck, the house is broken down, and it is taken out. The principal seasons for taking the Musquash are the autumn before the snow falls, and the spring after it has disappeared, but while the ice is still entire. In the winter time, the depth of snow prevents the houses and breathing holes from being seen. One of the first operations of the hunter is to stop up all the holes, with the exception of one, at which he stations himself to spear the animals that have escaped being struck in the houses, and come thither to breathe. In the summer, the Musquash burrows in the banks of the lakes, making branched canals many yards in extent, and forming its nest in a chamber at the extremity, in which the young are brought forth. When its house is attacked in the autumn, it retreats to these passages; but in the spring they are frozen up.

The Musquash is a watchful, but not a very shy animal. It will come very near to a boat or canoe, but dives instantly on perceiving the flash of a gun. It may be frequently seen sitting on the shores of small muddy islands, in a rounded form, and not easily to be distinguished from a piece of earth, until, on the approach of danger, it suddenly plunges into the water. In the act of diving, when surprised, it gives a smart blow to the water with its tail. Hearne states, that it is easily tamed, soon grows fond, is very cleanly and playful, and smells pleasantly (!) of musk.*

The fur of this animal is used in the manufacture of hats. Between four and five hundred thousand skins are annually imported into Great Britain from North America.

---

* It is singular that Hearne, who must have seen vast numbers of these animals, should describe the hind-feet as webbed. There is no vestige of a web; although the marginal row of long hairs fits the feet to act as oars.

FIBER ZIBETHICUS, var. B, NIGRA. *Black Musquash.*

It is not uncommon to find a Musquash of a very dark brown colour, approaching to black; but one covered with fur of a pure black colour is rare, though of occasional occurrence.

FIBER ZIBETHICUS, var. C, MACULOSA. *Pied Musquash.*

I have seen, in the possession of an Indian, the skin of a Musquash which was ariegated with dark, blackish-brown patches, on a white ground.

FIBER ZIBETHICUS, var. D, ALBA. *White Musquash.*

Fiber Zibethicus—albus. SABINE, *Franklin's Journ.*, p. 660.

An albino variety of the Musquash is not very unfrequent: I have seen several.

[35.] 1. ARVICOLA RIPARIUS. (Ord ?) *Bank Meadow Mouse.*

GENUS, Arvicola. LACEPEDE. CUVIER.
Arvicola riparius. ORD, *Journ. Acad. Sciences Phil.*, vol. iv. p. 305 ?
Marsh Campagnol. GODMAN, *Nat. Hist.*, vol. ii. p. 67.

A (riparius ?) supra hepatico-brunneus (ex fusco fuligneoque mixtis), subter plumbeus, auriculis mediocribus pilis obvelatis, cauda longitudine capitis, pedibus albidis.
Bank Meadow-mouse of a dull brown colour, intimately mixed with black ; beneath bluish-gray : ears of a moderate size, nearly hid by the fur ; tail the length of the head ; feet white.

It is so difficult to discriminate the different Meadow-mice by mere descriptions, that I have much hesitation in referring any of my specimens to those which have been named by authors. I have not had access to a museum containing many species ; and their forms and colours differ so little, that figures, unless very accurate, tend rather to mislead. Five species are common in the Hudson's Bay countries, exclusive of the lemmings ; and there are doubtless others which did not come under our notice. A considerable number have been described as inhabitants of the United States by Rafinesque, Ord, and others ; but the American naturalists are by no means agreed about the species, and have applied the names variously. The Meadow-mouse which I have referred to Mr. Ord's *riparius*, corresponds with the short account by that author in the Journal of Science ; but the description of Mr. Ord's specimen by Dr. Harlan, in his Fauna Americana, under the name of *Arvicola palustris*, differs in several particulars, and agrees more nearly with the *Arvicola xanthognathus* of the following article.

The animal which I am now about to describe was procured by Mr. Drummond, near the foot of the Rocky Mountains. Its manners are analogous to those of the common water-rat (*Arvicola amphibius*), with which it may be easily confounded, although the shortness of its tail may serve as a mark of distinction. It frequents moist meadows amongst the mountains, and swims and dives well, taking at once to the water when pursued. It is distinguished from the other American species of this genus which have come under my notice, by the length and strength of its incisors, which are twice the size of those of the *Arvicola xanthognathus*, although the latter is the largest animal of the two.

DESCRIPTION.

Shape—The *head* is rather large, and not easily distinguishable from the neck ; the *incisors* are much exposed, and project beyond the nose ; the upper ones are flattish ante riorly are marked with some scarcely perceptible perpendicular grooves, and have a somewhat irregular and rather oblique cutting edge; the lower incisors are twice the length of the upper ones, narrower, slightly curved, and rounded anteriorly. *Nose* thick and obtuse. *Whiskers* black; scarcely of the length of the head. *Eyes* small, much concealed by the fur. *Ears* moderately large, oval, rounded at the tip, covered on the outside with fur similar to that on the neighbouring parts, and on that account not easily distinguishable until the fur is blown aside. *Body* more slender posteriorly, the hind-legs not being so far apart as the fore ones. *Tail* about the length of the head, somewhat flat horizontally, tapering, and thinly covered with short hairs, which at the end form a small pencil-like point. *Fore-legs* short; feet rather small, with four, slender, well-separated toes, and the rudiment of a thumb, which is armed with a minute nail. Claws small, white, compressed, and pointed. The third toe nearly equals the middle one, which is the longest; the first is shorter than these two, and the outer one, which is the shortest of all, is half the length of the middle one. The hairs of the toes project over the claws, but do not conceal them. The toes of the *hind-feet* are longer than those of the fore ones, and their claws are also somewhat longer. The inner one is the shortest, the second is longer than the third, and the third than the fourth; but the difference between the three is but just perceptible. The first and fifth are considerably shorter than the others, and are situated further back. The first, or inner one, which is the smallest, from its shortness and position, resembles a thumb. The hind toes are turned a little inwards, as is usual in the meadow-mice; but there is no provision either of webs or the arrangement of the hairs to give them much power in swimming.

The *fur* on the back is about eight lines long, but not so soft and fine as in some others of the genus: it is nearly as long on the crown of the head and cheeks; but it is shorter and thinner on the chest and belly. The specimen described was killed in summer; in the winter perhaps the fur may be of a better quality.

Colour—Incisors yellow. The whole dorsal aspect, including the shoulders and outside of the thighs, is of a dull, dusky, dark brown, proceeding from an intimate mixture of yellowish-brown and black. These colours are confined to the tips of the hairs, and are so mingled as to produce a nearly uniform shade of colour without lustre. From the roots to near the tips the fur has an uniform shining blackish-gray colour. The ventral aspect is bluish-gray. The margin of the upper lip, the chin and feet, are dull white. The tail is dark-brown above and whitish beneath, the two colours meeting by an even line.

DIMENSIONS.

	Inches.
Length of head and body	7
„ tail	2

The depressed or flat tail may cause this animal to rank with the *Mynomes* of Rafinesque; but it is certainly not the same with the Meadow Mouse figured in Wilson's Ornithology, plate 50, with which Desmarest (Mamm. p. 286) unites it.

R

[36.] 2. ARVICOLA XANTHOGNATHUS. (Leach.) *Yellow-cheeked Meadow-Mouse.*

A. xanthognathus. LEACH. *Zool. Misc.,* vol. i. p. 60. t. xxvi.

A. (*xanthognathus*) *badio nigroque varius* (*nec tamen maculosus*) *ventre argenteo-cinereo, malis fulvis, pedibus fuscescentibus subtus albidis.*
Yellow-cheeked Meadow-Mouse, with a brown and black dorsal aspect, silvery-gray belly, dull orange-coloured cheeks, and brown feet.

DESCRIPTION.

Teeth corresponding in number with the rest of the genus. *Incisors,* pale-yellow, exteriorly. Lower ones longer, paler, and nearly round. Upper ones shorter, stronger, slightly rounded, with even cutting edges. Of the *upper molar* teeth the posterior one is the largest, and has three grooves on its sides. The two anterior ones have two grooves each, making, in all, ten ribs or projecting angles, in the upper molar teeth of each side. Of the *lower molars,* the anterior one is the largest, and has four grooves : the other two have two each, forming, in all, eleven ribs, which correspond to the angles of so many triangles on the grinding surfaces.

Form.—The body is nearly cylindrical, of the size of the water-rat, legs short, nose rather obtuse, its tip on a line with the incisors. *Ears,* nearly circular, rather large when compared with those of other meadow-mice, sparingly hairy within, well covered exteriorly with fur of the same colour with the rest of the superior parts. *Whiskers,* about the length of the head. *Tail,* shorter than the head, tapering, well covered with hairs, lying smoothly and coming to a point at the end. *Extremities*—legs, covered with short hairs, lying closely and smoothly. The *fore-feet* have naked palms, and four toes, with a callus protected by a very minute nail, in place of a thumb; the first toe is a little shorter than the third ; the second the longest, and the fourth the shortest. The toes are well covered with smooth hair above, and are naked below. The hair of the wrist projects a little over the palms. The claws are small. The *hind-feet* have five toes, of which the three middle ones are nearly equal in length ; the outer one is considerably shorter, and the inner one is the furthest back and the shortest. The posterior half of the sole is covered with hairs, which curve inwards. The soles of the hind-feet are narrower and longer than the palms of the fore-ones. *Fur,* soft and fine ; about four lines and a half long on the head, and nine on the posterior part of the back.

Colours.—The colour of the fur, from the roots to near the tips, is shining grayish-black. On the dorsal aspect of the head and body, the tips of the hairs are yellowish-brown or black, the black-pointed hairs being the longest. The colour resulting is a mixture of dark-brown and black, without spots, and appearing of different shades when moved in the light. The sides are a little paler than the back. The under parts are of a silvery bluish-gray, darkening into blackish-gray on two large patches anterior to the shoulders. There is a blackish-brown stripe along the centre of the nose. On each side of it there is a reddish-brown patch, which extends from the mouth to the orbit. Whiskers black. Tail brownish-black above, whitish beneath. Extremities, dark-brown, exteriorly; whitish, interiorly.

DIMENSIONS.

	Inches.	Lines.		Inches.	Lines.
Length of head and body . from 5½ to 8		0	Length of middle fore-toe and claw .	0	3
,, head	1	10	,, ,, hind-toe and claw .	0	3½
,, tail	1	6	,, hind-foot from heel to point of the		
,, ears (breadth or height) . .	0	7	claw of middle toe	0	10

This species makes long canals under the mossy turf on the dry banks of lakes and rivers, and also in woods, but does not burrow deep into the earth. It is plentiful in some quarters, but shews no disposition to enter the houses of the traders, and domesticate itself, like the following species. It is common in the immediate vicinity of Fort Franklin; and Mr. Drummond found it in abundance on the Rocky Mountains in latitude 56°, in places where the woods had been destroyed by fire. It has about seven young at a birth. It was first described by Dr. Leach; and an indifferent figure, half the natural size of a specimen, which he obtained from Hudson's Bay, was published in the Zoological Miscellany. Mr. Say, in the narrative of Long's expedition, mentions a Meadow-Mouse, which he terms the *Arvicola xanthognathus*, as an inhabitant of the banks of the Ohio, but gives no description; and Godman, who speaks of an animal under the same name, as common in the United States, and doing great injury to the banks, alludes to its *diminutive* size, and evidently refers to some other species. The description quoted by Mr. Sabine, in Franklin's Journey, under the title of *Arvicola xanthognatha*, does not belong to this animal, but to a much smaller species, which I have referred to the *Arvicola Pennsylvanicus* of ORD, in the following article.

[37.]   3. ARVICOLA PENNSYLVANICUS. (Ord.) *Wilson's Meadow-Mouse.*

Short-tailed Mouse. FORSTER, *Phil. Trans.*, vol. lxii. p. 380.   No. 18.
Meadow-Mouse. PENNANT's *Arctic Zool.*, vol. i. p. 133 ? *   WILSON, *Am. Ornith.*, vol. vi. p. 59. t. 50. f. 3.
Arvicola Pennsylvanioa.   " ORD, *Guthrie's Geography*" (quoted from Harlan.)   HARLAN, *Fauna*, p. 145.

*A. (Pennsylvanicus) rostro obtuso, auriculis vellere subcelatis, caudâ benè vestitâ obtusâ dimidium capitis longitudine
æquanti, corpore fusco subter griseo-albo.*
Wilson's Meadow-Mouse, with an obtuse snout, a blunt hairy tail, half the length of the head, back brown, belly
nearly white.

This campagnol was considered by Forster and Pennant to be specifically the
same with the Meadow-Mouse of the old continent (*Mus agrestis*, LINN. *Mus
arvalis*, PENN. *Arvicola vulgaris*, DESM.), which it greatly resembles both in
appearance and habits.   It was first described by Wilson, whose specimen, however, was half an inch longer than I have ever seen it in the Hudson's Bay
countries.   This little animal is very abundant from Canada to Great Bear Lake,
and multiplies with rapidity in the neighbourhood of the trading posts.   It seeks
shelter in the barns and out-houses, where it makes hoards of grain and of the seeds
of leguminous plants.   It is said to be very fond of the bulbous roots of the
Philadelphia lily; and it does much mischief in gardens by burrowing under
the drills, and carrying off the seeds.   This is the species which is described in
Captain Franklin's Journey (p. 660), under the name of *Arvicola xanthognatha*.

DESCRIPTION.

The body and head have conjointly a short oval shape.   The *head* is large, with an obtuse
nose, and the lips are clothed with very short hairs.   Margins of the nostrils and the septum
naked; upper lip very slightly cleft; a hairy patch on the inside of the mouth.   *Incisors*
yellow, dentition precisely similar to that of the *Arvic. xanthognathus*.   *Eyes* small; whiskers
about as long as the head, of a brownish-colour approaching to black at their roots.
Ears large, with a wide auditory opening, protected by a large rounded tragus.   External ear
erect, oval, rounded above, thin and membranous, clothed with a few short hairs; it is
nearly six lines high, but is hid by the fur.   The *tail* is cylindrical, and is thickly
clothed with short adpressed hairs, a few of which project beyond the obtuse extremity,
but they can scarcely be said to form a pencil or tuft.   The *extremities* are short, and

---

* There seems to be some mistake in Pennant's having ascribed the dimensions of six inches to this Meadow-Mouse.
He quotes Buffon, who describes his specimen as little more than half that size.

much concealed by the broad fleshy body. The palms of the fore-feet are naked, and have five little callous tubercles, of which one is common to the two middle toes, one to each of the other two toes, and two lie in contact with each other at the posterior part of the palm; one of the latter two is larger than any of the others, and supports the rudiment of a thumb, consisting of a small papilla, protected by a minute and rather obtuse entire nail. The two middle toes spring together from the extremity of the palm, and are nearly equal to each other in length. The inner toe arises higher up, and is next to the middle ones in length. The outer toe is opposite to the inner one, but is still shorter. All the toes are covered above with short adpressed hairs, some of which project beyond the claws. The *claws* are slender, pointed, very slightly arched, and have a lanceolate-shaped groove underneath. The feet are the only part of the fore-extremities which project beyond the fur of the body. The *hind-feet* have a pretty long and rather slender tarsus, clothed with short adpressed hairs. They have four toes similarly arranged with those of the fore-feet, but of greater length; and an inner toe or thumb of the same form with the outer one, but situated further back. The hind-feet are longer than the fore-ones, and part of the leg projects beyond the fur of the body.

Colour of all the upper parts, including the sides of the head, a hair-brown, or what is termed mouse-colour, without spots or mottling; there is no reddish spot on the cheek or face, but there is a very slight reddish-brown tinge on the hairs about the ears. The under parts, including the chin and part of the neck, are light-gray. The brown of the sides and gray of the belly mingle without any well defined line of junction. The fur is fine and long, and when blown aside appears over the whole body of a dark bluish or blackish gray, the colours proper to the back and belly being confined to the tips of the hair. The tail is of the colour of the back above and of the belly underneath.

DIMENSIONS.

	Inches.	Lines.
Length of head and body	3	6
,, tail	1	1
,, head alone	2	3

Many of these mice were killed at Carlton-house in the beginning of March, when the preceding description was drawn up. I should have hesitated in describing this animal as specifically distinct from the common meadow-mouse of Europe, had it not already received a distinct name.

[38.]   4. ARVICOLA NOVOBORACENSIS.   *Sharp-nosed Meadow-Mouse.*

Lemmus Novoboracensis. " RAFINESQUE, *Ann. of Nature*," (quoted from DESMAREST, *Mamm.* p. 286).

A. (*Novoboracensis*) *naso gracili acuto, auriculis prominulis, caudâ squamatâ nudiusculâ caput mediocre longitudine excedenti, corpore super obscurè fusco; subter sordidè murino.*
Sharp-nosed Meadow-Mouse, with ears slightly overtopping the fur, a slightly hairy scaly tail, more than half the length of the head, the body above dark brown; beneath soiled brownish-gray.

A Meadow-Mouse was observed by Mr. Drummond on the Rocky Mountains, inhabiting dry places along with the *Arvicola xanthognathus,* and having similar habits with that animal.   It answers to the short description given of Mr. Rafinesque's *Lemmus Novoboracensis;* and although that is not sufficient to prove their identity, I have adopted his specific name, to avoid the hazard of loading the science with another synonym.

DESCRIPTION.

Shape.—The *body* is thick; the head of a moderate size, tapers from the ears to the end of the nose; the nose is slender and acute when compared with other species of this genus, and it projects a little way beyond the incisors.   *Ears* rounded, rising slightly above the surrounding fur, but they are not very conspicuous, as the hairs on their margins have the same colour with those of the head and back.   The *tail* is covered with very short adpressed hairs, not close enough to hide the scales; a few of the hairs converge to a point at the end of the tail.   The *legs* are very short, the feet small, and the claws weak and compressed; a very minute nail occupies the place of a thumb.   The *fur* is less fine than that of the *Arvicola Pennsylvanicus.*   On the back it is grayish-black from the roots to near the tips, which are reddish-brown, terminated by black.   The resulting colour is an intimate mixture of brown and black, appearing in some lights dark reddish-brown, in others umber-brown, mixed with black.   The superior parts of the head have the same colour with the back, except that there is an obscure rufous spot beneath the ear.   The ventral aspect is yellowish-gray, which mingles on the sides with the colour of the back.   The feet are dark-gray.   The upper surface of the tail is liver-brown, the under one grayish-white.

DIMENSIONS.

	Inches.	Lines.		Inches.	Lines.
Length of head and body	4	3	Length of tail	1	5
„ head	1	4	„ longest fur on the back	0	8

Described from a summer specimen.

[39.] 5. Arvicola borealis. (Rich.). *Northern Meadow-Mouse.*

Mouse, No. 15. Forster, *Phil. Trans.*, 62. p. 380 ?
Arvicola borealis. Richardson, *Zool. Journ.*, No. 12, April, 1828, p. 517.
Awinnak. Dog-rib Indians.

A. (borealis) pentadactylus, auriculis vellere conditis, caudâ caput subæquanti, corpore villosissimo badio nigroque subter cinereo.
Northern Meadow-Mouse, with a strong thumb-nail, ears concealed in the fur ; tail about as long as the head ; fur very long and fine ; on the back chestnut colour mixed with black, on the belly gray.

This animal was found in abundance at Great Bear Lake, living in the vicinity of the *Arvicola xanthognathus,* and having similar habits. It very much resembles the *Arvicola Novoboracensis* in size and general appearance ; but, on comparing them with each other, the *A. borealis* is seen to have a rounder and smaller head, a less prolonged upper jaw, shorter ears, and a shorter and differently clothed tail. It may also be distinguished not only from the *A. Novoboracensis,* but also from the *A. xanthognathus* and *A. Pennsylvanicus,* by the form of its thumb-nail, which, instead of being thin, obtuse and rounded, lying closely on one side of a little tubercle, is larger, strap-shaped, and projects from the extremity of a minute rudimentary thumb. It has its outer and inner surfaces alike in being rather convex, and a small obtuse point projects from its truncated end. The form of the thumb-nail allies this animal very closely to the Norway lemming, and to one or two species of American lemming ; but its claws are smaller and more compressed, and apparently not so well calculated for scraping earth as the broader claws of the lemmings. It may, however, be considered as an intermediate link between the two sub-divisions of the genus *arvicola,* and may without inconvenience be ranked either as a true meadow-mouse or as a lemming.

DESCRIPTION.

The Northern-Meadow Mouse has the dentition and usual form of the campagnols, with a moderately large head, a convex forehead, and a short but acute nose projecting beyond the incisors. The eyes are small, and the ears, which, toward their margins, are thinly clothed with hairs of the same colour with the adjoining parts, are low, rounded, and shorter than the surrounding fur. The body and head are clothed with fur, which is very long in proportion to the size of the animal. The fur on the upper parts is shining blackish-gray, from the roots to the tips, some of which are yellowish or chestnut-brown, some black. The hairs with black tips are the longest, and are equally distributed amongst the others, giving the whole a dark umber or liver-brown colour, but producing no spots.   There is a rufous mark under the ears. The fur on the back is about ten lines long, that on the crown of the head is three or four. The fur on the under parts (including the chin and lips) has a lead-gray colour, and is shorter than that covering the back and sides.   The tail is round, well clothed with short stiff hairs, which do not permit any scales to be seen.   It is clove-brown above, and grayish-white beneath.   The hairs at the extremity of the tail are of the same length with the others, but converge to a point.   The fore and hind extremities are clothed with short hairs of a clove-brown colour, mixed on the toes and hind-parts of the fore-feet, with some longer white hairs.   The hind-toes are more slender and scarcely longer than the fore ones ; they turn obliquely inwards.   The fore-claws are small, whitish, much compressed, arched, and acute, with a narrow, elliptical excavation underneath.   The hairs of the toes reach to the points of the nails.   The claws of the hind-feet resemble the fore ones, but are not so strong. The thumb of the fore-feet consists merely of a small strap-shaped nail, slightly convex on both sides, and having an obtuse point projecting from the middle of its extremity.

DIMENSIONS.

	Inches.	Lines.		Inches.	Lines.
Length of head and body . . .	4	6	Breadth of the ear . . .	0	3
„ tail . . .	1	0	Length of fore-feet to end of middle claw	0	4½
„ head alone . . .	1	3	„ hind-feet, including heel and claw	0	7½
Height of ear . . .	0	4	„ fur on the back . .	0	10

Described from spring specimens, after the snow had melted.

[40.]   6.   ARVICOLA (GEORYCHUS ?) HELVOLUS.   (Richard son.
Tawny Lemming.

GENUS, Arvicola. CUVIER. *Sub-genus*, Georychus. ILLIGER. CUVIER.
[Arvicola (Lemmus) helvolus. RICHARDSON, *Zool. Journ.*, No. 12, 1828, p. 517.

A. GEORYCHUS (*helvolus*) *naso pallido obtuso, palmis pentadactylis, capite fulvo nigroque, corpore helvolo subter vix pallidiori.*
Tawny Lemming, with a pale blunt nose , a thumb ; tawny and black head ; and a reddish-orange coloured body, a little paler beneath.

The Lemmings are, by some authors, thought to possess characters which entitle them to form a distinct genus ; but I am inclined to agree with those who range them merely as a subdivision of the genus *arvicola,* characterised principally by the shortness of the ears and tail, and the larger and stronger claws, more fitted for digging. In the short character of the species given above, the insertion of the word *Georychus* is intended to indicate the presence of the characteristic features of the Lemmings.

This animal was found by Mr. Drummond, inhabiting alpine swamps, in latitude 56°; but he could not learn any thing of its habits. From the great similarity of its form, and the strong resemblance in the shape of its claws to the Norwegian Lemming, we may infer that its habits do not differ much from those of that animal.

DESCRIPTION.

Size of the Lapland Lemming. Body low, head oval, nose short, blunt, and nearly on a line with the incisors. Eyes small; ears broader than high, shorter than the fur, clothed with hair near the edges. Tail very short, clothed with stiff hairs, which are longest near its extremity, and there converge to a point.

The *fur* of the body has a reddish-orange colour, palest on the ventral aspect. On the back and sides there are interspersed a number of longer hairs tipped with black, but they do not produce any spotting. On the upper part of the head, round the eyes, and on the nape of the neck, the black hairs are more numerous, and the fur of those parts has a mixed black and orange colour. The nose is grayish-brown, the sides of the face are pale orange, and the margins of the upper lip white. The tail is coloured like the body. The feet are brownish. The fur on the body is about nine lines long, that on the nose and extremities is very short. The fur of the head is pretty long.

The cutting edge of the upper incisors is obliquely excavated in a lunated form, arising from their outer edges being inclined backwards, so that these teeth do not appear so flat anteriorly as those of the *Arvicola xanthognathus*, which have straight cutting edges. The claws of both extremities are much alike, greatly compressed, with sharp points, and an oblong, narrow excavation underneath. They are larger than those of any of the meadow-mice described in the preceding pages, although the *A. palustris* and *A. xanthognathus* are more than twice the size of the largest specimen of this animal. The thumb of the fore-feet consists almost entirely of a thick, flat, strap-shaped nail, resembling that of the Norway Lemming, and having, like it, an obliquely truncated summit. In the Tawny Lemming this summit presents obscurely two obtuse points.

DIMENSIONS.

	Inches.	Lines.		Inches.	Lines.
Length of head and body	4	6	Length of fore-feet and claws	0	4½
„ tail	0	7	„ hind-feet, from heel to end of claw	0	8
„ head	1	6	„ fur on the back	0	9

[41.]　7. ARVICOLA (GEORYCHUS) TRIMUCRONATUS.　(Richardson.)
Back's Lemming.

Arvicola trimucronatus.　RICHARDSON, *Parry's Second Voyage, App.* p. 309.

A. GEORYCHUS (*trimucronatus*) *auriculis vellere sub-conditis, rostro nigro obtusiusculo, palmis pentadactylis unguibus* (4) *lanceolatis curvis ; ungue pollicari ligulato tricuspidato, corpore super obscurè castaneo, laterè ferrugineo subter cinereo.*

Back's Lemming, with ears somewhat shorter than the fur; a blunt, black nose; four claws on the fore-feet, of a lanceolate form, and a strap-shaped thumb-nail, with three small points at the end; body, dark chestnut-colour above, reddish-orange or rust-colour on the sides, and gray beneath.

This animal was discovered by Captain Back on the borders of Point Lake, in latitude 65°, on Captain Franklin's first expedition.　Mr. Edwards, the Surgeon of the Fury, on Captain Parry's second expedition, brought a specimen from Igloolik, in latitude 69⅓°; and specimens were obtained on Captain Franklin's last expedition on the shores of Great Bear Lake.　At the latter place it was found in the spring, as soon as the ground began to thaw, burrowing under the mossy turf.　In the winter it travels under the snow in a semicylindrical furrow, very neatly cut to the depth of two inches and a half, in the mossy turf.　These hollow ways cross each other at various angles, but occasionally run to a considerable distance in a straight direction.　From their smoothness, it was evident that they were not merely worn by the feet, but actually cut by the teeth.　Their width is sufficient to allow the animal to pass with facility.　The food of this Lemming seems to consist entirely of vegetable matters.　It inhabits woody spots.　A female killed on Point Lake, June 26, 1821, contained six young, fully formed, but destitute of hair.

DESCRIPTION.

Size, a little inferior to the Hudson's Bay Lemming, or nearly about that of the Norwegian Lemming.　Head flat, covered by moderately long fur.　*Ears* shorter than the fur, inclined backwards, thinly clothed.　*Eyes* smaller than those of the English domestic mouse.　Upper lip deeply cleft.　The *nose* is obtuse, with a small, naked, but not pointed or projecting tip, and covered above with hairs of a deep black colour.　*Whiskers* numerous, black at the roots, brownish or white at the extremities; some entirely white.　Inside of the mouth hairy, the hairs springing from projecting glandular folds.　*Teeth*—incisors, somewhat yellowish; upper ones presenting a conspicuous, but shallow groove, with an obliquely notched, cutting edge. Grinders, three on a side in each jaw.

The *Body* is broad and rather flat, and is everywhere covered with a beautifully fine and soft fur, which is about nine lines long on the back, but rather shorter on the belly. The colour of the head and dorsal aspect of the neck and shoulders is a mixed reddish-gray, formed from the mingling of the clove-brown, yellowish-brown, and black tips of the hairs in nearly equal proportion. The back is chestnut-brown, but many of the longer hairs are tipped with black. The sides are reddish-orange, and the belly, chin, and throat, gray, intermixed with many orange-coloured hairs. The colours of this animal very strongly resemble those of the Tawny Lemming; but its nose is deep black, whilst the nose of the latter is pale. The tail projects a few lines beyond the fur, is clothed with stiff hairs converging to a point, dark above, grayish-white below.

The *fore-legs* are short, but the feet are moderately large, and turned outwards, like the feet of a turnspit dog. They are of a dark clove-brown colour above, and are clothed with longer white hairs posteriorly. The (4) toes are naked underneath, and are armed with moderate-sized strong nails, curved downwards, and inclined outwards. They are of an oblong form, convex above, not compressed, are excavated underneath more broadly than the nails of any of the other American lemmings I have met with, and have sharp edges fitted for scraping away the earth. The thumb is almost entirely composed of a strong nail which has two slightly convex surfaces, a strap-shaped outline, and a truncated extremity, from which three small points project. The palms are narrow.

The *posterior extremities* are considerably longer than the fore ones, the thighs and legs being tolerably distinct from the body. The sole is narrow, long, and somewhat oblique, having its inner edge turned a little forwards. The toes are longer, and the claws as long, but more slender than those of the fore-feet, and they are much compressed. In the tawny lemming (No. 40.) the claws of both the hind and fore feet are compressed.

DIMENSIONS
Of a male killed at Fort Franklin.

	Inches.	Lines.		Inches.	Lines.
Length of head and body . . .	5	0	Length of fur on the back . .	0	9
,, tail	0	6	,, palm and claw of middle toe, nearly	0	6
,, head	1	5	,, claw of middle toe . .	0	2
,, ears	0	4	,, sole, and middle claw of hind-foot	0	9
,, whiskers	1	3			

A female was only 4¾ inches long.

Since this animal was described in the Appendix to Captain Parry's Voyage, above quoted, a second American species (*A. helvolus*), armed with a thumb-nail, has been discovered, and the trivial name of Five-fingered American Lemming being no longer distinctive, I have given it a new English appellation, after the officer who first procured a specimen.

[42.] 8. Arvicola (Georychus) Hudsonius.   *Hudson's Bay Lemming.*

Mus Hudsonius. Forster, *Phil. Trans.*, lxii. p. 379. Pallas, *Glir.*, p. 208. Lin. Gmel., 137.
Hudson's rat. Pennant, *Quadr.*, vol. ii. p. 201. *Arctic Zoology*, vol. i. p. 132.
Hare-tailed Mouse. Hearne, *Journ.*, p. 387.
Lemmus Hudsonius. Captain Sabine, *Parry's First Voy., Suppl.*, p. clxxxv. Mr. Sabine, *Franklin's Journ.*, p. 661. *Dict. des Sciences Nat.*, tom. viii. p. 566. Harlan, *Fauna*, p. 146.
Arvicola Hudsonia. Richardson, *Parry's Second Voy., App.*, p. 308.
Hudson's Bay Lemming. Godman, *Nat. Hist.*, vol. ii. p. 73.
Spec. 107 a. British Museum.

A. Georychus (*Hudsonius*) *exauriculatus, unguibus duobus anticis intermediis maximis compressis bi-mucronatis,* (*mucrone uno super alterum*).
Hudson's Bay Lemming, earless, with two middle claws of the fore-feet unusually large, compressed, their very blunt extremity being rendered double by a deep transverse notch.

This curious animal was first described by Forster, from a mutilated specimen, and afterwards more fully by Pallas, who received a number of its skins from Labrador, one of which he sent to Pennant. A specimen, preserved in the *Museum du Roi*, at Paris, is described in the *Dict. des Sciences,* and there is an excellent specimen in the British Museum.

We did not meet with this lemming in the interior of America, and I believe it has hitherto been found only near the sea. It inhabits Labrador, Hudson's Straits, and the coast from Churchill to the extremity of Melville Peninsula, as well as the islands of the Polar sea, visited by Captain Parry. Its habits are still imperfectly known. In summer, according to Hearne, it burrows under stones, in dry ridges, and Captain Sabine informs us that in winter it resides in a nest of moss on the surface of the ground, rarely going abroad. The former author likewise acquaints us that it is very inoffensive, and so easily tamed, that if taken even when full grown, it will in a day or two be perfectly reconciled, very fond of being handled, and will creep of its own accord into its master's neck or bosom.

DESCRIPTION.

The body is thick, the head short and rounded, nose obtuse, eyes very small, and there are no exterior ears. The legs are short, and the tail is so short, that only the stiff hairs of its end project beyond the fur of the hips. The upper *incisors* are whitish, curved, flat anteriorly, and have even cutting edges. The lower ones are a little longer and more slender. The *fur* is remarkably fine and pretty long, blackish-gray from the roots to the tips, which are on the dorsal aspect white, dark brown and black. The result is a beautiful mottling of these colours, in which the dark brown predominates on the crown of the head and dorsal line: there is more white towards the sides. On the under parts of the cheeks, on the chest, about the ears, and on the sides, a bright rust colour prevails. The ventral aspect is grayish-white, more or less tinged with the rust colour. The extremely short tail is closely covered with stiff white hairs that converge to a point at its end.

The *feet* are clothed with long white hairs. On the fore-feet there are four toes, with a minute rudiment of a thumb, not armed with any nail whatever. The two middle toes are of equal length, and are each furnished with a disproportionately large claw, which is compressed, deep, very blunt at the extremity, and is there separated into two layers by a transverse furrow. The upper layer is thinner, the lower one has a blunt rounded outline. The latter has been described as an enlargement of the callus which exists beneath the roots of the claws of the lemmings and meadow-mice; it appears to me, however, to be of the same substance with the superior portion of the nail. The outer and inner toes have curved, sharp, pointed claws. The hind-feet have five toes, armed with slender, curved claws, like those of the other lemmings. The two middle claws, however, in full-grown individuals, shew some approach to the peculiar form of those on the fore-feet.

In the females and young, the subjacent production of the claws is less conspicuous. The description is drawn up from a summer specimen. In the winter, the tips of the hairs are white, but Hearne says the white colour of their fur never appears so pure as that of the ermine.

DIMENSIONS.

	Inches.	Lines.		Inches.	Lines.
Length of body and head	5	4	Length of tail	0	5
„ head	1	4	„ middle fore-claw	0	4½

[43.] 9. ARVICOLA (GEORYCHUS) GRŒNLANDICUS. *Greenland Lemming.*

Mouse. *Sp.* 15. FOSTER, *Phil. Trans.*, lxii. p. 379 ?
Hare-tailed Rat ? PENNANT, *Arct. Zool.*, vol. i. p. 133 ?
Mus Grœnlandicus. TRAILL, *Scoresby's Greenl.*, p. 416.
Arvicola Grœnlandicus, RICHARDSON, *Parry's Second Voy.*, *App.*, p. 304.
Owinyak. ESQUIMAUX.

A. GEORYCHUS (*Grœnlandicus*) *exauriculatus, rostro acuto, palmis tetradactylis hirsutis ; unguibus apice cylindrico producto, lineâ dorsali nigrâ.*
Greenland Lemming, earless, with a sharp nose ; fore-feet hairy beneath, with four toes, armed with claws, having sharp cylindrical points ; a dark stripe along the middle of the back.

Foster, in the Philosophical Transactions, notices a skin brought from Churchill, evidently of a lemming, but in too imperfect a state to enable him to determine the species. Both he and Pennant were inclined to refer it to the *Mus lagurus* of Pallas, because it agreed with that animal in having a dark line along the back ; it is more probable, however, that the skin belonged to the species which forms the subject of this article, which has also a dark dorsal stripe, and is certainly an inhabitant of Hudson's Bay. It was first described, and the specific name affixed, by Dr. Traill, from an individual procured by Captain Scoresby, on the east coast of Greenland ; and on Captain Parry's second expedition a considerable number were caught in Repulse Bay, and are described in the Appendix above quoted. They were found in similar situations with the Hudson's Bay lemming, and were considered to be the females of that species, by the officers of the expedition, and as such noted in their journals. A number of them being put into a cage, fought until they destroyed each other.

DESCRIPTION.

Size—rather less than the water-rat (*Arvicola amphibius*). In general form they resemble the other lemmings. *Head* rounded, narrower than the body, tapering slightly from the auditory openings to the eyes, but from the latter the acumination is more sudden, and it terminates in an acute nose. The general colour of the superior and lateral parts of the head, is the same with that of the back. There are no external *ears*, but the site of the auditory opening is denoted by an obscure transverse brownish streak in the fur. The *eyes* are near each other, and small. The fur on the *cheeks* is a little puffed up, has a rufous tinge, and is bounded posteriorly by an obscure blackish semicircular line, which commences at the anterior angle of the eye. The *nose*, covered with short black hairs, intermixed poste-riorly with some hoary ones, is rendered prominent by a depression on each side, anterior to the cheeks. Its acute apex is covered with black hairs disposed in a circular manner, and no naked space can be discovered above the nares in the dried specimen. The upper lip is deeply divided. *Incisors* slightly yellowish, inferior ones twice the length of the upper ones. *Whiskers* long, partly black, partly white. *Body* thick, having a smooth dense covering of long and soft fur. The colour on the dorsal aspect is dark grayish-brown, arising from an intimate mixture of hairs tipped with yellowish-gray and black; the black tips are the longest, and, predominating down the centre of the back, produce a distinct stripe. The ventral aspect of the throat, neck, and body, exclusive of some rusty markings before the shoulders, is of an unmixed yellowish-gray colour, which unites with the darker colour of the back by an even line running on a level with the tail and inferior part of the cheek. The fur, both above and below, presents, when blown aside, a deep blackish-gray shining colour from the tips to the roots. The *tail* is very short, and is of the same colour with the body at the root, but the part which projects beyond the fur of the rump is only a pencil of stiff white hairs, four or five lines long.

The *fore-extremities* project very little beyond the fur; the palms incline slightly inwards, are small, and the toes are very short; both are covered thickly above and below, with strong hairs curving downwards, and extending beyond the claws. The only naked parts on the foot are a minute, flat, unarmed callus, in place of a thumb, and a rounded smooth callus at the extremity of each toe. These callosities do not project forwards under the claws, and have no resemblance to the large, compressed, horny, under portions of the claws of the Hudson's Bay lemming. The *claws* are long, strong, curved moderately downwards, and also inclining inwards to the mæsial line, with a more slight curvature. The second claw from the inside, which is considerably the longest, is nearly four lines in length. At the root it has a compressed, conical form, and is much deeper than broad; it is rounded above, and flat or slightly grooved near its root underneath, but its curved extremity is lengthened out in a slender cylindrical manner. The other fore-claws, though smaller, are similar to this one. The third from the inside is the next in size, and the two extreme ones are considerably shorter. The length of the whole palm and middle claw is only six lines. The claws are fitted for digging, but not for cutting roots. *Hind-feet.*—The soles are hairy, and the hairs project further beyond the claws than on the fore-feet. *Toes* five, of which the three middle ones are nearly of a length, the two extreme ones arise further back, and are shorter. The hind-claws are shorter than the fore ones, slightly arched, narrow, but not sharp at the points; they are thin, hollowed out underneath, and calculated to throw back the earth which has been loosened by the fore-claws.

The description was drawn up from a male, killed August 22, in Repulse Bay.

Mr. Scoresby's Greenland specimen differs solely in colour, which on the upper parts is a mixture of mottled ash-gray with blackish-brown and reddish-brown, and on the belly and inferior parts is rufescent.

DIMENSIONS

Of the Repulse Bay male specimen.

	Inches.	Lines.		Inches.	Lines.
Length of the head and body . .	6	3	Length of longest fore-claw . . .	0	4
„ tail	0	9	„ of palm and middle-claw . .	0	6
„ fore-leg from palm to the axilla	1	1	„ whiskers	1	4

Specimens from Repulse Bay and from Greenland are preserved in the Museum of the Edinburgh University.

Of the four lemmings described in the preceding pages, three of them closely resemble lemmings of the old continent, described by Pallas, and may be considered as their American representatives. Thus the *Tawny Lemming* approaches the *Lapland Lemming* (PALL. *Glir.* t. 12. B.) in size, and in the form of its thumbnail. The latter differs in the colour of its fur; which is more varied, and on the throat and abdomen is white, whilst (except on the top of the head, where there are some dark markings) the tawny lemming is of a rusty colour throughout, the under parts being merely a little lighter than the back.

Back's Lemming again may be said to represent the *Norway Lemming* (*Op. citat.* t. 12. A.) The latter, however, has the claws of its fore-feet much compressed, and there are only two points on the end of the thumb-nail, instead of the three small points which characterise the thumb-nail of the former. The colours of its fur are also more lively and more agreeably varied, and its nose is whitish.

With respect to the *Hudson's Bay Lemming*, I believe no animal has been discovered in the old world, possessing the singular production of the two middle fore-claws of that animal.

The *Greenland Lemming* is most allied to the *Ringed Lemming* of Siberia (*Op. citat.* t. 11. B.) The Siberian animal, however, is of a smaller size, and has an obtuse nose; and the brown ring round the neck, surmounted by a paler one, whence it derives the specific appellation of *torquatus*, does not exist in the American animal. The *hare-tailed mouse* (*Op. citat.* t. 13. A.) agrees with the Greenland lemming, and also with the ringed lemming, in having a dark, dorsal stripe; but it may be readily distinguished from the former of these two, by its smaller size, obtuse nose, truncated furry tail, a large callus in place of a thumb, and a remarkable moveable callus on the palm.

NEOTOMA DRUMMONDII.

[44.] 1. NEOTOMA DRUMMONDII. (Rich.) *Rocky-Mountain Neotoma.*

GENUS. Neotoma. SAY.
Rat of the Rocky Mountains. LEWIS and CLARK, vol. iii. p. 41.
Myoxus Drummondii. RICHARDSON, *Zool. Journ.*, No. 12, March, 1828, p. 517.

N. (Drummondii) brunescenti-cervina subter alba, caudá floccosá corpus longitudine excedenti.
Rocky-Mountain Neotoma, yellowish-brown above, white beneath, tail more bushy towards the extremity, longer than the body.

PLATE VII.

This animal inhabits the Rocky Mountains, in latitude 57°, and though specimens of it have from time to time reached England, but little is known of its habits. Mr. Drummond informs me that it makes its nest in the crevices of high rocks, and seldom appears in the day-time, but its place of abode may be detected by its excrement, which has the colour and consistence of tar, and is always deposited in one place. Its food most probably consists of herbage of various kinds, and of small branches of pine trees, because there is generally a considerable store of these substances laid up in the vicinity of its residence. It is very destructive. In the course of a single night, the fur traders who have encamped in a place frequented by these animals have sustained much loss, by their packs of furs being gnawed, their blankets cut in pieces, and many small articles carried entirely away. Mr. Drummond placed a pair of stout English shoes on the shelf of a rock, and, as he thought, in perfect security, but on his return, after an absence of a few days, he found them gnawed into fragments as fine as saw-dust. When I published a short notice of this animal, in the Zoological Journal, I made use of the grateful privilege of a first describer, in distinguishing it by the name of my fellow-traveller, whose zeal for the promotion of every branch of natural history was unbounded. Not having had at that time an opportunity of examining the molar teeth, I was induced to refer it to the genus *Myoxus,* on account of its general appearance; but having lately seen the scull of an individual obtained on the Rocky Mountains by Mr. David Douglas, I have been enabled to ascertain that it belongs to the genus *Neotoma,* founded by Mr. Say. It has a great resemblance, particularly about the feet, to his figure of the Florida-rat, published in the Philadelphia Journal of Science, but differs from that species in having a

T

bushy tail, densely hairy, instead of a round tapering one, scaly and thinly hairy. In the softness of its fur, and general arrangement of its colours, it has much similarity to the common wood-rat. It differs, however, from the genus *mus*, as now restricted, in the form of its teeth, approaching more nearly to the genus *arvicola* in that respect than to any other, but receding from it, on the other hand, in the length of its tail, limbs, and in its general light, active form. Besides the specimen brought home by the Expedition, and from which the accompanying very correct engraving by Landseer was executed, there is another good specimen in the Museum of the Hudson's Bay Company; a third, with a mutilated tail, in the Zoological Museum; and a fourth, without a tail, and much over-stuffed, in the British Museum, all said to have come from the same quarter. I have also a hunter's skin, of a larger, and perhaps a specifically distinct kind. procured on the Rocky Mountains in latitude 63°.

DESCRIPTION.

In *size*, this Neotoma equals the Norway rat, and it has a good deal of the character of that animal in its physiognomy. Its nose is compressed and narrow, but appears rather obtuse if viewed laterally. There is a very narrow, naked margin to the nostrils, the tip and sides of the nose being covered with short hairs. The upper lip is divided about three lines deep.

Dental formula, incisors $\frac{2}{2}$, canines $\frac{0-0}{0-0}$, grinders $\frac{3-3}{3-3} = 16$.

The *incisors* have precisely the form of those of the meadow-mice, and wear away in the same manner at their points. The upper ones are short, slightly rounded, and not grooved on their anterior surface. The lower ones are long, narrow, and rounded anteriorly and on the sides. The *molar teeth* also very much resemble those of a meadow-mouse. (*A. xanthognathus*). The grooves on their sides, however, instead of running to the base of the tooth, terminate abruptly, where it is immersed in the socket; and some little distance below this termination, most of the grinders divide into two fangs. The two anterior pairs of lower grinders have these fangs very distinct, the space between the fangs being deep and wide; but the upper grinders and the posterior pair in the lower jaw have them much shorter, and as it were coalesced. The first grinder in the upper jaw has the rudiments of three fangs. The grinders of both jaws have a slight inclination backwards, and they gradually decrease from before backwards in size, and in the height of the part which projects above the sockets, preserving however an even grinding surface. In the *upper jaw*, the first grinder has three grooves on its exterior side, and as many inside, with an equal number of rounded projecting columns or ribs of a side; the second and third grinders have each two grooves, with three ribs exteriorly, and one groove with two ribs interiorly. In all, there are nine ribs on the exterior sides of the upper rows of grinders, seven on the interior sides, and ten triangles, formed by the folds of enamel on their crowns. In the *lower jaw*, the first grinder has two grooves exteriorly, and three interiorly; the second has two grooves on each side, and the third, one on each side. In all, there are nine ribs on the outsides, and eight on the insides of the lower rows of grinders, and nine triangular folds of enamel on their crowns. These triangles are disposed in a single series, or at least present, very obscurely, the double alternate arrangement which exists on the crowns of the grinders of the meadow-mice. The grinders of a Neotoma further differ from those of the meadow-mice in the ribs on their sides being broader and more rounded, and in the first upper grinder, instead of the last one, being the largest. The *whiskers* are considerably longer than the head; the anterior ones are white; the posterior ones, which are longer and stronger, are black, more or less tipped with white. The *ears* are large, oval, and rounded, closely covered on the back with short, adpressed, blackish-gray hairs, and they have a very narrow and obscure white margin. Their anterior surface is more sparingly hairy above, and is quite naked near the auditory opening.

The *fur* is remarkably fine, soft, and long, and has considerable lustre. The upper parts, including the head and cheeks, back, sides, and outer surface of the fore and hind thighs and legs, have a nearly uniform, light, yellowish-brown colour, intimately mixed with black hairs; the resulting tint is between a hair-brown and a fawn colour. The black hairs are more abundant on the sides of the nose, down the middle of the head and back, and about the rump. The upper lip, chin, throat, all the under parts, the inside of the thighs, and the whole of the feet, from the wrist and ankle joints, are pure white. The fur is longest on the back and sides, a little shorter on the belly, and shortest about the nose, but the furry coat is close throughout, and is everywhere of a deep blackish-gray colour for two-thirds of its length from the roots.

The *tail* at its commencement is cylindrical, and clothed with shorter hairs; but the fur gradually lengthens towards its extremity, where it is upwards of an inch long, and is somewhat distichously arranged, particularly beneath. The whole of the fur of the tail is very close and woolly at bottom. For a short space next the rump, the tail is coloured above like the back; but for the greatest part of its length it is of a dark lead-gray, arising from an intimate intermixture of blackish-gray and whitish hairs. Underneath it is throughout of a white colour, and when the tail is spread out, the white hairs form to it an indistinct white tip.

The *feet* are thickly clothed above with fur, which conceals the claws. On the fore-feet there are four toes, which do not differ much from each other in length, the two middle ones being longer only in a very slight degree. There is a small callus in place of a thumb, which is situated behind the roots of the toes, and is protected by a minute adpressed nail. There are besides five callous eminences of considerable size on the palm; three arranged in a triangular form at the roots of the toes, one a little longer posterior to the thumb tubercle, and another of equal size opposite to it. The claws are white, short, much curved, and very acute. The hind-feet have five toes, the four anterior of which much resemble those of the fore-feet, but are a little stouter, and have more spread.   The inner one, representing a thumb, is nearly as long as the outer one, though it is situated further back.   The posterior half of the sole is hairy.   The claws are like those of the fore-feet.

DIMENSIONS.

	Inches.	Lines.		Inches.	Lines.
Length of the head and body . . 9	9	0	Length from the wrist joint to the end of the middle claw . . . 0	0	8
,, tail (vertebræ) . . 6	6	6	,, of the middle fore-toe and claw . 0	0	5
,, tail with the fur . . 7	7	6	,, from heel to the end of the middle claw . . . 1	1	6
,, whiskers . . . 3	3	6			
Height of the ears posteriorly . . 0	0	10	,, of the middle hind-toe and claw . 0	0	7
Breadth of ditto 0	0	9	,, of the fur on the back . . 1	1	0
Distance from the tip of the nose to the anterior angle of the orbit . . . 0	0	11	Height of the back of the prepared specimen standing on its palms and soles . . 3	3	9

---

† 1. Mus rattus. (Linn.)   *The Black Rat.*

Genus. Mus. Cuvier.
Black Rat. Pennant, *Arct. Zool.*, vol. i. p. 129. Godman, *Nat. Hist.*, vol. ii. p. 83.
Mus rattus. Harlan, *Fauna*, p. 148.

This Rat was, most probably, not originally an inhabitant of North America, but was brought thither by the early European visitors of that continent.   It seems to have multiplied exceedingly fast in its new quarters, until the introduction of the still more destructive brown rat thinned its numbers, and it has now become as rare as it is in Europe, from the same cause.   We did not observe the black rat in any part of the fur countries; and I may also venture to affirm, that it has not advanced farther north than the plains of the Saskatchewan.   Indeed, I have no other reason for supposing that it may have got so far, than that an animal resembling a Musk-Rat, with a long round tail, is mentioned by the Indians of that quarter, under the name of *meestăhæ appeceooshees.*

† 2. Mus DECUMANUS. (Linn.) *The Brown Rat.*

Brown Rat. PENNANT, *Arct. Zool.*, vol. i. p. 130. GODMAN, *Nat. Hist.*, vol. ii. p. 78.

This very destructive animal came, according to the accounts of historians, from Asia to Europe about the beginning of the seventeenth century; was unknown in England before 1730, and, according to Dr. Harlan, did not make its appearance in North America until the year 1775. Pennant, writing in 1785, says he has no authority for considering it to be an inhabitant of the new continent, although he thinks it probable that it must by that time have been carried thither in ships. It is now very common in Lower Canada; but I was informed that in 1825 it had not advanced much beyond Kingston in Upper Canada. We did not observe it in the fur countries; and if it does exist there, it is only at the mouth of the Columbia River, or at the factories on the shores of Hudson's Bay.

---

† 3. Mus MUSCULUS. (Linn.) *The Common Mouse.*

Mouse. PENNANT, *Arct. Zool.*, vol. i. p. 131.
Mus musculus. SAY, *Long's Exped.*, vol. i. p. 262.
Common Mouse. GODMAN, *Nat. Hist.*, vol. ii. p. 84.

I have seen a dead mouse in a storehouse at York Factory filled with packages from England, and it is probable that the species may have been introduced into all the posts on the shores of Hudson's Bay; but I never heard of its being taken in the fur countries at a distance from the sea-coast. Mr. Say informs us, that it was introduced at Engineer Cantonment, on the Missouri, by Major Long's expedition.

[45.]   4. Mus leucopus.   (Rafinesque.)   *American Field-Mouse.*

Mus sylvaticus. Forster, *Phil. Trans.*, vol. lxii. p. 380.
Field Rat, A, American. Pennant, *Hist. Quad.*, vol. ii. p. 185. *Arct. Zool.*, vol. i. p. 131.
Mus leucopus. "Rafinesque-Smaltz, *Am. Month. Mag.*, vol. iii. p. 444 ; 1818" (quoted from Desmarest, *Mamm.*)
        Harlan, *Fauna*, p. 151. Richardson, *Zool. Journ.*, No. 12. p. 518.
Mus agrarius. Godman, *Nat. Hist.*, vol. ii. p. 88 ?
Appecooseesh. Cree Language.

    M. (leucopus), caudâ longâ vestitâ, corpore griseo-lutescente subter abruptè albo, auriculis magnis.
    American Field Mouse, with a long hairy tail, hair-brown back, white belly and feet, and large ears.

No sooner is a fur-post established than this little animal becomes an inmate of the dwelling-houses ; whilst the meadow-mouse, described in p. 124, under the name of *Arvicola Pennsylvanicus*, at the same time takes possession of the outhouses and gardens. We observed it as far north as Great Bear Lake ; and if the synonyms prefixed to this article are correctly applied, it is not uncommon in the United States. It also extends from Hudson's Bay across the continent to the mouth of the Columbia River. The gait and prying actions of this little creature, when it ventures from its hole in the dusk of the evening, are so much like those of the English domestic mouse, that most of the European residents at Hudson's Bay have considered it to be the same animal, altogether overlooking the obvious differences of their tails and other peculiarities. The American Field-Mouse, however, has a habit of making hoards of grain or little pieces of fat, which, I believe, is unknown of the European domestic mouse ; and what is most singular, these hoards are not formed in the animal's retreats, but generally in a shoe left at the bedside, the pocket of a coat, a nightcap, a bag hung against the wall, or some similar place. It not unfrequently happened that we found barley, which had been brought from a distant apartment, and introduced into a drawer, through so small a chink, that it was impossible for the mouse to gain access to its store. The quantity laid up in a single night nearly equalling the bulk of a mouse, renders it probable that several individuals unite their efforts to form it. This mouse does considerable mischief in the gardens, and in a very few nights will almost destroy a plantation of maize, by tracing the rows for the purpose of collecting the seeds, and depositing them in small heaps under the loose mould, generally by the side of a stone, or piece of wood. From the facility with which it seems to transport the substances it preys upon, I suspected that it had cheek-pouches, but none were found on examination. The ermine is a most inveterate enemy to this species, and pursues it into the sleeping apartments.

The *Mus leucopus* may be considered as the American representative of the European field-mouse (*mus sylvaticus*, Linn.), which it greatly resembles, and perhaps Pennant is quite right in terming it only a variety. The *mus sylvaticus* appears to have generally a more tapering, acute tail. Dr. Godman's description of the *Mus agrarius* corresponds so exactly with our animal, that I have quoted it as a synonym; but the *Mus agrarius* of Pallas differs, in having small ears. Dr. Harlan mentions, that several varieties of the *mus sylvaticus* exist in the neighbourhood of Philadelphia; but his description, which seems to be a translation of Desmarest's account of the European animal, does not agree with any variety of the *mus leucopus* that I saw to the north. The varieties that I met with differed principally in the size of the body, and the length of the tail. Specimens from the mouth of the Columbia were considerably larger than those from Hudson's Bay.

DESCRIPTION.

The American Field-Mouse has a larger head than the English domestic mouse; but in its general form it is similar to that animal. On the other hand, its head is smaller than that of Wilson's meadow-mouse; its body less fleshy, and it weighs less. Its muzzle is rather sharp. The whiskers are much longer than the head, part of them are black, the rest white. Eyes moderately large. Ears large, erect, membranous, of an elliptical form, with rounded tips, and covered rather thinly with short adpressed hairs.

The *fur* of the body is very fine, but not long, and is throughout of a dark, bluish-gray colour from the roots to near the tips. The colour of the upper parts is hair-brown, darkest on the crown of the head, and along the back; the sides are of a lighter hue, approaching to yellowish-brown, or sometimes, together with the hips, to reddish-brown. The cheeks have a still more lively colour than the sides, being somewhat rufous. The upper lip, a space on each side of the mouth, the chin, all the under parts, the inside of the thighs, and the whole of the legs and feet, are white.

The *tail* is thickly clothed with short hairs lying pretty smoothly, no scales whatever being visible. Its upper surface is of a hair-brown colour, considerably darker than any other part of the animal, and contrasts strongly with the inferior surface, which is white; the line of contact of the two colours is straight and well defined.

Fore-feet, with four toes and six tubercles on the palm. Of the three anterior tubercles, one is seated at the common origin of the two middle toes, and one at the commencement of each of the other two toes, which arise farther back. The other three tubercles lie nearly in a line at the posterior part of the palm. The smallest of these is the interior one, and it occupies the usual site of the thumb, of which there is no other vestige,—not even a nail.

The palms are naked, and the toes short. The *hind-feet* are long, particularly the tarsal bones, the hind toes being likewise a little longer than the fore-ones. There are six tubercles on the soles,—three at the roots of the toes, and three farther back; of the latter three, the one next the inner toe or thumb is large, the posterior one is small, and the exterior one minute.

DIMENSIONS.

	Specimens procured at Carlton House.		Columbia River Specimens.	
	Inches.	Lines.	Inches.	Lines.
Length of head and body	3	7	4	3
„ head alone	1	1	0	0
„ tail	2	3	2	9
Height of the ears	0	6	0	6½
„ of the back, when the animal stands on its palms and soles	2	9	0	0

[46.] 1. MERIONES LABRADORIUS. *Labrador Jumping Mouse.*

GENUS. Meriones. ILLIGER. F. CUVIER.
Labrador Rat. PENNANT, *Arctic Zool.*, vol. i. p. 132.
Gerbillus Hudsonius. " RAFINESQUE-SMALTZ, *Am. Month Mag.*, 1818, p. 446."
Mus Labradorius. SABINE, *Franklin's Journ.*, p. 661.
Gerbillus Labradorius. HARLAN, *Fauna*, p. 157.
Labrador Jumping Mouse. GODMAN, *Nat. Hist.*, vol. ii. p. 97.
Katsĕ (the leaper). CHEPEWYAN INDIANS.

PLATE VIII.

Pennant, in Arctic Zoology, first described a specimen of this animal, sent from Hudson's Bay by Mr. Graham, to the Museum of the Royal Society. Afterwards, in the third edition of his History of Quadrupeds, he is inclined to consider it as identical with the *mus longipes* of Pallas, (the *dipus meridianus* of Gmelin,) an inhabitant of the warm, sandy deserts, bordering on the Caspian sea. This opinion, which can scarcely be correct, was formed from an imperfect inspection of the Hudson's Bay specimen, whilst it was suspended in spirits, and is opposed by differences in colour and other characters which he himself points out. From Pennant's time, until Mr. Sabine described an individual, brought from Cumberland-house on Captain Franklin's first journey, the Labrador Jumping Mouse

MERIONES LABRADORIUS.

Published by John Murray. January 1829

does not appear to have attracted the notice of naturalists. Pennant mentions a` yellow lateral line in his specimen which did not exist in the one Mr. Sabine described, but this difference I attribute solely to the season in which they were procured. Mr. Sabine's specimen had its tail mutilated, an accident very common to the whole family of rats. Pennant, under the name of Canada Jerboid rat, and Colonel Davies, under that of *Dipus Canadensis,* describes another Jumping Mouse, which seems to differ from this in having ears shorter than the fur, but in other respects to be very similar to it. The *Gerbillus Canadensis* of Dr. Godman agrees in description with Rafinesque-Smaltz's *Gerbillus soricinus,* (DESMAREST, *Mamm.,* p. 332.) but has larger ears than the Canada rat of Pennant; and a specimen in the Philadelphia Museum, described by Dr. Harlan, under the name of *Gerbillus Canadensis,* appears to be entirely similar to the Labrador species. It is evident, therefore, that the Jumping Mice, inhabiting different districts of America, require to be compared with each other before the true number of species and their geographical distribution can be ascertained.

The Labrador Jumping Mouse is a very common animal in the fur countries as far north as Great Slave Lake, and perhaps further, but I was not able to gain any precise information respecting its habits.

DESCRIPTION.

Dental formula, incisors $\frac{2}{2}$, canines $\frac{0-0}{0-0}$, grinders $\frac{4-4}{3-3}$ = 18.

Incisors of a deep orange colour. *Upper ones* short and strong, rounded anteriorly, each marked near its exterior margin with a deep and conspicuous furrow. The *lower ones* are longer and much more slender, but they taper very slightly towards their tips, and are not so acute as the lower incisors of the genus *mus.* The *grinders* very much resemble in form those of the squirrels. The anterior one in the upper jaw is round and very small. The other three have slightly hollowed crowns, with points, as in the squirrels, on their outward margins. The second grinder has three of these points, and is also marked with a furrow on its inner side. The third is very little larger than the second, and has four points on the outer margin of its crown, but no furrow on its inner side. The fourth is smaller than the two last-mentioned ones, but considerably larger than the first one; it has two points on its outer margin. In the *lower* jaw, the first and second grinders are nearly equal in size, and the third or last is smaller. The zygomatic processes are scarcely arched, the breadth of the scull being greater at their junction with the temporal bones than at their middles.

The *head* of this animal is narrow, and the nose, which is also narrow, but with a small obtuse tip, projects about a line and a half beyond the incisors. The tip of the nose is covered with short, erect hair, and beneath it the minute round openings of the nostrils face sideways, and are protected anteriorly by a slight ventricose arching of their naked inner margins.

U

The whole naked space at the nostrils is not above a line wide. The mouth is small and far back. The *whiskers* are fine, black, and longer than the head. The eyes are small. The ears are nearly five lines high, of a semi-oval form, broadly rounded at the tips. They are clothed behind, and also on the inside near the margins, with short black hairs, mixed with some yellow ones. Their edges are pale.

Colour.—The back and the upper parts of the head are covered with hairs of a dark liver-brown colour, mixed with a few brownish-yellow ones. The sides are brownish-yellow, slightly sprinkled with black hairs. The margin of the mouth, the chin, the throat, and all the lower parts of the body are white. The yellowish-brown of the sides joins the white of the belly by a straight line extending between the fore and hind extremities. In some specimens this yellowish-brown colour occupies as much space as the darker colour of the back; in others, the latter encroaches so much on the sides as to leave merely a narrow yellowish line next the white; whilst in autumn specimens, when the animal has just acquired a new coat of fur, the dark colour of the back adjoins the white of the belly. The fur of this animal is not so long or so fine as that of the common or meadow-mice.

Extremities.—The *fore* ones are small and very short, the tip of the middle claw not projecting above half an inch from the body. They are covered above with short, whitish hairs. The palms are naked, and there are four slender toes armed with small, nearly straight, compressed acute nails. There is also a minute rudiment of a thumb, protected by a rounded nail, and situated considerably behind the root of the inner toe. The *hind-legs* are long and very slender, and there are five hind-toes, each with a very long slender tarsal bone. The inner-toe is the smallest and furthest back; the outer one is the next smallest, and the three middle ones are longer. They are all armed with small nails, not quite so much compressed as the fore ones. The upper surface of the feet and toes is covered with very short grayish hairs. The soles are naked to the heels. The *tail* is very long, tapers slightly, and is scaly, and thinly set with short hairs. It has no tuft at the end, but the hairs there are a little longer than elsewhere. It is of a dark brown colour above, and white beneath.

DIMENSIONS.

	Inches.	Lines.		Inches.	Lines.
Length of head and body	4	6	Length from knee to ankle joint	1	0
„ head	1	2	„ from heel to end of middle claw	1	4
„ tail	5	3	Height of the ears posteriorly	0	4½
„ from wrist joint to end of middle claw	0	4½			

PLATE 9.

ARCTOMYS EMPETRA.

Published by John Murray, January 1829.

[47.] 1. ARCTOMYS EMPETRA. (Schreber.) *The Quebec Marmot.*

GENUS Arctomys. GMELIN. CUVIER.
Quebec Marmot. PENNANT, *Hist. Quadr.*, 1st ed., No. 259'; 3d ed., No. 321. *Arctic Zool.*,
  vol. i. p. 111. BEWICK'S *Quad.*, 1st ed., p. 346. *Fig.* . 2d ed., p. 369. *Fig.*
Mus empetra. PALLAS, *Glir.*, p. 75. An. 1778.
Glis Canadensis. ERXLEBEIN, *Syst.*, p. 363.
Arctomys empetra. SCHREBER, *Quad.*, p. 743; pl. 210.
Common Marmot. LANGSDORFF's *Travels*, vol. ii. p. 75 ?
Arctomys empetra. SABINE. LINN., *Trans.*, vol. xiii. p. 24. HARLAN, *Fauna*, p. 160.
Quebec Marmot. GODMAN,*Nat. Hist.*, vol. ii. p. 108.
Weenusk. CREE INDIANS. Kath-hillœ-kooay. CHEPEWYANS.
Thick-wood Badger. HUDSON'S BAY RESIDENTS.
Siffleur. FRENCH CANADIANS, who apply the same name to the other species of marmot, and
  to the badger.
Tarbogan. RUSSIAN RESIDENTS ON KODIAK ?

A. (*empetra*) *super ex spadiceo nigroque canescens: subter helvolus, capite pedibusque nigrescenti-brunneis, genis albescentibus, auriculis mediocribus planis rotundatis, caudâ ex fuscâ canescenti apicem versus nigrescenti dimidium corporis vix superanti.*
Quebec Marmot, on the upper parts hoary, with an intermixture of black, and bright wood-brown shining through ; on the inferior parts reddish-orange ; and on the head and feet, blackish-brown ; cheeks, whitish ; flat, round ears, of a moderate size ; tail, about half the length of the body, brown and hoary, with a black tip.

PLATE IX.

This animal was first described by Pennant, under the name of Quebec Marmot, from a specimen kept alive in Mr. Brooks's menagerie. Pallas afterwards noticed what was supposed to be an animal of the same species, giving it the name of *mus empetra ;* and Mr. Sabine, in the Linnean Transactions, has given a good description of a specimen presented by the Hudson's Bay Company to the British Museum. The animal mentioned by Forster, in the Philosophical Transactions, as the Quebec Marmot, is not this species, but the *Arctomys Parryi,* to be afterwards noticed.

The Quebec Marmot inhabits the woody districts from Canada to latitude 61°, and perhaps still further north. I was able to collect but little information respecting it. It appears to be a solitary animal, inhabits burrows in the earth, but ascends bushes and trees, probably in search of buds and other vegetable matters, on which it feeds. Mr. Drummond killed two individuals,—one, on some low bushes, and the other upon the branch of a tree. Pennant says, that the one which he saw was very tame, and made a hissing noise. Mr. Graham mentions, that this Marmot burrows in the earth, in a perpendicular manner, selecting dry spots at some distance from the coast, and feeds on coarse grass, which it gathers by the river-sides. The Indians take it by pouring water into its holes. When fat, its flesh is considered to be a delicacy. Its fur is of no value. It very much resembles the *bobac* of Poland, in its form and general appearance.

DESCRIPTION.

Dental formula, incisors $\frac{2}{2}$, canines $\frac{0-0}{0-0}$, grinders $\frac{5-5}{4-4} = 22.$

Incisors exserted, strong, white ; *upper ones* rounded anteriorly, marked near their inner sides by an almost obsolete groove, and having even cutting edges. *Lower ones* longer, nearly linear, and rounded anteriorly. Of the *upper grinders* the anterior one is the smallest, the posterior one the widest, and the other three are nearly equal to each other in size. The crowns of the four posterior grinders are widest exteriorly, and exhibit a duplicature of enamel, folded as it were from the outer side in such a manner, that the inner crest of each tooth forms a single rounded eminence, a very little higher than the rest of the crown. The outer crest or edge of the tooth, consists of three lower and more acute points. The smaller anterior grinder has an oval crown divided into two sloping surfaces by a transverse ridge. In each of the *grinders of the lower jaw,* the bounding ridge of enamel forms an anterior and a posterior pair of points, of which the anterior pair is considerably higher, and particularly the inner point of that pair. The inner point of the posterior pair of each tooth wears away in a cup-shaped form. The area of a section of one of the lower grinders is obliquely quadrangular ; whilst the areas of the four posterior upper ones are more nearly triangular. The lower molars increase slightly in size from the anterior to the posterior one, which is the largest. The frontal bone is flat and depressed between the orbits, and its nasal process rises, to form with the nasal bones an oblong arch.

The *body* is thick and low, the head oblong, flat on the crown and between the eyes, with a slightly arched obtuse nose, covered with short hairs. Septum and margins of the *nostrils* naked. There is a duplicature or depression on the inside of the cheek, forming the rudiments of a pouch, and capable of containing a small bean. *Whiskers* shorter than the head, entirely black. There are some black *setæ* on the eyebrows, a tuft containing about eight black hairs as strong as the whiskers, at the back part of each cheek, and a similar tuft between the posterior angles of the lower jaw. *Eyes* moderately large. *Ears* low, flat, and rounded, the anterior edge only doubling in, to form a helix ; they are well covered with short, adpressed, hoary hairs on the inside ; posteriorly, they are clothed with hairs similar to those on the adjoining part of the head. The ears are conspicuous enough, unless when the fur is in prime order, and consequently long. The upper surfaces of the head and feet are covered with a thick, smooth coat, of rather short hair, having a shining dark umber-brown colour, which on the feet approaches to black. The end of the nose is, in some specimens, hoary ; in others brown.

The *fur* on the *back* is of two kinds,—one, a fine wool or down, which, for half its length from the roots upwards, is of a blackish-gray colour, and, for the other half, is of a pure, shining, yellowish, or wood-brown. Intermixed with the down there are many longer hairs, which are brownish-black for two-thirds of their length, and are tipped with white. Some of them, however, are merely ringed with white, and are tipped with black. The resulting colour of the back is grizzled or hoary *, the white predominating over the black, and the light-brown of the shorter wool being also more or less seen, according as the fur is in good or bad order. The sides of the upper-lip, the point of the chin, the cheeks, and the sides of the neck, are soiled, reddish-white, which mixes gradually with the dark colour of the head. The *under parts*, including the throat, breast, belly, and fore and hind legs, are of a reddish-orange-colour, without mixture. The fur there is thinner and rather shorter than on the back, of the same colour, for its whole length, and there is very little of the fine wool amongst it. On the sides the soft wool has reddish-orange tips, and is mixed with a few long, reddish-white hairs, forming a gradation betwixt the colours of the back and belly.

The *tail* is flattish, but not distichous; it is nearly linear, and is rounded at the tip. It is well clothed with hair, a little longer than that on the back, and is dusky above, from an intimate mixture of brownish-white and blackish-brown hairs, and is brownish-black underneath, and at the tip. In some specimens the tail is almost entirely dark-brown, with a very slight sprinkling of hoary hairs. *Legs*, very short, and muscular. *Fore-feet*, with four toes, and the minute rudiment of a thumb. The toes are well separated, not being connected by the skin for more than a third part of the first joint. The palms are naked, and have three tubercles at the roots of the toes, and two much larger ones further back, of which the largest has a minute rudiment of a thumb on its inner side, covered by a small triangular nail. The middle toe is the longest; the first and third, which are equal to each other, are but a little shorter, and the outer one is rather more than the length of its nail shorter than these. All the toes are covered with a smooth coat of hair above, and are perfectly naked underneath. The *fore-claws* are slightly arched, and rather obtuse; they are so much compressed, that there is merely a slight excavation near the tip underneath, their edges from thence to their roots being in contact; and their size, as compared with each other, is proportional to the toes to which they belong. The soles of the *hind-feet* are long and naked to the heel; the callous tubercles are not so conspicuous as in the fore-feet. There are five hind-toes, which are about the same size with the fore-ones. The middle claw projects a very little beyond the one on each side of it. The outer toe is more than the length of its claw shorter, and the inner one is still shorter and further back.

When the fur is out of season it loses its lustre, and the down and most of the hair on the belly falls off, so that the animal can scarcely be recognised.

DIMENSIONS.

	Inches.		Inches.
Length of head and body . . from 17 to 20		Distance between the ears	1¾
„ the head alone	4	Length of tail (vertebræ)	5½
„ from end of nose to anterior part of orbit	1½	„ including fur . . .	7
„ „ „ to auditory opening	3	„ middle-toe and its claw . .	1¼
„ from posterior part of the orbit to the audi-		„ middle claw alone . . .	0⁴⁄₁₂
tory opening	1	„ palm, middle-toe, and claw .	1¾
Height of ear (posteriorly)	0½	„ from heel to tip of middle-claw, hind-foot	2¾
Breadth of ear	0¾	„ of middle hind-toe and its claw .	1¼
Distance between the eyes	1½		

* Mr. Graham likens the colour of the back to a mixture of pepper and salt.

[48.] 2 ? ARCTOMYS ? PRUINOSUS. (Pennant.) *The Whistler.*

The hoary marmot. PENNANT, *Hist. Quadr.*, vol. ii. p. 130 ; *Arct. Zool.*, vol. i. p. 112.
Ground hog. MACKENZIE, *Voy.*, p. 315.
Whistler. HARMON's *Journ.*, p. 427.
Arctomys ? pruinosus. RICHARDSON, *Zool. Journ.*, No. 12, p. 518, March, 1828.
Quisquis-su. CREE INDIANS. Deh-dehie. CHEPEWYANS.
Skwey-kwey. ATNAH INDIANS. Thidnu. NAGAILERS.
Souffleur or Mountain Badger. FUR-TRADERS.

A. (pruinosus) vellere corporis antioè rudi canescenti posticè fuscescenti, caudâ pilosissimâ badiâ nigrâgue.
Hoary Marmot, with long coarse fur, particularly on the chest and shoulders, where it is hoary ; hind parts dull
yellowish-brown ; tail blackish-brown, bushy.

DESCRIPTION.

" Tip of the nose black ; ears short and oval ; cheeks whitish ; crown dusky and tawny ; hair, on all parts rude and long ; on the back, sides and belly cinereous at the bottom, black in the middle, and tipped with white, so as to spread a hoariness over the whole ; legs black ; claws dusky ; tail full of hair, black and ferruginous. Size of the Maryland Marmot."

The above is Pennant's description of a specimen which was preserved in the Leverian Museum, and said to have been brought from Hudson's Bay. That specimen is now lost, and the species does not appear to have come under the notice of any other naturalist. If I am correct in considering it as the same with the Whistler of Harmon, we may soon hope to know more of it, for the traders who annually cross the Rocky Mountains from Hudson's Bay to the Columbia and New Caledonia, are well acquainted with it. I failed in obtaining a specimen, as I did not visit the Rocky Mountains myself ; and one which was procured for me by a gentleman was so much injured, that he did not think it fit to be sent.

The Whistler inhabits the Rocky Mountains from latitude 45° to 62°, and probably further both ways ;—it is not found in the lower parts of the country. It burrows in sandy soil, generally on the sides of grassy hills, and may be frequently seen cutting hay in the autumn ; but whether for the purpose of laying it up for food or merely for lining its burrows, I did not learn. While a party of them are thus occupied, they have a sentinel on the look out upon an eminence, who gives the alarm on the approach of an enemy by a shrill whistle, which may be heard at a great distance. The signal of alarm is repeated from one to another as far as their habitations extend. According to Mr. Harmon, they feed on roots and herbs, produce two young at a time, and sit upon their hind-feet when they give their young suck. They do not come abroad in the winter.

Mr. Macpherson describes one killed in the month of May on the south branch of the Mackenzie as follows :—" It was 27½ inches long, of which the head 2¼, and the tail 8½. It is, I think, of the same genus with the Quebec Marmot. In the fore-teeth, and in the shape of the head and body, it resembles a beaver. The hair, especially about the neck and shoulders, is rough and strong. The breast and shoulders, down to the middle of the body, is of a silver-gray colour; the rest of the body, and the brush, are of a dirty yellowish or brown. The head and legs are small and short in proportion to the body."

Mr. Harmon represents them as about the size of a badger, covered with a beautiful long silver-gray hair, and having long bushy tails. Mr. Drummond says they resemble the badger of the plains (*Meles Labradoria*) in colour, but are of a rather smaller size. The Indians take the Whistlers in traps set at the mouths of their holes, consider their flesh as delicious food, and by sewing a number of their skins together, make good blankets.

[49.] 3. Arctomys brachyurus. (Harlan.) *Short-tailed Marmot.*

Burrowing squirrel. Lewis and Clark, vol. iii. p. 35. (but not of vol. i.)
Anisonyx brachyura. Rafinesque-Smaltz, *Am. Month. Mag.*, 1817, p. 45. Desmarest, *Mamm.*, p. 329.
Arctomys brachyura. Harlan, *Fauna*, p. 304.

A. (*brachyurus*) *auriculis obtusiusculis, corpore super xerampelino rubro tincto et sub-maculato, naso ventre pedibusque lateritiis, caudâ depressâ ellipticâ fulvâ albo marginatâ ; subter griseâ.*
Short-tailed Marmot, with short obtusely pointed ears, the head and body above of a brownish-gray colour, tinged with red, and speckled with a lighter colour; nose, feet, and under surface of the body, brick-red; a flat oblong oval tail, fox-red above, with a white margin, and iron-gray colour on the under surface.

This animal inhabits the plains of the Columbia. It is known to us only by the description quoted below from the narrative of Captains Lewis and Clark. M. Rafinesque, evidently from a misapprehension of the account of its feet, has constituted for its reception the genus *Anisonyx*, the characters of which are fictitious *;

* The *Sewellel*, which is also provisionally referred to *Anisonyx*, by M. Rafinesque, belongs to a distinct genus to be hereafter described under the name of *Aplodontia.*

no character being assigned to it in the original description that can separate it from the marmots.

DESCRIPTION.

" The burrowing squirrel (of the Columbia) somewhat resembles those found on the Missouri; he measures one foot and five inches in length, of which the tail comprises two and a half inches only : the neck and legs are short ; the ears are likewise short, obtusely pointed, and lie close to the head, and the aperture is larger than will generally be found among burrowing animals. The eyes are of a moderate size, the pupil black, and the iris of a dark sooty brown ; the whiskers are full, long and black ; the teeth, and indeed the whole contour, resemble those of the squirrel; each foot has five toes; the two inner ones (thumbs) of the fore-feet are remarkably short, and are equipped with blunt nails ; the remaining toes on the front-feet are long, black, slightly curved, and sharply pointed ; the hair of the tail is thickly inserted on the sides only, which gives it a flat appearance, and a long oval form : the tips of the hair forming the outer edges of the tail are white, the other extremity of a fox red ; the under part of the tail resembles an iron-gray; the upper is of a reddish-brown ; the lower part of the jaws, the under part of the neck, legs and feet, from the body and belly downwards, are of a light brick red ; the nose and eyes are of a darker shade of the same colour ; the upper part of the head, neck and body, are of a curious brown-gray, with a slight tinge of brick-red ; the longer hairs of these parts are of a reddish-white colour at their extremities, and falling together, give this animal a speckled appearance."

" These animals form in large companies, like those on the Missouri, occupying with their burrows sometimes two hundred acres of land ; the burrows are separate, and each possesses, perhaps, ten or twelve inhabitants. There is a little mound in front of the hole, formed of the earth thrown out of the burrow, and frequently there are three or four distinct holes, forming one burrow, with their entrances around the base of these little mounds. The mounds, sometimes about two feet in height and four in diameter, are occupied as watch-towers by the inhabitants of these little communities. The squirrels, one or more, are irregularly distributed on the tract they thus occupy, at the distance of ten, twenty, or sometimes from thirty to forty yards. When any one approaches, they make a shrill whistling sound, somewhat resembling tweet, tweet, tweet, the signal for their party to take the alarm, and to retire into their intrenchments. They feed on the roots of grass, &c."

The specific name of *brachyurus* is not particularly happy in its application to this animal; the *A. Richardsonii* has a tail equally short; the tail of the *A. Ludovicianus* is even shorter ; that of the *A. citillus* both American and Siberian specimens is shorter still ; and that of the *A. mugosaricus* of Lichtenstein is fully twice as short in proportion to the length of the body.

† 4. ARCTOMYS MONAX. (Gmelin.) *The Wood-Chuck.*

Bahama Coney. CATESBY, *Carolina*, vol. ii. p. 79. An. 1743.
Monax. EDWARDS, *Birds*, pl. 104.
Maryland Marmot. PENNANT, *Arctic Zool.*, vol. i. p. 111. GODMAN, *Nat. Hist.*, vol. ii. p. 100.
Arctomys Monax. SABINE. *Linn. Trans.*, vol. xiii. p. 585. HARLAN, *Fauna*, vol. i. p. 158.
GRIFFITHS, *An. King.*, vol. iii. p. 170. Cum figura, vol. v. No. 633.

A. (monax) auriculis conspicuis rotundatis, corpore ex ferrugineo cinerascenti, vultu plumbeo, caudâ fuscâ mediocri.
Wood-Chuck, with prominent rounded ears ; fur on the body rust-coloured, tipped with gray; bluish-gray face ; a moderately long, dark-brown, rather bushy tail.

To render the list of American marmots, given in this work, as perfect as our present knowledge permits, I shall insert here short *compiled* accounts of two species, which inhabit parts of North America, lying to the southward of the district to which this work more particularly relates. Of these the Wood-chuck, or Maryland Marmot, has been longest known to Naturalists. It is common in all the middle states, and is described, by Drs. Harlan and Godman, as living in society, and forming burrows in the sides of hills, which extend to great distances under ground, and terminate in various chambers, according to the number of inhabitants. The chambers are lined with dry grass, leaves, or other similar materials, and the animals pass the winter in them in a torpid state, after having closed the entrance. They feed on vegetables, are particularly fond of red-clover, and often prove injurious to the farmer, by the extent of their depredations. They sally forth in a body on their marauding excursions, generally at mid-day, and, having placed sentinels, proceed to fill their mouths. On the approach of danger, the sentinel gives the alarm by a clear, shrill whistle, and they betake themselves to their burrows with their utmost speed. If one of them is intercepted by a dog, it boldly offers battle, and bites severely. They are capable of being tamed, and become very playful, and fond of being handled. They are cleanly animals, removing all fragments of food, and even loose earth, from the mouths of their burrows, and carefully burying their excrement. The female produces six young at a litter. Dr. Godman, from whom chiefly the above account of the habits of this animal is borrowed, informs us that Edwards's figure is very unlike, and that the only good representation is that given in Griffith's Animal Kingdom, which is copied from a print done in America by Lesueur. He likewise mentions that it has *ample cheek-pouches,* and an extension of the skin between the toes, rendering the feet, especially the hind ones, distinctly semi-palmated. He also observes, that the width of the auditory opening seems, at first sight, ill adapted for the subterranean life which the Wood-chuck leads, but that it possesses the power of closing it accurately.

DESCRIPTION.

[Extracted from GODMAN'S Nat. Hist.]

" The *body* of the Maryland Marmot is about the size of that of a rabbit, and is covered by long, rusty-brown hair, generally gray at the tips; the face is of a pale, bluish ash-colour. The *ears* are short, but broad, and as if they had been cropped at their superior edges; the *tail* is about half the length of the body, and is covered with dark-brown hairs, somewhat bushy at its extremity. The feet and claws are black; the claws are long and sharp." Warden says that, in Vermont, the largest weigh eleven pounds; but that in the southern states they attain a greater size.

---

† 5. ARCTOMYS (SPERMOPHILUS?) LUDOVICIANUS. (Ord.)
The Wistonwish.

GENUS Arctomys. GMELIN. *Sub-genus* Spermophilus. F. CUVIER.

Prairie dog. GASS, *Journal*, p. 50. An. 1807.

Prairie dog, or Wistonwish. PIKE, *Journey*, p. 207. An. 1811.

Petit chien, Prairie dog, Barking squirrel, Burrowing squirrel. LEWIS and CLARK, vol. i. pp. 93, 95, 254, &c.

Barking squirrel. IDEM, vol. iii. p. 38 (but not the " burrowing squirrel," mentioned in the same volume).
 An. 1814.

" Arctomys Ludovicianus. ORD, *Guthrie's Geog.*, vol. ii. p. 302. An. 1815."

Cynomys socialis et cinereus. " RAFINESQUE-SMALTZ. *Am. Month. Mag.* An. 1817."
					DESMAREST, *Mamm.*, p. 314.

Monax Missouriensis. WARDEN, *United States*, vol. i. p. 225. An. 1819.

Arctomys Ludovicianus. SAY, *Long's Journey*, vol. ii. p. 334. HARLAN, *Fauna*, p. 160.

Arctomys latrans. HARLAN, *Fauna*, p. 306.

The Prairie Marmot. GODMAN, *Nat. Hist.*, vol. ii. p. 114.

A. SPERMOPHILUS ? (*Ludovicianus*) *super cervinus pilis nigris interspersis : subter sordidè albus, ungue pollicari̦conico majusculo, caudâ brevi apicem versus fusco torquatâ.*

Wistonwish, having cheek-pouches? back reddish-brown mixed with gray and black; soiled white belly; a rather large conical thumb-nail; and a short tail banded with brown near the tip.

This animal, which has acquired so many appellations since the year 1807, inhabits the banks of the Missouri and its tributaries. The best account of its habits are given by Lieutenant Pike, and Captains Lewis and Clark. M. Rafinesque, considering the *petit chien*, briefly noticed by Lewis and Clark, in their first volume, to be distinct from the *barking squirrel*, more fully described in their third volume, drew up from their notices the characters of his *Cynomys socialis*, and *C. cinereus*. Dr. Harlan has given the name of *Arctomys latrans* to the *Cynomys socialis*, at the same time treating of the *Arctomys Ludovicianus* as a separate species. An attentive perusal of Lewis and Clark's narrative, however, has led to the conclusion, that, in the passages cited above, these travellers speak only of one species of marmot under a variety of names *; and Mr. Say seems, also, to have been of this opinion. Lewis and Clark, vol. i., page 246, mention a small animal, about one-third of the size of their Missouri burrowing squirrel, but otherwise closely resembling it. They could not obtain a specimen, and its characters, therefore, have not been recorded by them; but from their vicinity at the time to the plains of the Saskatchewan, from the general colour of the animal, and from their description of its earths, it most probably was the *tawny marmot* (No. 52) of this work. The genus *Cynomys* of M. Rafinesque corresponds to the *Spermophilus* of M. F. Cuvier; but the characters given by the latter author are more precise and more skilfully drawn up. The following account of the Wistonwish is quoted from Mr. Say, whose description was taken chiefly from a well-prepared specimen, presented by Lewis and Clark to the Philadelphia museum. It seems to differ from other American marmots, in the length of its thumb-nail, and to approach in that respect to the *A. fulvus* of Lichtenstein.

"This interesting and sprightly little animal has received the name of Prairie dog, from a fancied resemblance of its warning cry, to the hurried barking of a small dog. The sound may be imitated, by the pronunciation of the syllable, 'chek, chek, chek!' in a sibilated manner, and in rapid succession, by propelling the breath between the tip of the tongue and the roof of the mouth. As particular places are in general occupied by the burrows of these animals, such assemblages of dwellings are denominated *Prairie-dog villages*, by the hunters. They vary widely in extent,—some being confined to an area of a few miles, others are bounded by a circumference of many miles. Only one of these villages occurred between the Missouri and the Prairie towns; thence to the Platte they were much more numerous. The entrance to the burrow is at the summit of the little mound of earth, brought up by the animal during the progress of the excavation below. These mounds are sometimes inconspicuous, but generally somewhat elevated above the common surface, though rarely to the height of eighteen inches. Their form is that of a truncated cone, on a base of two or three feet, perforated by a

* The Burrowing Squirrel of the Columbia (*Arctomys brachyurus* of Dr. Harlan, and of this work) is described by them, in their third volume, as different from the Missouri animal, mentioned also by the name of burrowing squirrel in their first volume.

X 2

comparatively large hole or entrance at the summit, or in the side. The whole surface, but more particularly the summit, is trodden down and compacted, like a well-worn pathway. The hole descends, vertically, to the depth of one or two feet, whence it continues in an oblique direction downward. A single burrow may have many occupants. We have seen seven or eight individuals sitting upon one mound. The burrows occur usually at intervals of about twenty feet. They delight to sport about the entrance of their burrows in pleasant weather. At the approach of danger they retreat to their dens, or when its proximity is not too immediate, they remain, barking and flourishing their tails, on the edge of their holes, or sitting erect, to reconnoitre. When fired upon in this situation, they never fail to escape, or, if killed, instantly to fall into their burrows, where they are beyond the reach of the hunter. As they pass the winter in a lethargic sleep, they lay up no provision of food for that season, but defend themselves from its rigours by accurately closing up the entrance of the burrow. The further arrangements, which the Prairie dog makes for its comfort and security, are well worthy of attention. He constructs for himself a very neat globular cell, with fine dry grass, having an aperture at top, large enough to admit the finger, and so compactly formed, that it might almost be rolled over the floor without injury."

DESCRIPTION.

"The animal is of a light, dirty reddish-brown colour above, which is intermixed with some gray, also a few black hairs. This coating of hair is of a dark lead colour, next the skin, then bluish-white, then light reddish, then gray at the tip. The lower parts of the body are of a dirty white colour. The *head* is wide and depressed above, with large eyes; the iris is dark brown; the *ears* are short and truncated; the *whiskers* of moderate length, and black; a few bristles project from the anterior portion of the superior orbit of the eye, and a few also from a wart on the cheek; the nose is somewhat sharp and compressed; the hair of the anterior legs, and that of the throat and neck, is not dusky at the base. All the *feet* are five-toed, covered with very short hair, and armed with rather long, black nails: the exterior one of the fore-foot nearly attains the base of the next, and the middle one is half an inch in length; the thumb is armed with a conic nail, three-tenths of an inch in length; the *tail* is rather short, banded with brown near the tip, and the hair, excepting near the body, is not plumbeous at the base."

DIMENSIONS
Of the above specimen.

	Inches.	Lines.
Length of head and body	16	0
„ the tail	2	9
„ the tail, including the fur	3	4

Lieutenant Pike's description, as far as it goes, agrees nearly with the above. "They have a dark brown colour, except their bellies, which are white; their tails are not so long as those of the gray squirrels, but are shaped the same." In page 93 of the first volume of Lewis and Clark's narrative, where the animal is termed *petit chien*, it is stated that " The head resembles the squirrel in every respect, except that the ear is shorter; the tail like that of the ground squirrel; the toe-nails are long; the fur fine, and the long hair is gray." In the third volume, where it is called barking squirrel, the following particulars are mentioned:—" This animal commonly weighs three pounds; the colour is an uniform, bright brick-red and gray, and the former predominates; the under side of the neck and belly are lighter than other parts of the body; the legs are short, and the breast and shoulders wide; the head is short and muscular, and terminates more bluntly, wider, and flatly than the common squirrel; the ears are short and have the appearance of amputation; the jaw is furnished with a *pouch* to contain his food, but *not so large as that of the common squirrel**; each foot has five toes, and the two outer ones are much shorter than those in the centre. The two inner toes of the fore-feet are long, sharp, and well adapted to digging and scratching. From the extremity of the nose to the end of the tail, this animal measures one foot and five inches, of which the tail occupies four inches."

Of the five preceding Marmots, the *Arctomys Empetra* has a slight folding of the lining of the mouth, forming the rudiment of a cheek-pouch; the *A. pruinosus* has not been examined; the presence or absence of cheek-pouches in the *A. brachyurus* is not noted by its describers; the " ample" cheek-pouches of the *A. monax* rest on the authority of Dr. Godman; and those of the *A. Ludovicianus* are mentioned by Lewis and Clark alone, whilst their having escaped the notice of so accurate an observer as Mr. Say, excites some doubt of their existence. The Spermophiles, described in the following pages, have all cheek-pouches, which, indeed, furnish the only character that distinguishes the sub-genus from the other marmots. The solitary mode of life attributed to the Spermophiles, and some other peculiarities, apply principally to *A. citillus*, and so many species have been added since M. F. Cuvier first described the genus *Spermophilus*, that its characters require to be re-modelled.

---

* It is not easy to divine what the " Common Squirrel" is which has ample cheek-pouches.

[50.]   6. Arctomys (Spermophilus) Parryi.   (Richardson.)
Parry's Marmot.

Genus Arctomys. Gmelin. Cuvier. *Sub-genus* Spermophilus. F. Cuvier.
Ground squirrel. Hearne's *Journey*, pp. 141 and 386.
Quebec marmot. Forster, *Phil. Trans.*, lxii. p. 378.
Arctomys alpina. Parry's *Second Voy.*, p. 61, *Narrative.*
Arctomys Parryi. Richardson, *Parry's Second Voy.*, App., p. 316.
Seek-Seek. Esquimaux. Thœ-thiay (Rock badger.) Chepewyans.

A. Spermophilus (*Parryi*) *auriculis brevissimis, corpore super griseo nigrove creberrimè albo guttato ; subter helvolo, vultu badio, caudâ pedes posticos extensos tertiâ parte superante planâ versus apicem nigrâ margine extimo albescenti subtus helvolâ.*

Parry's Marmot, with cheek-pouches, very short ears, body thickly spotted above with white on a gray or black ground, pale rust-coloured beneath, face chestnut-coloured, the tail one-third part longer than the hind-feet, stretched out flat, black at the extremity, with a narrow white margin, rust-coloured beneath.

Plate x.

This spermophile inhabits the barren grounds skirting the sea-coast from Churchill in Hudson's Bay round by Melville Peninsula, and the whole northern extremity of the continent to Behring's Straits, where specimens precisely similar were procured by Captain Beechey.   It abounds in the neighbourhood of Fort Enterprise, near the southern verge of the barren grounds, in latitude 65°, and is also plentiful on Cape Parry, one of the most northern parts of the continent.   It is found generally in stony districts, but seems to delight chiefly in sandy hillocks amongst rocks, where burrows, inhabited by different individuals, may be often observed crowded together.   One of the society is generally observed sitting erect on the summit of the hillock, whilst the others are feeding in the neighbourhood. Upon the approach of danger, he gives the alarm, and they instantly betake themselves to their holes, remaining chattering, however, at the entrance until the advance of the enemy obliges them to retire to the bottom.   When their retreat is cut off, they become much terrified, and seeking shelter in the first crevice that offers, they not unfrequently succeed only in hiding the head and fore-part of the body, whilst the projecting tail is, as is usual with them when under the influence of terror, spread out flat on the rock.   Their cry, in this season of distress, strongly resembles the loud alarm of the Hudson's Bay Squirrel, and is not very unlike the sound of a watchman's rattle.   The Esquimaux name of the animal *seek-seek* is an attempt to express this sound.   According to Hearne, they are easily tamed, and are very cleanly and playful in a domestic state.   They never come abroad during

ARCTOMYS (SPERMOPHILUS) PARRYI.

Published by John Murray, January 1829.

the winter. Their food appears to be entirely vegetable; their pouches being generally observed to be filled, according to the season, with tender shoots of herbaceous plants, berries of the alpine arbutus, and of other trailing shrubs, or the seeds of bents, grasses, and leguminous plants. They produce about seven young at a time.

The accompanying figure was drawn from a specimen procured on the banks of the Mackenzie.

DESCRIPTION.

Dentition the same as in the *A. Richardsonii* hereafter described. *Forehead* flat, straight; *nose* short, thick, and very obtuse, projecting a little beyond the upper incisors, and covered with a close coat of very short, pale, yellowish-brown hairs. The *face* is clothed with short brownish-orange or reddish-brown hairs, mixed with a few coarser black ones. There are some short black *whiskers* on the upper lip, also a few black hairs over the eye and on the posterior part of the cheeks, none of them exceeding half the length of the head. The *eyes* are large and prominent. The *ear* consists merely of a low, much rounded, hairy flap, not above two or two and a half lines high, and situated above the auditory opening, which is large. The *cheeks* are of a paler red than the face, and in some specimens exhibit a considerable intermixture of gray. The *cheek-pouches* are pretty large, and open into the mouth immediately anterior to the grinders. The *body*, when the animal is fat, is thick, and flattish on the back, with a considerable breadth posteriorly. It is covered above with a dense coat of short soft fur, consisting of a fine down, which has a dark smoke-gray colour at the roots, pale French-gray in the middle, and yellowish-gray at the summits; and of longer hairs, of which the greater part are tipped with white, but many have lengthened black summits. The colours are so disposed as to produce a crowded assemblage of somewhat quadrangular white spots, margined and separated by black and yellowish-gray. The spots are nowhere well-defined, but they are most so on the posterior part of the back. On the upper aspect of the neck, and towards the sides, the white hairs, although numerous, do not produce spots. The throat, sides of the neck, outside of the shoulders, fore and hind legs, and the whole inferior aspect of the body have a colour intermediate, between brownish-red and brownish-orange, which is generally most intense on the sides of the neck, but varies in brightness with the season of the year. The hair on the belly and thighs is longer, and not so close as that of the back, and has less down intermixed with it.

The *tail* is flat, and rounded at the tip; its hair, particularly that inserted on the sides, being capable of a distichous arrangement. In this state it presents on its upper surface a mixture of gray, brown, and black in the centre, then a black border, becoming much broader towards the tip of the tail; and, lastly, a narrow margin of soiled brownish-white. Underneath it has an unmixed brownish-red colour to near the tip, where the black border and pale margin appear. The hairs of the tail become longer towards its extremity, and there they are blackish at the roots, then yellowish-brown for a short space, then brownish-black for more than half their length ; and, lastly, tipped with pale brownish-white.

Extremities.—There is a tuft of four or five longer hairs on the posterior part of the fore-leg. The toes are well separated, and are covered above with short adpressed hairs. They are naked beneath, and have a callous enlargement at the roots of the claws. The palms are also naked, and have similar callous eminences to those mentioned in the description of *A. Richardsonii.* The very small thumb is protected and almost entirely covered by a short, convex, rounded nail, and is situated on the inner side of a large tubercle at the posterior part of the palm. The toes and claws have the same relative length to each other as those of the *A. Richardsonii,* but the claws are larger in proportion. The hind-feet are also similar in form to those of the animal just-mentioned. About half an inch of the sole, next the heel, is well clothed with hair, the remainder is naked.

DIMENSIONS.

	Inches.	Lines.		Inches.	Lines.
Length of head and body .	12 to 14	0	Length of the palm and middle fore-toe .	1	4
,, head . . .	3	0	,, the middle fore-claw . .	0	7
,, tail (vertebræ) . .	3	0	,, from heel to point of middle hind-		
,, tail including fur . .	4	6	claw	2	2
,, from the orbit to end of the nose	1	0	,, the middle hind-claw . .	0	4
,, orbit . . .	0	5			

Arctomys Parryi, var. β, Erythrogluteia.

This variety, procured by Mr. Drummond on the Rocky Mountains, near the sources of the Elk River, in latitude 57°, differs from the preceding merely in being of somewhat smaller size, with a proportionally shorter head, longer tail, smaller claws, and an ear about as high again, and of a more ovate form. There is an obscure brownish streak down the centre of the back, the black hairs predominate on the upper part of the neck, and the space between the ear and eye is occupied by an intimate intermixture of black and white hairs. The feet and posterior surface of the hips and thighs are of a bright brownish-red colour. In all other respects the resemblance to var. α, is very close.

DIMENSIONS OF VAR. β.

	Inches.	Lines.		Inches.	Lines.
Length of head and body	11	0	Length of middle fore-claw	0	5½
,, head	2	4	,, from heel to end of middle hind-claw	1	9
,, tail (vertebræ)	3	6			
,, tail, fur included	5	3	,, middle hind claw	0	4
,, palm and middle fore-claw	1	2			

Arctomys Parryi, var. γ, Phæognatha.

There is a specimen of a third variety in the Museum of the Zoological Society, which was also brought from Hudson's Bay, but the particular district not mentioned. It is characterized chiefly by a well-defined, deep, chestnut-coloured mark under the eye.

[51.] 7. ARCTOMYS (SPERMOPHILUS) GUTTATUS? *The American Souslik.*

Mus citillus, var. guttata. PALLAS, *Glir. tab.* 6. B. ?
Spermophilus guttatus. TEMMINCK, *Tab. Meth.* ?

A. SPERMOPHILUS (*guttatus Americanus*) *auriculis nullis, corpore super xerampelino creberrimè albo guttato subter ochreo, caudâ abbreviatâ corpore concolori, naso convexo ferrugineo, palpebris labiisque albidis.*
American Souslik, without external ears, having the upper parts of a clove-brown colour, varied by small crowded white spots, the under parts and feet ochre-coloured, a short tail coloured like the body, convex reddish-brown nose, and whitish eyelids and lips.

Mr. Douglas brought a small marmot from the western side of the Rocky Mountains, and several injured specimens of the same species exist in the Museum of the Hudson's Bay Company. I can detect no external characters (except that the spots on its fur are more crowded and indistinct) to distinguish it from the *mus noricus* of Agricola or Hungarian Souslik, which I know only from the descriptions and figures given by authors ; but a scull of the latter preserved in the College of Surgeons, although of the same size with the American animal, differs from it in having a more arched facial line, and in possessing an uniform degree of curvature from the occiput to the end of the nose. The American Souslik has a convex nose, with the frontal bone depressed between the orbits as in the *A. Richardsonii.* It resembles the *A. Parryi* very closely in the colours and markings of its fur, though it has not, when recent, one-third of the weight of that animal, and its feet and claws are much smaller, being less than those even of the *A. lateralis.* I have been able to collect no particular information respecting its habits. It seems to be confined to the western declivity of the Rocky Mountains. Buffon mentions that the name of Souslik given to the *A. guttatus,* on the Wolga, is intended to express the great avidity that animal has for salt, which induces it to go on board vessels laden with that commodity, where it is often taken.

DESCRIPTION.

Dentition precisely similar to that of *A. Richardsonii.* *Incisors* slightly yellow. Cheek-pouches. Nose obtuse ; facial line slightly arched. *Whiskers* black, not strong. No external *ears,* the auditory opening being surrounded merely by a thickish margin, having the appearance of the cicatrix of an ear that has been cut off. *Colour.*—The upper surface of the nose is reddish-brown, mixed with a few black and some white hairs. The upper lip, the upper and under eye-lids, and the whole under jaw, are white. The cheeks and upper aspect of the head are of a mixed gray colour, produced by the fur being dark brown, ringed towards the tips with white and frequently tipped with black. The whole of the back and upper surface of the tail has a motled gray colour, produced by numerous small, somewhat quadrangular white spots, spread over a dark ground, of a colour intermediate between clove and liver-browns. The white nearly equals the brown in quantity, and the spots are less distinct towards the sides. The *fur* is short and not very fine, dark towards the roots, ringed with white above, and intermixed with longer hairs, having black tips. The throat, breast, belly, under surface of the tail, and the extremities, are of a pale ferruginous colour, approaching to ochre-yellow.

The *tail* is rather slender, and, excluding the fur at its tip, is about the same length with the posterior extremities when stretched out, or about one-seventh of the length of the body. The *extremities* are shaped like those of the other spermophiles, but the toes and claws are more slender than usual. The *palms* are naked, and are entirely occupied by five tubercles, viz. three at the roots of the toes, and two behind them of a larger size. At the base of the inner posterior tubercle there is a small convex obtuse nail, which is the only vestige of a thumb. As is usual in the spermophiles, the second fore-toe and claw is longer than the others, the third is scarcely longer than the first, and the fourth is the length of its claw shorter than the third. The fore-claws are slender, much compressed, and slightly curved. The soles are naked, but in some measure protected by the hairs which curve in from the margins of the tarsus. The three middle toes are nearly equal in length, the fifth is considerably shorter, and the first is shorter than the fifth. The hind claws are much shorter, not so much compressed, less curved, and more excavated beneath than the fore ones. All the claws are black.

DIMENSIONS.

	Inches.	Lines.		Inches.	Lines.
Length of the head and body . .	8	6	Length from heel to end of middle hind-claw	1	3
„ body . .	6	6	„ of the middle hind-toe and claw .	0	6
„ head . .	2	0	„ middle hind-claw .	0	1¼
„ tail (vertebræ) .	1	6	„ cranium from end of the		
„ tail, including fur .	2	3	nasal bones to the occipital spine .	1	7½
„ from wrist joint to end of middle fore-claw . . .	1	0	„ nasal bones . .	0	6½
„ middle fore-claw .	0	3½	Breadth of the frontal bones between the orbits	0	3½

DIMENSIONS
Of a second specimen.

	Inches.	Lines.		Lines.
Length of the head and body . .	9	6	Length from the anterior part of the orbit to	
„ tail . . .	1	6	the end of the nose . . .	7½
„ tail, including the fur .	2	0		

[52.]　8. Arctomys (Spermophilus) Richardsonii.　(Sabine.)
The Tawny Marmot.

Arctomys Richardsonii.　Sabine, *Linn. Trans.*, vol. xiii. p. 589. t. 28.　Idem, *Franklin's Journ.*, p. 662.
　　　　　　Griffith, *An. Kingd.*,vol. v. p. 246.　No. 639.　British Museum, *Spec.* 110.
Tawny American Marmot.　Godman, *Nat. Hist.*, vol. ii. p. 111.

A. Spermophilus (*Richardsonii*) *super cervinus pilis nigris interspersis subter pallidior, caudâ brevi corpore concolori margine pallido, auriculis brevissimis.*

Tawny Marmot, with cheek-pouches ; back yellowish-gray, interspersed with black hairs; belly pale grayish-orange ; a short tail coloured like the body with a pale margin ; very short ears.

Plate XI.

This animal inhabits the grassy plains that lie between the north and south branches of the Saskatchewan River, living in deep burrows, formed in the sandy soil.　It is very common in the neighbourhood of Carlton-house, its burrows being scattered at short distances over the whole plain.　. It can scarcely be said to live in villages, though there are sometimes three or four of its burrows on a sandy hummock, or other favourable spot.　The burrows are proportionable to the size of the animal, generally fork or branch off near the surface, and descend obliquely downwards to a considerable depth ; some few of them have more than one entrance. The earth scraped out in forming them is thrown up in a small mound at the mouth of the hole, and on it the animal seats itself on its hind-legs, to overlook the short grass, and reconnoitre before it ventures to make an excursion.　In the spring, there are seldom more than two, and most frequently only one individual seen at a time at the mouth of a hole ; and although I have captured many of them at that season, by pouring water into their burrows, and compelling them to come out, I have never obtained more than one from the same hole, unless when a stranger has been chased into a burrow already occupied by another.　There are many little, well-worn pathways diverging from each burrow, and some of these roads are observed, in the spring, to lead directly to the neighbouring holes, being most probably formed by the males going in quest of a mate.　The males fight when they meet on these excursions, and it not unfrequently happens that the one which is worsted loses a part of its tail as he endeavours to escape.　They place no sentinels, and there appears to be no concert between the tawny marmots residing in the neighbourhood, every individual looking out for himself.　They

PLATE. II.

ARCTOMYS (SPERMOPHILUS) RICHARDSONII.

Published by John Murray, January 1829.

never quit their holes in the winter; and I believe they pass the greater part of that season in a torpid state. The ground not being thawed when I was at Carlton-house, I had not an opportunity of ascertaining how their sleeping apartments were constructed, nor whether they lay up stores of food or not. About the end of the first week of April, or as soon as a considerable portion of the ground is bare of snow, they come forth; and, when caught on their little excursions, their cheek-pouches generally contain the tender buds of the *anemone nuttalliana*, which is very abundant, and the earliest plant on the plains. They are fat when they first appear, and their fur is in good condition; but the males immediately go in quest of the females, and in the course of a fortnight they become lean, and the hair begins to fall off. They run pretty quick, but clumsily, and their tails at the same time move up and down with a jerking motion. They dive into their burrows on the approach of danger, but soon venture out again if they hear no noise, and may be easily shot with the bow and arrow, or even knocked down with a stick, by any one who will take the trouble to lie quietly on the grass near their burrow for a few minutes. Their curiosity is so great, that they are sure to come out to look around. As far as I could ascertain, they feed entirely on vegetable matters, eating in the spring the young buds and tender sprouts of herbaceous plants, and in the autumn the seeds of grasses and leguminous plants. Their cry, when in danger, or when angry, so nearly resembles that of the *Arctomys Parryi*, that I am unable to express the difference in letters. Several species of falcon, that frequent the plains of the Saskatchewan, prey much on these marmots; but their principal enemy is the American badger, which, by enlarging their burrows, pursues them to their inmost retreats. Considerable parties of Indians have also been known to subsist for a time on them, when the larger game is scarce, and their flesh is palatable when they are fat. I have no precise information respecting the range of this animal. It inhabits sandy prairies, is not found in thickly wooded parts, and nowhere, I believe, further north than latitude 55°. It is mentioned in the Appendix to Captain Franklin's Journey, that it was found on the shores of the Arctic Sea; but incorrectly, as I have since ascertained that I had mistaken the preceding species for this one. It is one of the animals known to the residents of the fur-countries by the name of *ground squirrel*; and to Canadian voyagers, by that of *siffleur*;—it has considerable resemblance to the squirrels, but is less active, and has less sprightliness and elegance in its attitudes. It is most readily distinguished from the squirrels by the smallness of its ears; the shape of its incisors, which are larger, but not so strong, and much less compressed; the second, and not the third fore-toe, being the largest; and its comparatively long claws, and less bushy tail.   It seems to be the American representative of the *A. (spermophilus) concolor*, or *jevraska* of Siberia.   The Tawny Marmot has been hitherto known only by Mr. Sabine's account, in the Linnean Transactions, of one obtained on Captain Franklin's first expedition.   That description, owing to the imperfections of the specimen, is incorrect, in ascribing to the animal "a tapering, sharp nose," instead of a thick one, fully as obtuse as that of either the preceding or following species, and considerably more so than that of the Hudson's Bay squirrel.   The figure, also, in the Linnean Transactions, is incorrect, in the shape of the ears, and does not give so good an idea of the form of the animal, as Landseer's excellent etching in this work.

DESCRIPTION.

Dental formula ; incisors $\frac{2}{2}$, canines $\frac{0-0}{0-0}$, molars $\frac{5-5}{4-4} = 22.$

The *incisors* are straw-coloured, rounded anteriorly, without the vestige of a groove, and not so much compressed as the incisors of a squirrel, being fully as broad transversely as they are from before backwards.   The upper ones have even cutting edges, the under ones rounded edges.   In general form and structure, the upper *grinders* are similar to those of the *Arct. empetra*, but they are rather more compressed, the duplicature of the plate of the enamel from without inwards being more acute in proportion.   The lower grinders have also the same general form, but the points are more distinct, and the anterior pair of points on each tooth rise in a more remarkable manner above the posterior pair.

Skull.—The os-frontis is flat between the orbits as in the Quebec marmot ; the nasal process, however, does not rise, but forms with the nasal bones part of a flat, elliptical arch, that extends from the occipital ridge nearly to the end of the nose, when it drops rather suddenly.   The margin of the orbits is a little raised.   The distance between the orbits is only about 4 lines or 4$\frac{1}{2}$ lines, being less than in the *Arctomys Franklinii*, and not above half the space that exists between the orbits of the *Sciurus Hudsonius*.   The capacity of the cavity for containing the brain, is less in proportion than in the *A. Franklinii*.   The zygomatic process is broader than in the latter animal, and has a large, hollow surface, for the lodgment of the muscles.

Body, a little shorter, but thicker than that of the Hudson's Bay squirrel.   Head roundish, depressed ; nose obtuse ; the naked septum and margin of the nostrils is of a blackish-brown colour.   The end of the nose is covered with very short, grayish hairs ; the rest of the face, and dorsal aspect of the head, is coloured like the back, but has sometimes a darker yellowish-brown tinge.   *Whiskers*, black, shorter than the head.   *Cheek-pouches*, capable of containing a chestnut.   *Eyes*, large.   *Ears*, small, rounded, about a line high, situated above and behind the auditory opening, thick, and clothed with short hairs.   There is no part of the auricle anterior to the auditory canal.

The colour of the *back* is yellowish-brown, verging towards gray, intermixed with black hairs ; the fur is short and fine.   On the *sides* the fur is a little longer, and has more of a yellowish-gray hue, with few of the black hairs. The fur on the *belly* is longer but thinner than that on the back, and its colour is between pale rufous and yellowish-gray. The cheeks, throat, and inside of the thighs are very pale ash-gray, verging towards white. The buttocks and under surface of the tail have generally more or less of a rufous tinge. The fur throughout the body is shining pale ash-gray for the greater part of its length, the brown tints being confined to the tips. The black hairs which are intermixed, are longer than the others, and are of one colour their whole length. The tail is flat or depressed, nearly linear, and rounded at the end. It is less than one-fourth of the length of the body and head, and is clothed with hairs longer than the fur of the body, and capable of a distichicous arrangement. The upper surface of the tail is darker than the back, the central parts, when the hairs are spread out, being a mixture of black and rusty brown, in nearly equal proportions, but not banded or spotted, being only clouded. The extremities of the hairs, and, consequently, the margin of the tail, have a rusty colour, becoming paler towards the tips, which are almost white. The hairs of the tail at its extremity are the longest, being an inch in length, but the tail is by no means bushy.

There are four toes and a minute thumb on the *fore-feet*. The toes are covered above with a close smooth coat of hair. On the naked palms there are five callous tubercles, one small one at the root of the inner toe, a similar one at the root of the outer toe, and one a little larger common to the two middle ones. There is a pretty large one adjoining to the thumb, and one nearly as large and of a conical form opposite to it. The first and third toe are of the same length, the middle one is the longest, and the outer one is the shortest and furthest back. The thumb has a very short joint, and is armed with a small convex obtuse claw. The claws of the toes are long and much compressed, their edges being in contact beneath, nearly to the tips, where they separate to form a narrow groove. *Hind-feet* with five toes. Sole naked, its heel alone being protected by hairs which grow on its sides and curve over it. It has four smooth tubercles, one of which is common to the middle and fourth toe, and the other three are proper to the inner, second, and outer toes. The toes are slender and distinct, a slight duplicature only of naked skin appearing at their bases when they are pulled apart. The three middle toes differ little in length, and arise together; the other two are considerably smaller, and have their origin further back. The claws are shorter than those of the fore-feet, though of similar shape, except that the edges are not in contact beneath. All the claws are dark brownish-black.

DIMENSIONS
Of a recent full-grown specimen.

		Inches.	Lines.		Inches.	Line
Length of body and head	.	9	8	Length from heel to tip of middle hind-claw	1	10
,, head . .	.	2	4	,, of middle hind-claw . .	0	3
,, body . .	.	7	4	,, of the cranium, from the end of the		
,, tail (vertebræ) .	.	2	0	nasal bone to the occipital ridge .	1	10¼
,, tail, including fur	.	3	3	,, of the nasal bones . .	0	8
,, middle fore-claw .	.	0	6	Distance between the orbits in the scull .	0	4½

The females are generally smaller than the males.

[53.]   9. Arctomys (Spermophilus) Franklinii.   (Sabine.)
Franklin's Marmot.

Arctomys Franklinii.   Sabine, *Lin. Trans.*, vol. xiii. p. 19.   *Franklin's Journ.*, p. 662.   Harlan, *Fauna*, p. 167.
Franklin's Marmot.   Godman, *Nat. Hist.*, vol. xi. p. 109.

A. Spermophilus (*Franklinii*) *corpore super cervino ferrugineove creberrimè nigro maculato subter albido, vultu ex nigro canescenti, caudâ elongatâ cylindricâ pilis albis nigro ter quaterve torquatis vestitâ.*
Franklin's Marmot, with cheek pouches ; the upper surface of the body spotted thickly with black, on a yellowish-brown ground, under surface grayish-white ; face black and white, intimately and equally mixed ; tail long, cylindrical, and clothed with hairs which are ringed alternately with black and white.

Plate XII.

This animal was seen only in the neighbourhood of Carlton-house, where it lives in burrows dug in the sandy soil, amongst the little thickets of brushwood that skirt the plains.   It is about three weeks later in its appearance in the spring than the *Arctomys Richardsonii*, probably from the snow lying longer on the shady places it inhabits than on the open plains frequented by the latter.   It runs on the ground with considerable rapidity, and never, as far as I could learn, ascends trees.   It has a louder and harsher voice than the *A. Richardsonii*, more resembling that of the *Sciurus Hudsonius* when terrified.   Its food consists principally of the seeds of leguminous plants, which it can procure in considerable quantity as soon as the snow melts and exposes the crop of the preceding year.

DESCRIPTION.

Franklin's Marmot has somewhat the shape of the Hudson's Bay Squirrel, but is larger. It is more slender than the *Arctomys Richardsonii*.   Its nose is not so obtuse as that of the latter, but the difference is not great. The septum, naked margins of the nostrils, and margins of the lips are of a light flesh-colour.   In the *Arctomys Richardsonii* these parts are dark, approaching to black.   The ears are longer than those of the *A. Richardsonii*, having a more conspicuous erect rounded flap, covered with hairs similar to those on the crown of the head ; they resemble in form the ears of the Hudson's Bay Squirrel, but are not so large. *Eye* larger than that of *Sciurus Hudsonius*. *Cheek-pouches* of a moderate size. *Whiskers* mostly black.

The *fur* is coarser than that of *A. Richardsonii ;* it is about four or five lines long.   The *colour* of the back is pale reddish-brown, minutely and regularly speckled with black.   The tips of all the hairs are brown ; the black forms a ring beneath the brown ; below the black

ARCTOMYS (SPERMOPHILUS) FRANKLINII.

Published by John Murray January 1829.

the hair is brownish-gray, and at the bottom bluish-black. Black specklings occur on the crown of the head, cheeks, and shoulders, but the tips of the hairs covering those parts are white. On the top of the head the hair is short, and there is a large proportion of black in that part. The eyelids are white, sometimes the brownish tints are very pale, and the animal is grizzly on all the upper parts as in the individual described by Mr. Sabine. The throat, chin, lower parts of the cheeks, inside of the thighs, and all the under parts are of a soiled white without spots. The fur on the belly is rather thin, but of the same length with that on the back.

The hairs on the *tail* are longer than those on the back, and are barred with black and white, which, when the hairs are distichously arranged, produces an indistinct appearance of longitudinal stripes. But when the animal is pursued, the tail is cylindrical, the hairs standing out in every direction. In this state, which was the only one in which I had an opportunity of observing the living animal, the black and white colours of the tail are intermixed, and in nearly equal proportions. The white forms the tips of the hairs, and when they are spread out, the tail consequently appears to be bordered with white. There is no difference of colour between the upper and under surface of the tail, in which respect this species of marmot differs from all the others I have seen except the *A. Beecheyi* and *A. Douglasii*. The scrotum is large and prominent in the spring, but not pendulous. The *feet* are formed like those of *A. Richardsonii*, and are covered with short hairs, black at the roots and white at the tips. The thumb has one joint, and is larger than those of the latter animal, but has a smaller nail, which is white. The *Sciurus Hudsonius* has a shorter thumb than either of these marmots, but it is armed with a more conspicuous nail. The hind-feet when stretched out reach to the middle of the tail. The palms are naked. The hind soles are hairy for about two-thirds of their length from the heels. The claws are dark at the base, and pale-brown at their points.

On comparing the skull of this marmot with that of *A. Richardsonii*, the cavity for containing the brain appears greater in proportion, and there is a considerably greater breadth between the orbits. The margins of the orbits are not elevated as in the latter, but the bone lying between them forms a regular arch in a longitudinal direction, though it is flat transversely. The space between the orbits is not, however, so great as that between the orbits of *Sciurus Hudsonius*. The teeth do not differ from those of *A. Richardsonii*.

DIMENSIONS
Of a recent specimen full grown.

	Inches.	Lines.		Inches.	Lines.
Length of body and head	10	6	Length of middle fore-claw	0	6½
,, head	2	2	,, from heel to end of middle hind-claw	2	2
,, body	8	4	,, middle hind-claw	0	3½
,, tail (vertebræ)	5	3	,, height of ear	0	3½
,, including fur	6	3			

z

† 10. Arctomys (Spermophilus) Beecheyi. (Richardson.) *Beechey's Marmot.*

Quauhtecallotlquapachtli, ant Coztiocotequallin. Fernandez, *Quad. Nov. Hisp.*, p. 8. ? nec tamen " *Le Coquallin*" du Buffon.

A. Spermophilus (*Beecheyi*), *auriculis conspicuis, corpore super rufescenti-albo fuligneoque minutè maculato undulatove subter cervino, caudâ elongatâ e nigro canescenti.*

Beechey's Marmot, with cheek-pouches, conspicuous ears ; body above minutely spotted or waved with reddish-white marks on a blackish-brown ground, under parts pale brownish-yellow ; an unusually long, round tail of a mixed black and white colour.

Plate XII B.

Mr. Collie, surgeon of His Majesty's ship Blossom, informs me that this kind of Spermophile " burrows in great numbers in the sandy declivities and dry plains in the neighbourhood of San Francisco and Monterey, in California, close to the houses. They frequently stand up on their hind legs when looking round about them. In running, they carry the tail generally straight out, but when passing over any little inequality, it is raised, as if to prevent its being soiled. In rainy weather, and when the fields are wet and dirty, they come but little above ground. They take the alarm when any one passes within twenty or thirty yards of them, and run off at full speed till they reach the mouth of their hole, where they stop a little and then enter it. They soon come out again, but with caution, and if not molested will proceed to their usual occupations of playing or feeding. Artemisias and other vegetable matters were found in their stomachs."

I have not met with any description or notice of this animal by preceding writers unless my quotation of Fernandez be correct. In colour, size, appearance of the tail, and in general form, it approaches closely to the *Arctomys Franklinii ;* its most evident distinctive character being the greater size of its ears. The specific name has been adopted in honour of the able and scientific Commander of the Blossom. The *Arctomys Beecheyi* is an inhabitant of more southern districts than that to which this work is confined ; but it is introduced here for the purpose of giving as complete a list as possible of the American marmots, which, until very lately, have not received their due share of attention.

ARCTOMYS (SPERMOPHILUS) BEECHEYI.

Published by John Murray. January 1829

DESCRIPTION.

Dentition precisely similar to that of the *A. Richardsonii*. Incisors orange-coloured. The shape of the scull resembles that of *A. Franklinii*, and in the size of its body it also corresponds with that animal. *Head* broad, depressed; nose very obtuse, covered with short brownish hairs. *Cheek-pouches* moderate sized. *Whiskers* strong, black. *Eyes* large; eyelids whitish. *Ears* flat, semi-oval, and thin like those of a squirrel, covered with short adpressed hairs, which project a little beyond the margin at the apex. At the base both the anterior and posterior margins of the ear fold in a little and are hairy.

The *fur* covering the ear behind is brownish-black, fading towards the posterior margin into pale brown. The hairs lining the inner surface of the ear are pale brown. The upper aspect of the head is clothed with short yellowish-brown hairs. A stripe of a darker brown colour, slightly sprinkled with white, is continued from the hind head to the back, on each side of which, from the ears to the shoulders, the fur is hoary. The whole dorsal aspect of the body is coloured by a mixture of blackish-brown and very pale wood-brown or brownish-white, so disposed that the whitish parts appear in small but not very distinct spots, which cover more space than the blackish tints which separate them from each other. These markings occupy only the tips of the fur, which is short, close, lies smoothly, and has considerable lustre. When blown aside, the fur presents an uniform brownish-black tint, from the roots to near the tips. The upper parts of the cheeks are hoary; their lower parts, the margins of the mouth, the chin, the throat, and all the under parts, the insides of the thighs and shoulders, and the fore and hind legs and feet, are of an unspotted very pale brownish-yellow colour. The hair covering the belly is, as in the rest of the spermophiles, thinner and somewhat coarser than that on the back. The colours of the back and belly mingle a little at their line of junction on the sides. The *tail* is linear, covered throughout with hair an inch and a half long, which is capable of a somewhat distichous arrangement. In this state it presents three longitudinal brownish-white stripes, and two brownish-black ones on each side of the vertebræ; one of the white stripes forms the margin of the tail, and the black one next it is the broadest. These stripes originate in the hairs being of a brownish-white colour at the root, then black, and so on in alternate broad rings to the tips, which are whitish. The animal when pursued resembles the *A. Franklinii* in carrying its tail horizontally, and it is most probable that the hairs will then stick out in every direction, when it will appear variegated and hoary but not striped. The rings of alternate black and white on the hairs of the tail of *A. Franklinii* are more numerous, smaller, and less distinct.

The shape of the *extremities* is precisely the same as in the other American spermophiles. The *fore-ones* are proportionably larger than those of *A. Richardsonii*, or *A. Franklinii*. Palms, naked; thumb, larger than even that of *A. Parryi*, and armed with a thin, convex, obtuse nail, adhering closely to its phalanx. Claws black, and similar to those of *A. Richardsonii*. The soles of the *hind-feet* are covered with hairs posteriorly, and furnished with naked tubercles anteriorly; and the hind-claws are similar to those of the two spermophiles just mentioned.

Z 2

DIMENSIONS.

	Inches.	Lines.		Inches.	Lines.
Length of head and body	11	0	Height of the ear	0	6
„ head	2	3	Width of ditto at its base . . .	0	6
„ body	8	9	Length of middle fore-claw . .	0	6
„ tail (vertebræ) . . .	5	0	„ from heel to tip of middle hind-claw	2	2
„ tail, including fur . .	6	6	„ of middle hind-toe and claw .	1	0
„ whiskers	2	0	„ middle hind-claw . . .	0	$3\frac{1}{2}$

The following *anatomical notices* were furnished by Mr. Collie:—This spermophile has an epiglottis; a firm, bony clavicle; a large, simple stomach, resembling that of man in form, and equalling, in bulk, the whole stomach and liver. The intestinal canal is five times the length of the body, and it is not furnished within with *valvulæ conniventes*. The *cæcum* is a large, curved, membranous pouch, three inches and a half long. The liver is of a dark-red colour, and has a large *lobulus spigelii*, but its left lobe is small. The gall-bladder is deeply imbedded in the liver. The spleen is oblong and purplish. There is no well-marked pancreas. The kidneys are situated close to the liver; and there are no vestiges of *capsulæ renales* *.

[54.] 11. ARCTOMYS? (SPERMOPHILUS?) DOUGLASII. *Douglas's Marmot.*

A. SPERMOPHILUS? (*Douglasii*), *auriculis conspicuis, corpore super anticè pruinoso lineâ interscapulari nigrescenti; posticè pallidè brunnescenti maculis fuligneis interstincto; subter sordide albescenti, caudâ elongatâ cylindricâ pilis albis nigro torquatis vestitâ.*

Douglas's Marmot, with cheek-pouches, conspicuous ears, upper surface of the body hoary anteriorly, with a black stripe betwixt the shoulders; pale-brown posteriorly, with many indistinct transverse dark marks; tail long, cylindrical, and clothed with hairs, which are ringed alternately with black and white.

Through the kindness of Mr. David Douglas, I have received from the banks of the Columbia, a hunter's skin of an animal, which very much resembles the preceding one. The skull and teeth are wanting, neither is it possible now to ascertain whether cheek-pouches existed or not, so that, until more perfect speci-mens are procured, some doubt must remain as to its place in the system. The

* Fernandez gives the following account of the Coztiocotequallin:—" Quauhtecallotl-quapachtli, aut Coztiocote-quallin à luteo alvi colore dictus, in duplam ferè crescit magnitudinem (*Sciuri Mexicani*) alboque, nigro et fusco colore promiscuè tegitur, si ventrem excipias qui pallens est, aut fulvus quemadmodum attigimus, et caudam gerit prælongam, pilosamque quâ se interdum operit; vivit in terræ foraminibus, et antris inclusus, in quibus quoque educat prolem: vescitur indico frumenti, quod raptum ab arvis in hyemem recondit. Versutus est velut et reliqui, nec unquam cicuratur, aut congenitam deponit feritatem."

form of its claws, however, the second fore-toe being the longest, together with the shortness of its tail and ears, and the quality and colours of its fur, induce me to think that it is a true spermophile, nearly allied to *A. Franklinii* and *A. Beecheyi*. It agrees with these two, in the length, form, and colours of its tail ; and the colours exhibited by its fur have such a general resemblance, that, although they can be readily distinguished by any one who has compared them, it is not easy to convey a distinct idea of the differences by description. The *A. Douglasii* is larger than either of the other two referred to ; and its claws are shorter. Its ears are less than those of *A. Beecheyi ;* but considerably larger in proportion than those of *A. Franklinii.*

DESCRIPTION.

The *fur*, as in the other marmots, is of two kinds,—a short down, and longer and coarser hairs. The longer hairs are slender at their roots, become thicker upwards, and then taper suddenly near the points, which are acute. They are not so long, nor do they produce so fine and close a covering as the fur of any of the North American squirrels which have come under my notice. On the back the *down* has a blackish-brown colour, deepening into black over the spine ; on the sides, and also on the belly, it has a clove-brown colour ; but the skin being apparently a summer one, there is very little down on the belly. The longer hairs are, for about two-thirds of their length, of a brownish-black colour, then brownish-white for a space, and lastly, terminated by fine black tips of various lengths. On the shoulders the hairs near their tips are pure white, instead of brownish-white, and the black tips are slender, and not conspicuous, except in the hairs covering the spine.

Colours of the surface of the fur.—The sides of the mouth, and a narrow space round the eyes, are of a soiled, white hue. The tip of the nose is covered with very short, brownish hairs. The upper surface of the head is hoary, with a slight tinge of brown ; the hairs covering this part are short, and their black tips are much less conspicuous than their brownish-white parts. The ears are clove-brown posteriorly, deepening into blackish-brown at their margins ; they are of a paler brown anteriorly. The superior surface of the neck and anterior part of the back appear hoary from an intimate mixture of pure white and blackish-brown, in which the former greatly predominates except over the spine, where there is a stripe of blackish-brown, varied by a very few of the hairs being ringed with white. The predominating colour of the surface of the fur on the posterior part of the back, is brownish-white, on which there are many small, transverse, blackish specks, not distinctly marked. The whole under parts are of a soiled white colour, with a brownish tinge on the throat, on the inside of the thighs, and close to the tail. The extremities are whitish, with more or less of a brownish tinge.

The *tail* is long for an animal of this genus, and exactly resembles in form and colour that of *A. Franklinii*. It is clothed with long hairs, white or brownish-white at the roots, then ringed alternately with black and white, and, lastly, tipped with white. There are three black rings on each hair, and the one next to the white tip is considerably broader than the others. The hairs are equally long on all sides of the tail, which has, therefore, a cylindrical form; but if the animal is capable of giving them a distichous direction, there will then appear four white stripes, and three black ones on each side of its vertebræ, of which the exterior of the black ones will be the broadest, and the whole tail will have a white border. The *whiskers* are black, and are shorter than the head. The external *ears* have a semi-ovate form, and are well clothed on both sides with short hair. The margins fold in at their base like the ears of a squirrel, or like those of *A. Beecheyi*.

The feet are shaped like those of the other spermophiles; the hind soles, for more than half their length from the heels, are thickly clothed with hair. The claws are black. The thumb-nail very much resembles that of *A. Beecheyi*, but the thumb is not quite so large and distinct as in that animal.

DIMENSIONS.

	Inches.	Lines.		Inches.	Lines.
Length of the head and body	13	6	Length of middle fore-claw	0	4½
,, head	2	7	,, from the heel to the tip of the middle		
,, tail (vertebræ)	5	9	hind-claw	2	1
,, ,, with fur	7	3	Height of the ear, measured posteriorly	0	6

[55.]   12. ARCTOMYS (SPERMOPHILUS) LATERALIS.   *Say's Marmot.*

Small Gray Squirrel. LEWIS and CLARK, vol. iii. p. 35.
Sciurus lateralis. SAY, *Long's Exped.*, vol. ii. p. 235 (vol. ii. p. 46. *Amer. edit.*) HARLAN, *Fauna*, p. 181.
The Rocky Mountain Ground Squirrel. GODMAN, *Nat. Hist.*, vol. ii. p. 144.
Arctomys (Spermophilus) lateralis. RICHARDSON, *Zool. Journal*, vol. ii. No. 12, p. 519. April, 1828.

A. SPERMOPHILUS (*lateralis*), *lineâ in utroque latere luteo-albâ nigro marginatâ.*
Say's Marmot, with a yellowish-white stripe, bordered with black, on each flank.

PLATE XIII.

This animal is an inhabitant of the Rocky Mountains, and the first notice of it occurs in Lewis and Clark's memorable expedition to the Pacific Ocean. Mr. Say first described it, and placed it among the squirrels in the sub-genus *tamias*. I have, however, removed it to M. Frederick Cuvier's sub-genus *spermophilus*, on

PLATE 13

ARCTOMYS (SPERMOPHILUS) LATERALIS.

Published by John Murray. January 1829.

account of the form of its claws and incisors. It is, indeed, intermediate between these very nearly allied sub-genera, with respect to its claws and teeth; and its fur also is finer than that of the *spermophilus Hoodii*, but less so than the fur of *tamias Lysteri*. Its incisors are stronger and shorter, in proportion to its size, than those of the other marmots, but less compressed, and more slender than those of the squirrel. The claws likewise are rather more curved, and deeper at the base than those of the marmots; but considerably larger, and not so sharp as the claws of a squirrel. The second toe from the thumb of the fore-feet is the longest, as in the spermophiles, and not the third, as in the squirrels. Its ears very much resemble the ears of a ground squirrel, but are not so much pointed. I have been able to collect no certain information respecting the manners of this little animal. Mr. Drummond obtained several specimens on the Rocky Mountains, in latitude 57°, and noticed that it burrowed in the ground. Lewis and Clark say that it is common to every part of the Rocky Mountains, where wood abounds.

DESCRIPTION.

Form of body, that of a spermophile or squirrel; head, rather large; legs, shorter than those of a squirrel. *Incisors* yellowish, flattened anteriorly, and narrower behind, but not shewing the fine and numerous grooves which are visible when the incisors of the *Sciurus Lysteri* or *quadrivittatus* are viewed with a lens. Mouth situated about as far back as that of *Arctomys Hoodii*. Forehead convex; nose obtuse, covered with very short hairs, except a naked space round the nostrils. *Whiskers* black, shorter than the head. A few long black hairs over the eye, and on the posterior part of the cheek. *Eyes*, moderately large; *ears*, rather larger in proportion than those of *A. empetra*; but smaller than the ears of *Sciurus Hudsonius* or *quadrivittatus*, consisting of a somewhat triangular flat flap, much rounded at the apex, placed on the upper or mæsial side of the auditory opening, thickly clothed on both surfaces with short hairs, and having a small doubling of the anterior margin to form a helix, which, where it approaches the auditory canal, is covered with longer hairs. The *fur* on the back is dark at the roots, then pale-smoke-gray for a space, then brown, and, lastly, its tips are barred with white and dark hair-brown. The colour of the surface, when the fur lies smooth, may be termed a hoary brownish-gray. There is no vestige of a dorsal line. A yellowish-white streak commences close behind each ear, and running backwards along the sides, terminates at the hip. It is widest in the middle, being there three lines broad, and in some specimens it is very faint on the neck, though its commencement at the ear is always distinct. The white streak is bounded above and below between the shoulder and the hip by a pretty broad border of brownish-black. The sides under the lower black border, all the ventral aspect, the inner surfaces of the extremities, and the breast and throat, are of a soiled yellowish-white, sometimes tinged with brown. The cheeks, sides of the neck, and exterior parts of the fore and hind extremities, have more or less of a chestnut-brown hue. The crown of the head is brown, mixed with a little gray, and is darker on the mæsial line. The ears are brown on the margins, but pale elsewhere. There is a white circle round the eye. The nose and forehead are pale yellowish-brown, and the upper lip and chin are nearly white.

The *tail* is depressed or distichous, nearly linear, being only very slightly broader towards the tip. It is black above, with an intermixture of brownish-white hairs, and is bordered with the latter colour; yellowish-brown beneath, and margined with black and brownish-white.

The feet are shaped like those of the four preceding spermophiles. The claws are much longer and better fitted for digging than those of the *Sciurus Lysteri* or *quadrivittatus*. The thumb-tubercle is far back, and has a small, obtuse nail. The hind soles are naked to the heel, as are also the palms, and under surface of the toes. The upper surfaces of the feet are covered with short, yellowish-white hairs, which scarcely reach beyond the roots of the black claws.

DIMENSIONS

Of prepared skins.

	Inches.	Lines.	Inches.	Lines.		Inches.	Lines.
Length of body and head, from 7		9	to 8	6	Length of palm and middle fore-claw . . 0		11
„ head alone . . .			2	2	„ hind-sole and middle-claw . 1		6
„ nose to auditory opening . .	1			8	„ hind middle-claw . . . 0		2¼
„ tail (vertebræ) . . .			2	9	Height of ear 0		4
„ „ including fur . . .	3			9	Breadth of base of external ear . . . 0		5
„ middle fore-claw . .	0			4½			

PLATE.14.

ARCTOMYS) SPERMOPHILUS) HOODII.

Published by John Murray January.1829.

[56.]   13. ARCTOMYS (SPERMOPHILUS) HOODII.   (Sabine.)
The Leopard-Marmot.

Leopard ground-squirrel.   SCHOOLCRAFT, *Travels*, p. 313 and Index. An. 1821.
Sciurus tridecem-lineatus.   "MITCHELL, *Med. Repository*, An. 1821.   Described from Mr. Schoolcraft's specimen."
Arctomys Hoodii.   SABINE, *Lin. Trans.*, vol. xiii. p. 590, An. 1822.   IDEM. *Franklin's Journ.*, p. 663.
Striped and spotted ground squirrel.   SAY, *Long's Exped.*
Spermophile rayé.   F. CUVIER, *Histoire Naturelle des Mamm.* cum figura.   I never saw this Marmot assume the rounded form given to it in this figure.
Arctomys tridecem-lineata.   HARLAN, *Fauna*, p. 164.
Hood's Marmot.   GODMAN, vol. ii. p. 112.

A. SPERMOPHILUS (*Hoodii*), *dorso occupato lineis octo pallidè rufescenti-croceis cum lineis novem fuscis alternantibus quarum quinque latioribus serie guttarum crocearum notatis et quatuor (duobus nempe utrinque) inferioribus interruptis, caudâ gracili elongatâ.*

The Leopard Marmot, with cheek-pouches, having its back striped with eight pale brownish-yellow lines, which alternate with nine broader chocolate-brown ones, of which the two inferior ones on each side are interrupted, and each of the other five is marked with a row of pale spots ; a long slender tail.

PLATE XIV.

This, the most beautiful of the marmots, inhabits, in considerable numbers, the open parts of the plains in the vicinity of Carlton-house, on the Saskatchewan. Its burrows are interspersed among those of the *A. Richardsonii*, but may be distinguished by their smaller entrances and more perpendicular direction. Some of them will admit a stick to be thrust straight down to the depth of four or five feet.   The manners of the Leopard-marmot are similar to those of *A. Richardsonii,* but it is a more active animal, and of a bolder and more irritable disposition.   When it has been driven to take shelter in its burrow, it may be heard expressing its anger in a shrill and harsh repetition of the syllable *seek-seek.*   This Marmot makes its appearance in spring, about the same period with the *A. Franklinii,* the depth of burrow evidently preventing the warmth of the sun from reaching it so early as it does the *A. Richardsonii.*   The males very soon after coming abroad go in quest of their mates, and from their boldness at that period, they are easily captured by the many beasts and birds of prey which frequent the plains.   The males fight when they meet, and in their contests their tails are often mutilated.   I observed several individuals which had been recently injured in this way, and it is rare to meet a male which has a tail equalling those of the females in length.   Mr. Sabine's figure, and that in the *Histoire Naturelle des Mammifères,* have both been made from mutilated specimens.

2 A

The most northern habitat of this animal is, as far as I know, latitude 55°, and, according to Mr. Say, they are not uncommon at Engineer Cantonment on the Missouri, and on the plains which extend from thence to the Arkansas. Mr. Schoolcraft mentions that they are numerous on the river St. Peter, a tributary of the Missouri, and have been found destructive to the gardens. They also carry away grain from the fields at Carlton-house. They appear to be confined to the level sandy country, and not to inhabit the rocky and more thickly wooded parts.

A female, killed at Carlton-house on the 17th of May, had ten young in the uterus.

DESCRIPTION
From recent Specimens.

Dentition the same as that of *A. Richardsonii.*

Form of the body much like a squirrel's. Top of the head convex, the forehead and nose forming a more remarkable curve than in *A. Franklinii,* and consequently considerably more than in *A. Richardsonii.* The nose is as obtuse in proportion as that of the former of these species, and of similar form. It is covered above and on the sides with very short pale-brownish hairs. The septum, and naked space round the nostrils, have a pale flesh colour. The mouth is farther back than that of *A. Richardsonii,* but not so much so as that of the *A. Franklinii,* having nearly the position of the mouth of the *Sciurus Hudsonius. Whiskers* black, tipped with yellowish-brown, shorter than the head. The *eyes* are rather larger in proportion than a squirrel's. The *ears* consist of a very low lobe behind and above the auditory opening, covered on both sides and on the margin with short hairs; it curves in anteriorly to form a minute helix, which is hairy. The inferior part of the auditory opening has a naked margin, which is not elevated, and appears as if a portion had been cut away.

Colour.—The end and sides of the nose, the lower part of the cheeks, the eyelids, the throat, belly, part of the sides, and the extremities, are covered with a moderately close coat of pale yellowish-brown hair, sometimes, especially on the shoulders and hips, tinged with rust-colour. The upper part of the cheeks and side of the head are covered with a mixture of pale yellowish-brown and black. The lower jaw is nearly white. On the back there are five stripes of a chocolate-brown colour, each stripe having down its middle a row of square spots nearly of the same colour with the fur on the belly. The central stripe, running from the crown of the head to the root of the tail, is a little broader than the others, and the pale spots in it are smaller. These chocolate stripes are separated from each other by narrower stripes of the same colour with the belly. There are also two narrower stripes of chocolate-brown on each side, less distinctly marked, and without spots, but separated by yellowish-brown stripes;—forming in all nine chocolate stripes and eight pale ones, five of the former being spotted with the pale colour.

The linear *tail* is narrower and longer than the tails either of *A. Franklinii* or *A.*

Richardsonii, having, when its hairs are distichously arranged, a pale chocolate-brown colour down its middle, bounded on each side by a deeper colour, approaching to black, and lastly the whole tail is margined by pale brownish-gray. The same colours occur on the under surface of the tail, but there is more of the pale-brown colour and less of the black.

The feet are formed like those of *A. Richardsonii.* The thumb is smaller, but it has a larger nail; it has one joint and its nail is obtuse.

DIMENSIONS
Of a recent Male Specimen.

	Inches.	Lines.		Inches.	Lines.
Length of head and body	7	6	Length of the palm, middle fore-toe, and		
„ head	1	9	claw	0	9½
„ tail (vertebræ)	3	4	„ middle fore-claw	0	3
„ tail including fur	4	1	„ sole, middle hind-toe, and		
Distance from tip of the nose to the anterior			claw	1	4
angle of the orbit	0	9	„ middle hind-claw	0	2
„ posterior angle of the orbit to the auditory opening	0	7½			

The largest individual I saw was a male, which measured nearly nine inches to the insertion of the tail. The females were smaller than the males.

In addition to the nine marmots described in the preceding pages from specimens either recent or prepared, and to the four of which I have given compiled notices, North America most probably possesses many others, among which may possibly be reckoned the *Techallotl* of FERNANDEZ*, which, like the *Coztiocotequallin* (referred to the *A. Beecheyi*) feeds on grain, and lives in burrows. The author has not mentioned whether either of these two animals has cheek-pouches or not.

The EUROPEAN species of the genus are:—

1. A. BOBAC (*Polish marmot*).
2. A. MARMOTTA (*marmot of the Alps*).
3. A. SPERMOPHILUS GUTTATUS (*souslik or marmot of the Wolga*).

* " Techallotl caudam ferè depilem gestat, ac breviorem, nec dodrantem vincit longitudine; non cicuratur, sed perpetuò mordet atrociter et corrodit oblata omnia; fusco et candenti colore promiscuè tingitur; et posterioribus quoque innixus pedibus oblatam edit alimoniam, sed precipuè maizii spicas apprehensas anterioribus; oculi sunt magni si illos cum ceteris partibus conferas; vivit in antris quæ unguibus facile excavat; consternitque lanâ, gossipiove et quovis alio molli stramento, ac passeres voce imitatur."—FERNANDEZ, *Quadr. Nov. Hisp.*, p. 9.

4. A. SPERMOPHILUS CITILLUS seu UNDULATUS (*Zizel, Suset or Hungarian marmot*).

5. A. SPERMOPHILUS CONCOLOR (*Jevraschka or Siberian marmot*).

The three last were described, by Pallas, as varieties of one species, which he named *mus citillus.* M. Lichtenstein has lately given the characters of the three following new ones, brought from Bucharia by M. Eversman :—

6. A. FULVUS, resembling the bobac, but having only eleven inches length, exclusive of the tail, which is three inches and one-third long. Its fur is of a shining, yellowish-brown colour, and is mixed with an ash-gray down. Its toes are slender, and much longer than those of the bobac, and its thumb-nail is peculiarly long.

7. A. LEPTODACTYLUS is nine inches long, exclusive of the tail, which is two inches and three-quarters. It is remarkable for the length of its toes, which is so great, that the distance from the heel to the root of the claw of the middle hind-toe is equal to one-fourth of the length of the body (whilst in the bobac and Siberian souslik, it does not exceed the eighth part). The sole is not naked, as in the other marmots, but is clothed, as well as the under surfaces of the toes, with the exception of the two middle ones, with close, coarse hairs. The thumb of the fore-feet is armed with a strong, obtuse nail, which curves inwards. The fur of the back is composed of long, crowded, silky hairs, of a gold-yellow colour, mixed with black down. The belly is white ; the crown of the head is grayish-brown, which colour forms sharp, angular projections towards the nose, and is intersected by a white stripe, occupying the space between the eye and the nose, and surmounted by a black streak reaching from the inner angle of the eye to the upper lip. The tail is coloured above like the back; beneath, it has a shining, black centre, with a white border.

8. A. MUGOSARICUS, is nearly nine inches long, with a tail a little exceeding an inch, and is without nails to the fore-feet. The soles of the hind-feet are broad and short, having only about one-tenth of the length of the body. In other respects it perfectly resembles the souslik.

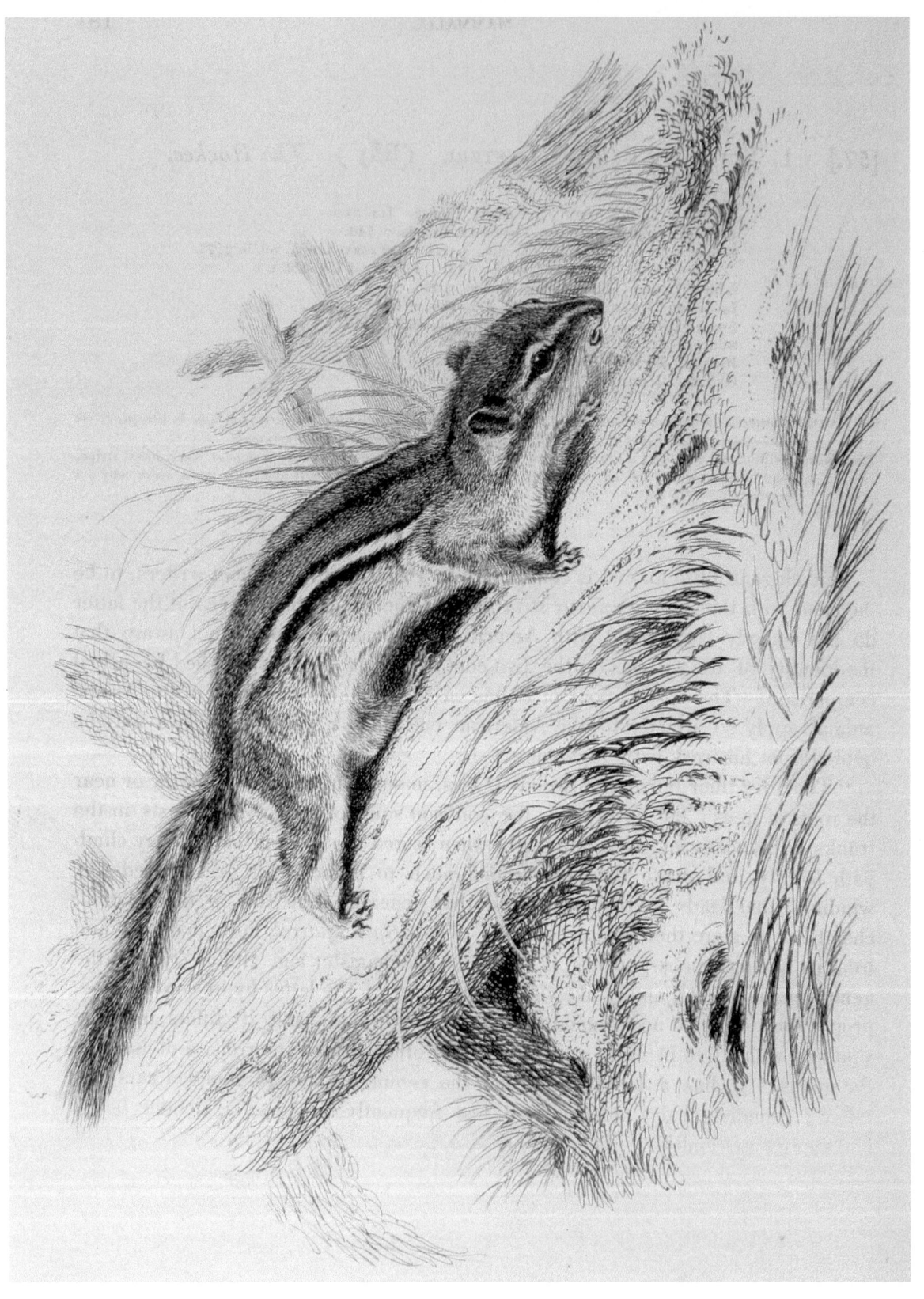

SCIURUS (TAMIAS) LYSTERI.

Published by John Murray. January 1829.

[57.]   1. Sciurus (Tamias) Lysteri.   (Ray.)   *The Hackee.*

GENUS, Sciurus. LINN. *Sub-genus*, Tamias. ILLIGER.
Escurieux Suisses. SAGARD-THEODAT, Canada, p. 746.
Ground squirrel. LAWSON, Carolina, p. 124. CATESBY, *Carol.*, vol. ii. p. 75.
                EDWARDS, vol. iv. t. 181. KALM, vol. i. p. 322. t. i.
Sciurus Lysteri. RAY, *Synops. Quadr.*, p. 216.
Le Suisse. CHARLEVOIX, *Nouv. Fr.*, vol. v. p. 198.
Striped Dormouse. PENNANT, *Arct. Zool.*, vol. i. p. 126.
Sciurus striatus. HARLAN, *Fauna*, p. 183.
Hackee. UNITED STATES.
Ohihoin. HURONS.

Sc. TAMIAS (*Lysteri*), *dorso brunnescenti-griseo posticè helvolo lineâ centrali nigrâ percurso, lineâque in utroque latere albâ breviori latiori super subterque nigro marginatâ, ventre albo, caudâ breviusculâ.*

The Hackee, with cheek-pouches ; a brownish-gray back, bright orange-brown buttocks, a slender black dorsal stripe, and a broader white one on each flank, with a broad black border above and below it ; a white belly ; a shortish tail.

PLATE XV.

This elegant little animal is considered, by Pallas and subsequent writers, to be the same with the Asiatic *sciurus striatus ;* but the descriptions given of the latter do not exactly correspond with American specimens, and I am not aware that the identity of the species on the two continents has been established by actual comparison.   The observations of Pallas on the manners and form of the Asiatic animal apply so exactly to the American one, that a passage or two may be quoted from his work with advantage :—

" They dig their burrows in woody places, in small hummocks of earth, or near the roots of trees ; but never, like the common squirrels, make their nests in the trunks or branches of trees, although, when scared from their holes, they climb with facility, and make their way from branch to branch with great speed.   A winding canal leads to their nest, and they generally form two or three lateral chambers, to store their winter food in.   The *striped squirrel,* in its manners, and from its having cheek-pouches, is allied to the hamster and citillus (type of the genus *spermophilus*), and is likewise connected with the latter by its convex nose, proper for an animal accustomed to dig.   In its whole habit it differs from the squirrels which live in trees, and forms, with other striped squirrels, a division of the genus.   It has a longer head than the common squirrel ; rounded ears, not tufted ; roundish, hairy tail, which it less frequently turns up ; a slender body, and shorter extremities.   The fur, likewise, is very short, and less fine.   Yet, in its diurnal habits, and in not becoming torpid in winter, it comes near the squirrels. It is difficult to tame."

The hackee is common on the north shores of Lakes Huron and Superior; but I do not believe that it exists in a higher latitude than the 50th parallel. Although very wild, it is fond of establishing its abode in the immediate vicinity of man, and multiplies greatly in cultivated places.

DESCRIPTION

Of a recent male specimen, killed in April at Penetanguishene.

Dental formula; incisors, $\frac{2}{2}$; can. $\frac{0-0}{0-0}$; grinders, $\frac{4-4}{4-4} = 20$.

Incisors of a deep yellowish-brown colour, and marked with a number of very fine longitudinal furrows. They are compressed, as is usual in squirrels, but they are not so strong in proportion as the incisors of the Hudson's Bay squirrel, though they are longer. The lower incisors are twice as long as the upper ones. The molars are nearly equal to each other in size, and their crowns have nearly circular slightly excavated areas, with a small notch exteriorly; they are surrounded by a thin plate of enamel, which acquires a black crust.

Form.—Body slender; the *head* tapers from the ears to the nose; the forehead is slightly convex, but the crown of the head is depressed; the nose is not very obtuse, and is clothed with short hairs. The nostrils open downwards, and their margins and septum are naked. The *whiskers* are fine, rather shorter than the head, and of a black colour. There are also some fine black hairs on the cheek, and one or two longer ones springing from the eyebrow. *Eyes,* large; *ears,* ovate, rounded, erect, covered with short hair, and without tufts on their margins. The *cheek-pouches,* which are of moderate size, and extend but a very short way behind the ear, open into the mouth between the incisors and grinders. There are nine transverse folds or plaits on the palate, of which the five posterior ones are divided by a mæsial ridge.

Colour.—The dorsal aspect of the head is covered with yellowish-brown hairs, which are mixed with a smaller number of black ones. There is a black spot near the tip of the nose. The eyelashes are black, the eyelids white; there is a dark brown streak between the eye and the ear, and a broad, yellowish-brown stripe extends from the nose, under the eye, to behind the ear, deepening in its middle to chestnut-brown. The anterior part of the back is hoary-gray, from a mixture of black and white hairs. The rump, hips, and exterior surfaces of the thighs are of a bright orange-brown colour, mixed with a few black hairs. A dark dorsal line commences at the occiput, and reaches to within an inch of the tail. This line is brownish at its commencement, but deepens to black posteriorly. There are also, on each flank, two black lines, which commence behind the shoulders, extend to the hips, and are separated by a moderately broad white stripe. All these stripes are more or less bordered with brown. The sides, beneath the stripes, present a mixture of gray and very light brown. The fur, covering the throat, chin, belly, and inner surface of the extremities, is longer and thinner than that on the dorsal aspect, and is white throughout its whole length. The fur on the upper parts of the body forms a smooth coat, and is blackish-gray at its roots. There is no defined line of separation betwixt the colours of the back and belly.

Tail, sub-distichous, not bushy, brown for a small space at its root, afterwards grayish approaching to black on its upper surface, the black hairs predominating over the whitish ones. Underneath it is reddish-brown, with a margin of hoary-black. When the hind-legs are stretched out, they reach within a quarter of an inch of the tip of the tail.

Extremities.—The fore-feet have four toes, and an imperfect thumb; the palm is naked, with five tubercles, three of which are situated at the roots of the toes, and two larger ones behind. On the inner side of one of these there is a minute wart in place of a thumb, entirely covered by a thin, roundish nail; the claws are curved, compressed, and sharp-pointed, convex above, and channelled underneath; they bear the same proportion to the size of the animal that those of the Hudson's Bay squirrel do; are much smaller than the claws of the spermophiles, and are partially concealed by the hairs of the toes. There are five toes on the *hind-feet,*—the three middle ones nearly of equal length, the outer and inner ones shorter; the hind part of the sole is hairy.

DIMENSIONS
Of a recent specimen.

	Inches.	Lines.		Inches.	Lines.
Length of head and body	6	0	Height of ears	0	4
„ head	2	0	Breadth of ditto	0	3
„ tail, including fur	3	8	Length of middle fore-claw	0	0¾
„ „ (vertebræ)	3	0			

[58.]   2. Sciurus (Tamias) quadrivittatus.   (Say.)
             Four-banded Pouched Squirrel.

Sciurus quadrivittatus. Say, *Long's Exped.*, vol. ii. p. 349. (vol. ii. p. 45, *Amer. Ed.*) Harlan, *Fauna*, p. 180.
Four-lined squirrel. Godman, *Nat. Hist.*, vol. ii. p. 137.
Four-banded squirrel. Griffith, *An. Kingd.*, vol. v. No. 665.
Sciurus (Tamias) quadrivittatus. Richardson, *Zool. Journ.*, No. 12, p. 519, April, 1828.
Sassacka-wappiscoos. Cree Indians.

Sc. Tamias (*quadrivittatus*), *lineis quinque nigrescentibus cum quatuor albis alternantibus dorsumque totum occupanti-*
     bus, lateribus ferrugineis, ventre cinereo, caudá gracili elongatá fuligneo spadiceoque variá.
Four-banded pouched Squirrel, having five blackish lines and four alternating white ones occupying the whole back ;
     reddish-brown sides and gray under parts ; with a long slender tail exhibiting dusky and light-brown colours.

Plate XVI.

This diminutive squirrel is common throughout the woody districts, as far north
as Great Slave Lake, if not farther.   It is found at the south end of Lake Winipeg,
in latitude 50°, and within that range it seems to replace the *Sciurus Lysteri*.
Mr. Say observed it on the Rocky Mountains near the sources of the Arkansas and
Platte ; and Mr. Drummond brought specimens from the sources of the Peace River,
which rises on the same ridge.   It is an exceedingly active little animal, and very
industrious in storing up provision, being generally observed with its pouches full
of the seeds of leguminous plants, bents, and grasses.   It is most common in dry
sandy spots, where there is much underwood, and is often seen in the summer
time sporting among the branches of willows and low bushes.   It is a lively,
restless animal, troublesome to the hunter, and often provokes him to destroy it
by the angry chirrupping noise that it makes on his approach, and which is a
signal of alarm to the other inhabitants of the forest.   During the winter it
resides in a burrow with several openings made at the root of a tree, and is never
seen on the surface of the snow at that season.   When the snow disappears, many
small collections of hazel nut shells, from which the kernel has been extracted
by a minute hole gnawed in the side, are to be seen on the ground near its
holes.   Mr. Say states its nest to be composed of an extraordinary quantity of
the burrs of *xanthium*, portions of the upright *cactus*, small branches of pine-trees,
and other vegetable productions, sufficient in some instances to fill a cart.   On
the banks of the Saskatchewan the mouths of their burrows are not so protected.
The four-banded squirrel is, in common with the *hackee*, named *Le Suisse* by the

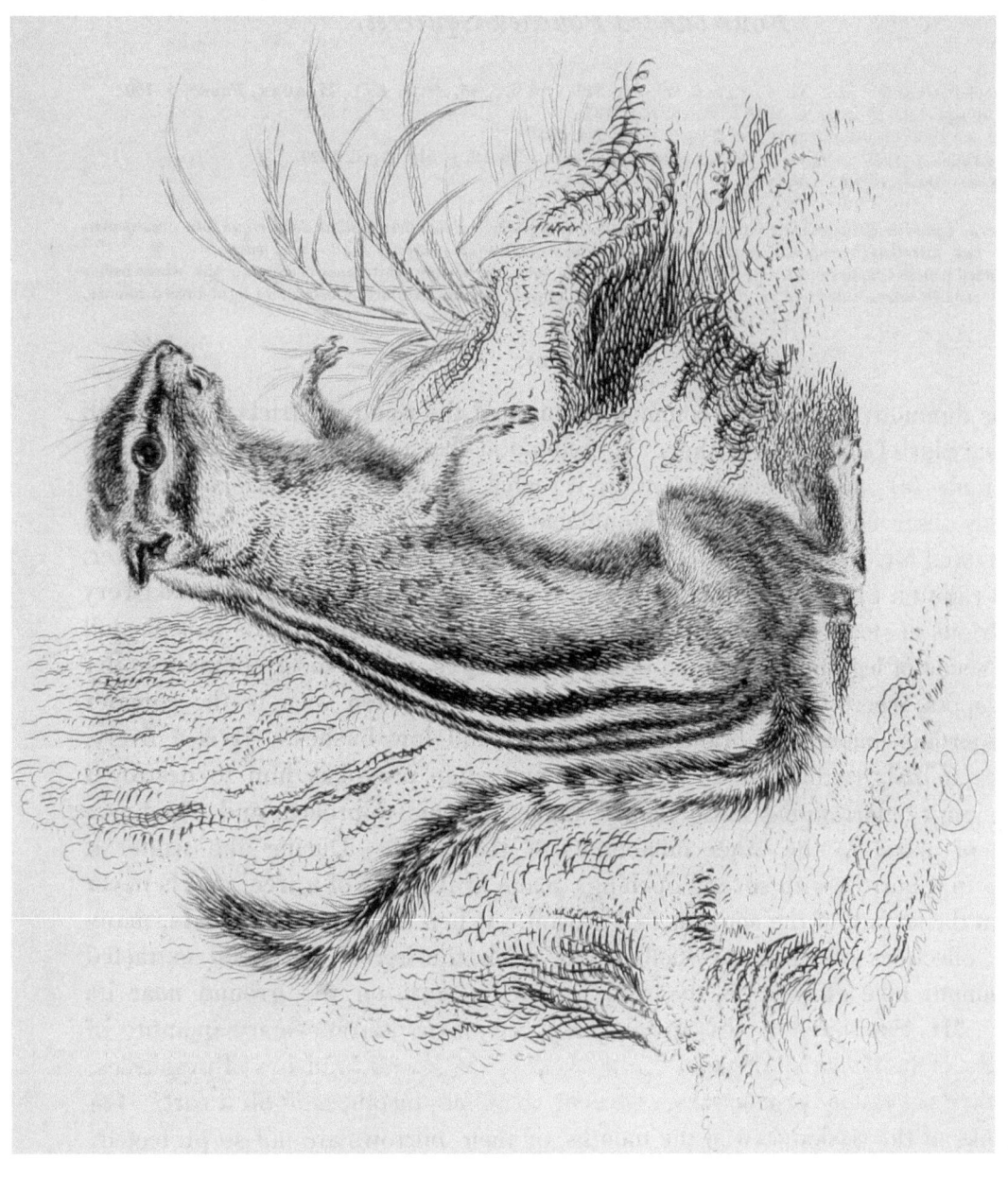

SCIURUS (TAMIAS) QUADRIVITTATUS.

Published by John Murray, January 1829.

French Canadians, an appellation which, according to Father Theodat, arose from their skins being rayed with black, white, red and gray, like the breeches of the Switzers who formed the Pope's guard. The same author informs us that they bite bitterly when taken.

DESCRIPTION.

Dental formula, incisors $\frac{2}{2}$, canines $\frac{0-0}{0-0}$, *grinders* $\frac{5-5}{4-4}$, $= 22$.

Incisors much compressed, like those of the squirrels. The upper ones are short, and their even cutting edges have an inclination backwards. Their anterior surfaces are of a deep yellow colour, flatly convex, and under the microscope they appear to be marked with longitudinal grooves. They are narrower behind. The anterior grinder of the upper jaw has a round crown, and is very much smaller than the others, which resemble those of the *Sciurus Hudsonius.* The inferior grinders are intermediate in form between the corresponding teeth of the squirrels and those of the spermophiles, their areas not being so much hollowed as in the former, nor presenting points so high as in the latter. The frontal bone is more arched between the orbits than in the *Sciurus Hudsonius*, and its proportional breadth is not so great. The scull has an uniform slight curvature from the occiput to the end of the nose ; the cavity for containing the brain is larger, the orbit is much smaller in proportion, and the zygomatic arch less projecting than in the Hudson's Bay squirrel, or than in any of the spermophiles noticed in this work.

Form.—The *head* is long, and tapers considerably from the eyes to the end of the nose, which is not, however, remarkably sharp. The mouth is situated far back. The *whiskers* are black and rather shorter than the head. The *eye* is small when compared with a squirrel's. The *ear* is erect, semi-ovate, obtuse and flat, except a slight duplicature at the base of the anterior margin ; it is covered on both sides with a coat of short hair. The cheek-pouches extend to the angle of the jaw. The *body* has a more slender form than that of the squirrels in general.

Colour of the head.—A narrow black line runs from the nostrils to the anterior part of the orbit, and is continued from behind the orbit to near the ear. The cartilaginous margins of the eyelids are black, but the eyelids themselves, both upper and under, are grayish-white. This white marking is continued from the ear to the end of the nose, in two lines, separated from each other by the black line above-mentioned. The upper white line reaches the end of the nose, where it is separated from its fellow on the opposite side of the face by a narrow mæsial line of a dark brown colour. The under white line, after passing over the under eyelid, is lost in the white of the upper and lower lip ; and there is a dark brown streak immediately below its posterior part. The upper aspect of the head is dark hair-brown, sprinkled with a few hoary specks. This colour is bounded by the white line, which passes over the upper eyelid, and it is continued forward of a darker hue till it ends acutely at the tip of the nose. This arrangement of colour gives a peculiar character of sharpness to the features, and causes the nose to appear more pointed than it really is.

2 B

Colour of the body.—A black or blackish-brown dorsal line commences between the ears and terminates in the dark colours of the upper surface of the tail. A similar but rather broader line begins at each shoulder and ends on the buttocks near the tail; and on each flank there is another line, which extends from the shoulders to the haunches; they are separated on each side of the back by two other lines of equal breadth and of a grayish-white colour, intermixed or bordered with reddish-brown hairs. These lines conjointly, viz. the five dark ones and four pale ones, occupy the whole back, and there is none of the beautiful gray ground which exists betwixt the lateral stripes of the *hackee*. The *sides* are bright reddish-brown mixed with chestnut colour. The thighs and buttocks are hair-brown. The upper and under lips, the throat, the belly, and the insides of the extremities, are pale smoke-gray.

The *tail* is long and narrow, linear, or perhaps rather thicker at the root than at the tip, covered above with hairs, which are light wood-brown at the roots, then blackish-brown, and lastly wood-brown at the tips. The hairs are capable of an obscure distichous arrangement, and the resulting colour is an ill-defined border of wood-brown, bounding a mixture of blackish-brown, with a little wood-brown. The under surface of the tail presents an unmixed reddish-brown colour in the centre, bounded by a black line which is faintly bordered by reddish-brown.

Extremities.—The *fore-feet* are shaped like those of the Hudson's Bay squirrel, and have four toes, with a small thumb of only one joint, armed with an obtuse nail. Palms naked. *Claws* black, compressed, curved, and sharp like those of the squirrels, better fitted for climbing than for digging. *Posterior extremities* long. Hind-feet with five slender toes divided to the base, and four naked callous eminences on the sole at their roots: the rest of the sole is well clothed with short hairs. Both the fore and hind-feet are covered above with a smooth coat of pale grayish-brown hair. The hind soles are longer and more slender than those of the spermophiles described in this work, and considerably more so than those of the *sciurus Hudsonius*. The hind-claws likewise differ very much from the curved, sharp, hind-claws of the last-mentioned animal, and have more resemblance to those of the spermophiles.

DIMENSIONS
Of a recent specimen killed at Carlton-house.

	Inches.	Lines.		Inches.	Lines.
Length of the head and body	5	6	Height of the ear	0	4½
„ tail	4	3	Breadth of the ear at its base	0	5
„ head	1	5	Length from the heel to the tip of the		
„ from end of the nose to the centre			middle hind-claw	1	3
of the orbit	0	7	„ of hind and fore-claws, about	0	1
„ from the end of the nose to the					
auditory opening	1	1			

The tails of this kind of squirrel, particularly of the males, are often mutilated in their contests with each other, and they are very liable to be broken off in the attempt to catch them, so that it is rare to obtain a specimen with a perfect tail.

SCIURUS HUDSONIUS.

Published by John Murray, January 1829.

[59.] 3. Sciurus Hudsonius. (Pennant.) *The Chickaree.*

Genus, Sciurus. Linn.
Escurieil commun ou Aroussen. Sagard-Theodat, *Canada*, p. 746.
Common Squirrel. Forster, *Phil. Trans.*, lxii. p. 378. An. 1772.
Sciurus vulgaris, var. E. Erxlebein, *Syst.* An. 1777.
Hudson's Bay Squirrel. Pennant, *Arctic Zool.*, vol. i. p. 116. *Hist. of Quadr.*, vol. ii. p. 147.
Common Squirrel. Hearne, *Journ.*, p. 385.
Red Barking Squirrel. Schoolcraft's *Journ.*, p. 273.
Red Squirrel. Warden, *United States*, vol. i. p. 330, No. 54.
Sciurus Hudsonius. Ejusdem, vol. i. p. 231, No. 56. Sabine, *Franklin's Journey*, p. 663. Harlan,
    Fauna, p. 185. (The *Sc. Hudsonius* of Gmelin is a *Pteromys*.)
Ecureil de la Baie d'Hudson. F. Cuvier, *Hist. Naturelle des Mammifères.*
Hudson's Bay Squirrel. Godman, *Nat. Hist.*, vol. ii. p. 138.
Chickaree. United States. Aroussen. Hurons. Annekcootchass. Cree Indians.

Plate XVII.

This squirrel is an inhabitant of the forests of white spruce, which cover a great portion of the surface of the earth in the fur countries. The limits of its range to the southward have not been mentioned by American writers, but they say that it is common in the middle states. It is found as far north as the spruce trees extend, that is, to between the sixty-eighth and sixty-ninth parallel of latitude, and it is one of the most numerous animals in the northern districts. It digs its burrows, generally at the root of one of the largest and tallest trees it can select, and forms four or five entrances, around which very large quantities of the scales of spruce-fir canes are in process of time accumulated. It does not come abroad in cold or stormy weather, but even in the depth of winter it may be seen, during a gleam of sunshine, sporting among the branches of its tree. On the approach of any one, it conceals itself behind a branch, but soon betrays its position by the loud noise it makes, somewhat like the sound of a watchman's rattle, and from whence it has obtained the expressive appellation of Chickaree. When pursued and harassed it makes great leaps from tree to tree, but as soon as it observes the way clear, it descends to the ground and seeks shelter in its burrow. It does not appear to quit the tree beneath which it burrows, by choice, unless when it makes an excursion in the spring in quest of a mate. In the fur countries it subsists chiefly, if not entirely, on the seeds and young buds of the spruce-fur. In the winter it collects the cones from the tree and carries them to the entrance of its burrow, where it picks out the seeds beneath the snow. Like the English squirrel, it makes hoards on the approach of severe weather.

The flesh of this squirrel is tender and edible, but that of the male has a strong murine flavour. The Indian boys kill many with the bow and arrow, and also take them occasionally with snares set round the trunks of the trees which they frequent. Hearne states that they are hard to tame. Their skins are of no value, have never formed an article of trade, and are not applied to any purpose even by the Indians.

DESCRIPTION.

Dental formula, incisors $\frac{2}{2}$, canines $\frac{0-0}{0-0}$, grinders $\frac{4-4}{4-4}$ or $\frac{5-5}{4-4}$ = 20 or 22.

Incisors, strong, very much compressed, deep at their roots from before backwards; flat on the sides; convex and of a deep orange colour anteriorly. The upper ones have an even chisel-shaped cutting edge. The lower ones are not much longer and are more pointed. *Grinders.*—The squirrels are said to have five grinders in the upper jaw when young. The Hudson's Bay squirrel loses the small anterior one very early, as, after examining a great many specimens, I found none with more than four on a side in the upper jaw. In the *tamias* and *spermophiles* the fifth grinder remains when the animal is full grown, but is proportionably larger in the latter than in the former. The inner surface of the upper grinders of the Hudson's Bay squirrel are more obtuse, and consequently their areas are less wedge-shaped than those of the grinders of the Spermophiles. They are likewise more excavated on the crowns, and have less elevated ridges of enamel. The lower grinders have also excavated crowns, and the two anterior points of each tooth do not form an elevated crest as in the spermophiles. The under jaw is shorter but rather stronger than in the latter genus, and the space for the lodgement of the brain is larger. There is also a much greater distance between the orbits; the frontal bone is flat; and the nose less arched than in the genera *tamias* and *spermophilus.*

Form.—Nose obtuse, forehead very slightly arched. Mouth rather far back. Whiskers black, longer than the head. *Ears* rounded, somewhat concave; the posterior margin doubles forwards to form a valve over the auditory opening, and the anterior one curves in to form a helix. Both sides of the ear are covered with hair; that which clothes the outside being longest, and when the fur is in prime order, projecting upwards beyond the margin; but there is not at any time a distinct tuft on the tip of the ear, like that which ornaments the common English Squirrel.

Colour.—There is a short blackish central stripe on the end of the nose; the sides of the nose are pale brown, sometimes almost white. A broad stripe of bright chestnut commences between the ears, and is continued down the back and along the tail nearly to its tip : this chestnut colour is intimately speckled with black, and mixes more or less gradually with the colour of the sides in different specimens. The forehead, cheeks, sides, and exterior surfaces of the extremities are of a grayish-brown speckled colour, resulting from minute black specks being equally distributed over a pale yellowish-brown or wood-brown ground. The upper and under eyelids, a space round the mouth and the throat, are white. The belly and inner sides of the extremities are smoke-gray. These colours vary with the season and condition of the animal. In some seasons the chestnut-coloured dorsal stripe commences behind the shoulders, is of a less bright hue, mingles more gradually with the colour of the sides, and is not continued so far down the tail. The outer surfaces of the extremities have occasionally an orange hue. The belly is in some instances nearly white, in others pretty dark gray, from the number of black hairs interspersed over it. In summer specimens, when the fur on the belly becomes thin, the colours of the upper and under parts are separated by a blackish-gray line, extending from the shoulders to the thighs. This line is produced by the roots of the fur being seen, and is not perceptible in the northern specimens, whose fur is finer and longer; nor have I noticed it in more southern specimens procured in the winter. The fur on the back is fine and of a blackish-gray colour, from the roots for half its length upwards; the remainder of its length is wood-brown, with two or three rings of black; the tips of the longest hairs are black. The fur on the belly is in the winter rather longer than that on the back, bluish-gray at the roots, then white, with a ring or two of black below the tips.

The *tail* is somewhat depressed, and is linear. It is full of long hair, but is not nearly so bushy as the tail of the English squirrel. Its hairs are capable of a somewhat distichous arrangement, and then it presents on its upper surface a bright chestnut centre, and a light, brown margin, separated from the chestnut by a black band, most distinct near the tip of the tail. Beneath, the tail exhibits an intimate mixture of light brown and black, the latter forming a band near the tip. The hairs on the upper surface of the tail carry the bright brown colour to their roots, many of those beneath are black throughout their whole length, but the majority of the under ones are brown, with a black tinge near their tips.

The *extremities* are covered with longer fur than those of the spermophiles; the limbs are robust; the fore-feet have four toes, with the rudiment of a thumb, covered by an obtuse, thin nail, closely applied; the third toe is rather the longest, the second is next in length; the first and fourth are shorter, and arise further back. In the spermophiles, on the contrary, the second toe is decidedly the longest, and the first does not arise so far back as in the Hudson's Bay squirrel. The *claws* are very much compressed, and so much curved, as to feel hooked; and they are very acute. The palms and under surfaces of the toes are naked. The *hind soles* are thickly hairy from the heel to the naked tubercles at the roots of the toes, which are five in number, are rather stout, and not long; the outer toe is longer than the innermost one; the hind claws are of the same form with the fore-ones, but are rather smaller. The long hairs on the toes reach to the points of the claws. The scrotum is large, and in the spring is rather pendulous.

DIMENSIONS
Of a recent specimen.

	Inches.	Lines.		Inches.	Lines.
Length of head and body	8	6	Length of fur on the back	0	10
„ head	2	4	„ „ at the tip of the tail	1	6
„ tail (vertebræ)	5	0	Height of the ears (measured posteriorly)	0	6½
„ „ including fur	6	6	*Measurements of the scull:—*		
„ palm and middle fore-claw	1	0	Distance between the orbits	0	7
„ sole and middle hind-claw	1	10	Length of nasal-bones	0	6

Sciurus Hudsonius, var. β.   *Columbia Pine-Squirrel.*

Small brown Squirrel.   Lewis and Clark, vol. iii. p. 37.

Lewis and Clark describe a small squirrel which inhabits the banks of the Columbia, and has similar habits with the Hudson's Bay one.   It is most probably a distinct species; but as all our knowledge of it is derived from the short account of it given by those authors, I have, as its discoverers have not bestowed on it a specific name, ranked it, for the present, merely as a variety of the *sciurus Hudsonius.*

The DESCRIPTION given of it is as follows :—

" The small brown squirrel is a beautiful little animal, about the size and form of the red squirrel (*Sc. Hudsonius*) of the Atlantic states, and Western lakes.   The tail is as long as the body and neck, and formed like that of the red squirrel; the eyes are black; the whiskers long and black, but not abundant; the back, sides, head, neck, and outer part of the legs are of a reddish-brown; the throat, breast, belly, and inner part of the legs are of a pale red; the tail is a mixture of black and fox-coloured red, in which the black predominates in the middle, and the red on the edges and extremity; the hair of the body is about half an inch long, and so fine and soft, that it has the appearance of fur; the hair of the tail is coarser, and double in length.   This animal subsists chiefly on the seeds of various species of pine, and is always found in the pine-country."

The *sciurus rufiventer* of Geoffroy, an inhabitant of the country around New Orleans, has much similarity in colours to the above animal; but neither the description given by M. Desmarest, nor that by Dr. Harlan of the New Orleans specimens, correspond exactly with the account of Lewis and Clark.   The *sciurus rufiventer* seems to have a shorter tail.

[60.]   4. SCIURUS NIGER.   (Linn.)   *Black Squirrel.*

Sciurus niger. SAY, *Long's Expedition*, vol. i. p. 262.
Otchitamon. ALGONQUINS.

So much confusion has crept into the accounts of the American squirrels, that great uncertainty, respecting the species alluded to by authors, must exist, until some resident naturalist favours the world with a good monograph of the squirrels of that country.   The black squirrels have been considered by some to be a variety of the *sciurus cinereus*, or of the *sc. vulpinus*, and by others have been referred to the *sc. capistratus*.   M. Desmarest describes a small black squirrel, which is distinguished from the large black variety of the masked squirrel, by the softness of its fur.   Pennant's black squirrel is evidently the *sc. capistratus* of later writers.

The squirrel, which is the subject of this article, is larger than the *écureil gris de la Caroline* of M. F. Cuvier (lesser gray squirrel, *Pennant, Hist. Quad.*), and rather smaller than the "large gray squirrel" of Catesby.   It is not an uncommon inhabitant of the northern shores of Lakes Huron and Superior, where the greater or smaller gray squirrels are never seen, and is by far the largest squirrel existing on the eastern side of the Rocky Mountains, to the northward of the Great Lakes. It does not extend further north than the 50th parallel of latitude, but its range to the southward cannot be determined until the species of American squirrels are better known.   It is probable that it is not rare in the United States.   There are at present two pairs of American gray squirrels in the menagerie of the Zoological Society, which differ from each other in size, and in the smaller kind (lesser gray squirrel) having a tawny-coloured belly.   Both these kinds have, as was pointed out to me by Mr. Vigors, a peculiar wideness in the posterior part of the body, and a fulness of the skin of the flanks, being an approach to the form of a *pteromys*.   In the *sciurus Hudsonius*, the hind quarters are as slender and distinct from the flanks, as in common European squirrels; and there does not appear to have been any peculiar extension of the skin of the flanks, in the specimen of a black squirrel procured for me at Penetanguishene, by Mr. Todd, surgeon to the naval depôt there, and from which the following description was drawn up.

DESCRIPTION
Of a full grown but young individual.

Dental formula, incisors, $\frac{2}{2}$, canines $\frac{0-0}{0-0}$, grinders $\frac{5-5}{4-4} = 22$.

Incisors, much compressed, very strong, and having a deep orange colour on their exterior surfaces. The first or deciduous *grinder* is round and very small, the others are precisely similar in form to the grinders of the Hudson's Bay squirrel.

Form.—Its head is somewhat narrower and its nose sharper than that of the large gray American squirrel. Its frontal bone is not so flat between the orbits as that of the Hudson's Bay squirrel, and has a nearly regular flat elliptical curve from the occipital ridge to the end of the nose. Its scull is about twice as big as that of a Hudson's Bay squirrel. *Ears* elliptical rounded at the tip, covered with short fur, and entirely without tufts.

Fur.—The whole fur is black, that on the back being particularly close and having a glossy hue. When blown aside it appears downy towards the roots, and has a grayish-black colour without lustre. The fur is much shorter and coarser than that of the gray squirrels. On the dorsal aspect of the head it is of a shining black colour, without any lighter coloured spots about the muzzle or behind the ear. On the cheeks and throat it is of a brownish-black. The *tail* is clothed with long hair, unmixed with down, except close to the body. The feet are clothed with a smooth coat of short black hair. The claws are curved, much compressed, and sharp, exactly resembling those of the gray squirrel. The thumb tubercle is armed with a rounded nail closely adhering to it. The claws of the hind-feet are somewhat sharper than the fore ones, but are similar in form and nearly of equal size.

DIMENSIONS
Of the Penetanguishene Specimen.

	Inches.	Lines.		Inches.	Lines.
Length of the head and body	13	0	Length from the thumb to the tip of the middle		
,, head	3	0	fore-claw	1	3
,, tail (vertebræ)	9	6	,, of the sole, middle hind-toe, and claw	2	8
,, tail including fur	13	0	,, middle hind-toe and claw	1	1
,, palm to the tip of the middle			,, longest claws	0	3
fore-claw	1	6	,, fur on the back	0	9
,, third or longest fore-toe and claw	1	1	,, fur on the sides of the tail	2	6
			,, fur at the end of the tail	3	6

DIMENSIONS
Of the Scull of the same Specimen.

	Inches.	Lines.
Smallest breadth of the os frontis between the orbits	0	9
Length of the nasal bones	0	$9\frac{1}{2}$

There is a specimen of rather larger dimensions, procured at Fort William, on Lake Superior, and presented to the Zoological Society by Captain Bayfield. It has a few white hairs scattered amongst the fur of the body and rather more in the tail.

Lewis and Clark mention their having met with gray squirrels on the Columbia; but from our ignorance of the species to which they belong, they cannot be admitted into this work.

[61.] 1. PTEROMYS SABRINUS. *Severn River Flying Squirrel.*

GENUS Pteromys. CUVIER. Sciuropterus. F. CUVIER.
Greater Flying-Squirrel. FORSTER, *Phil. Trans.*, vol. lxii. p. 379.
Severn-River Squirrel. PENNANT, *Hist. Quad.*, vol. ii. p. 153. *Arct. Zool.*, vol. i. p. 122.
Sciurus Hudsonius. GMELIN, *Syst.*, vol. i. p. 153.
Sciurus Sabrinus. SHAW, *Zool.*, vol. ii. pt. i. p. 157.
Pteromys Sabrinus. RICHARDSON, *Zool. Journ.*, No. 12. p. 519, April, 1828.

Pt. (Sabrinus), super ex rubescenti brunneus, caudâ planiusculâ corpus subæquanti dorsoque concolori, lobulo membranæ volitantis rotundato.
Severn River Flying-squirrel, pale reddish-brown above ; tail flattish, nearly as long as the body, and of the same colour with the back ; flying membrane having a small rounded projection behind the wrist.

This is a very distinct species from the much smaller Assapan, (*Pt. volucella*) which is common in the United States. It was first described by Forster, who saw a specimen brought from Severn River that falls into James's Bay, and was considered by him to be the same species with the European flying squirrel, which it much resembles. I have followed Pennant and Shaw in separating it from the latter, on account of its longer tail, different coloured fur, and the smallness of the rounded projection of the flying-membrane behind the wrist.

The Severn River Flying Squirrel does not extend its range further north than latitude 52° (unless the Rocky Mountain one prove to be only a variety of it.) Mr. Tod sent me a specimen from Penetanguishene on Lake Huron, and I have seen others from Moose Factory, at the bottom of James's Bay.

DESCRIPTION.

Dental formula, incisors $\frac{2}{2}$, canines $\frac{0-0}{0-0}$, grinders $\frac{5-5}{4-4} = 22$.

Head round, nose short and obtuse, covered above with a smooth shining coat of light gray hair. *Incisors* nearly even with the end of the nose, anteriorly of a deep orange colour. *Whiskers* black, longer than the head. *Eyes* large, surrounded by a blackish-gray marking in the fur. *Flying-membrane* extending from the wrist to the middle of the hind leg, nearly straight, having only a very slight rounded projection close to the wrist.

The *fur* is every where remarkably fine and soft. On the dorsal aspect of the head, body, and flying-membrane, it is of a deep blackish-gray colour from its roots to its tips, which are of a pale reddish-brown, and which form the colour of the surface when the fur lies smoothly. There is no different coloured stripe on the flying-membrane, but the dark colour of the roots of the fur is more easily seen there. The outer surfaces of the fore and hind-feet are

2 C

pale bluish-gray. The margins of the mouth, sides of the nose, cheeks, and whole ventral aspect of the body are white, with in some parts a slight tinge of buff colour, particularly on the under surface of the flying-membrane. The *tail* is depressed, slightly convex on its upper surface, but quite flat or even somewhat channelled beneath. It is broadest about an inch from the body, and then tapers gradually but slightly towards the extremity, which is rounded. The flattened form is given to the tail, not so much by the distichous arrangement of the hair, as by the fur on its sides being much longer than that on its upper surface. Its colour above is nearly that of the back, with an intermixture, however, of black hairs; beneath it has a bright buff colour.

The *extremities* are small. The fore-ones are connected with the flying-membrane down to the wrist, and the feet are hairy both above and below. There are four toes on the fore-feet, which are short, and the claws are small, compressed, curved, sharp-pointed, and white. Under their roots there is a compressed callus projecting from the end of each toe. The third toe is the longest, then the second, next to it the outer one, and lastly the inner one; but the difference of length betwixt them is not great. There is a flat callus in place of a thumb, armed with a very minute nail. There are five hind-toes, of which the inner one is the shortest, then the outer one, and the remaining three are nearly equal to each other. The claws resemble those of the fore-feet, and are almost concealed by the hair of the toes. The soles are covered with a dense brush of soft white fur like the feet of a rabbit or hare.

DIMENSIONS

Of the skin of a Lake Huron specimen.

	Inches.	Lines.		Inches.	Lines.
Length of the head and body . .	7	9	Length of the hind-sole, middle toe and		
,, head . . .	2	0	claw	1	4
,, tail (vertebræ) . .	4	0	Greatest breadth of the tail . .	1	9
,, tail including fur .	5	0	Width from margin of one flying-membrane		
,, fore palm, middle-toe, and			to the margin of the opposite one .	4	0
claw	0	9			

PTEROMYS ALPINUS.

Published by John Murray. January 1829.

Pteromys Sabrinus. var. β. alpinus. *Rocky Mountain Flying-Squirrel.*

Pteromys alpinus. Richardson, *Zool. Journ.* No. 12. p. 519. March, 1828.
Specimens in the Zoological Museum and Hudson Bay Co.'s Mus.

Pteromys (alpinus), super luteo-fuscus, caudâ planâ fuligneâ corpus longitudine excedenti, margine membranæ volitantis recto.
Rocky Mountain Flying-Squirrel, yellowish-brown above ; tail flat, longer than the body, blackish-gray ; flying membrane with a straight border.

Plate XVIII.

This animal was discovered by Mr. Drummond, on the Rocky Mountains, living in dense pine-forests, and seldom venturing from its retreats, except in the night. I have received specimens of it from the head of the Elk River, and also from the south branch of the Mackenzie. It approaches nearer to the *Pt. volans* of Siberia in the colour of its fur than to *Pt. Sabrinus,* but it has much resemblance to the latter in its form. It is entirely destitute of any rounded process of the flying membrane behind the fore-leg* ; and when its scull is compared with that of *Pt. Sabrinus,* the frontal bone between the orbits appears narrower. The size of its limbs and tail is also greater. These remarks were made on a comparison of the specimens of this animal, and of the *Pt. Sabrinus,* which I at first received, and I was induced to think that they were specifically distinct ; but having lately had an opportunity of examining a more complete suite of specimens from Hudson's Bay, doubts were excited on the subject, and although it is probable, from the distance between their respective localities, that they may prove eventually to be distinct, I think it better at present to describe them as mere varieties. Except that the size of both these species is considerably greater than that of *Pt. volans,* they might be united with that species, without any great inconvenience.

DESCRIPTION.

Dentition the same as in *Pt. Sabrinus.* Head and extremities larger than in the latter animal ; and its tail is also longer, flatter, and has a more elliptical form. The flying-mem-

* In the accompanying plate the artist has fore-shortened the tail, so that it does not appear to possess its relative length to the body, and the position he has given to the fore-foot has produced a slight rounding of the flying-membrane at the elbow. The true form of the membrane is given in the figure in the distance.

2 C 2

brane is not so full as in the latter, and its border is straight. The end of the nose is hair-brown, and the fur about the mouth and on the sides of the nose has a dark smoke-gray colour. The ears are thin and membranous in appearance, thinly covered on both sides with short adpressed hairs, but having some fur at their base posteriorly, similar to that on the adjoining parts of the head. Their form is semi-oval with rounded tips. The surface of the fur on the back has a yellowish-brown colour, without any tendency to the more red hue of the back of the *Pt. Sabrinus.* The fur of the throat and belly is a grayish-white, without any tinge of buff-colour; the tail has a flat, oblong, oval form, and has a blackish-brown colour above, and is merely paler beneath.

The *extremities* are shaped like those of *Pt. Sabrinus*, but are larger in proportion. The soles, palms, and under surfaces of the toes, are well covered with fur, except a small callous eminence at the end of each toe, five eminences on the palm of which the two posterior ones are the largest, and four on the soles situated at the root of the toes. The brush of soft fur near the outer edge of the soles is as conspicuous as in the *Pt. Sabrinus.*

DIMENSIONS.

	Inches.	Lines.		Inches.	Lines.
Length of head and body	8	0	Height of the ears posteriorly	0	6
,, head	2	2	*Dimensions of the scull.*		
,, tail (vertebræ)	5	3	Length from tip of nasal bones to occipital		
,, ,, including fur	6	3	ridge of nasal bones	1	6
,, palm, middle fore-toe and claw	0	10	,, of nasal bones	0	
,, sole, middle hind-toe and claw	1	6	Breadth at the posterior part of the zygo-		
,, whiskers	2	6	matic process	1	0
Breadth between the outer edges of the flying membrane	4	9	Breadth of frontal bone between the orbits	0	3

There is a specimen in the Hudson's Bay Museum, which measures nine inches from the end of the nose to the origin of the tail.

Geomys. (Rafinesque.) *Sand-Rat.*

Geomys. " Rafinesque-Smaltz, *Amer. Month. Mag.* for 1817, p. 45." Desmarest, *Mamm.*, p. 314.
Lesson. *Man. de Mammal.*, p. 260.
Ascomys. Lichtenstein ? Saccomys. F. Cuvier ?
Pseudostoma. Say ?

Plate xviii C. Fig. 1 to 6.

CHARACTERS.

Dental formula, incisors, $\frac{2}{2}$, canines $\frac{0-0}{0-0}$, grinders $\frac{4-4}{4-4}$ = 20.

Incisors strong ; linear and flattish anteriorly ; narrower posteriorly, and chamfered away evenly from their insertion into the sockets to their tips. The upper ones are generally marked with one or more grooves anteriorly; the lower ones have sometimes a faint groove on their exterior sides. The second and third pairs of grinders in each jaw are quite simple in their structure, each tooth consisting merely of a slightly curved cylinder of enamel, without roots, compressed from before backwards, with a longitudinal depression or shallow furrow on one side, which it renders more acute than the opposite one ; the acute side of the grinders faces outwards in the upper jaw, and inwards in the lower one. The crowns of these teeth are flat, and have a transversely pear-shaped area, composed of soft bone, enclosed by a rim of enamel, but there are no transverse ridges. The posterior pair of grinders in each jaw are not so much compressed as those just described, but are nearly cylindrical, and have a roundish, slightly angular crown. The anterior pair, above and below, differ still more widely from the rest in being double, each of them being composed of two cylinders, shaped like the other teeth, and connected with each other by a narrow neck : the anterior cylinder is smaller than the other, and the long diameter of its crown is parallel to the axis of the jaw, and consequently is at right angles to the transverse pear-shaped crown of the posterior cylinder, and to the crowns of the teeth which succeed it. The upper grinders incline slightly backwards, the lower ones have a similar inclination forwards, and the grinding surfaces of both are very even.

The *lower jaw* is particularly thick and strong, and its symphysis which slopes upwards nearly in the same direction with the incisors is about one-third of its whole length. The palate is very narrow, and in the scull exhibits a central longitudinal crest of bone, with a deep and partially covered furrow on each side of it for the passage of vessels.

The *head* is large and depressed ; the nose short. On examining the scull, the frontal and nasal bones are observed to be in the same plane, and the zygomatic arch is but a little depressed below the crown of the head.

The *nostrils* are small round openings, facing downwards and somewhat laterally, separated from each other by a naked furrowed septum, and surrounded by a small naked space ; their inner margins are a little arched or ventricose.

The *mouth* is small, being contracted by an union of the lips behind the upper incisors. The *cheek-pouches* are large and pendulous, thinly clothed with short hairs, or sometimes almost naked, opening into the mouth by the side of the molar teeth.

The *eyes* are small and far apart. There is no other external *ear* than a slightly raised margin to the auditory openings, which are large.

Body cylindrical. *Tail* of moderate length, round and tapering, more or less hairy.

Extremities short, with five short toes to each foot. The *palms* are naked, and have a remarkable callous protuberance, projecting like a heel at their posterior part. The second and fourth toes are united nearly the whole length of their first phalanges to the middle one by skin. The fifth toe is considerably smaller and much further back than these, and the thumb is the smallest of all, and is situated a little further back than the fifth toe. The *fore-claws* are long, compressed towards their roots, slender and awl-shaped near their points, acute and considerably curved ; the middle one is the longest, the thumb one is small and more blunt, and the others are of intermediate sizes, proportionable to the length of their respective toes. The *hind-feet* are more slender than the fore-ones, and their soles, which are entirely naked, are narrower than the palms. The outermost and innermost hind-toes are situated further back than the other three, of which the middle one is the longest. The *hind-claws* are much shorter and more obtuse than the fore-ones, are excavated underneath, and are but slightly compressed.

The *fur* resembles in quality that with which the meadow-mice are clothed. The tail and feet are covered with shorter and coarser hair.

HABIT.—The sand-rats burrow in sandy soils and feed on acorns, nuts, roots, and grass, which they convey to their burrows in their cheek-pouches. They throw up little mounds of earth like mole-hills, in the summer, but are not seen abroad in the winter, nor do they throw up earth during that season. Their pouches when full have an oblong form, and nearly touch the ground, but when empty they are retracted for three-fourths of their length. Their interior is very glandular, particularly round the orifice that opens into the mouth.

REMARKS.—M. Rafinesque-Smaltz, in 1817, founded his genus geomys on the *hamster of Georgia (geomys pinetis)*, described by Mitchill, Anderson, Meares, and others, and referred to it, as a second species, the *Canada pouched-rat (mus bursarius* of Shaw). He at the same time ranged under another genus, named by him *diplostoma,* some Louisiana or Missouri animals, known to the Canadian voyagers by the appellation of *gauffres,* and remarkable for their large cheek-pouches, which open forwards exterior to the mouth and incisors, to which they form a kind of hood. These two genera have been adopted by few naturalists; and the American systematic writers have either overlooked **M. Rafinesque's** species entirely, or referred them all to the *mus bursarius*. In the latter case, they are undoubtedly wrong, for there are at least six or seven distinct species belonging to one or other of these genera, which inhabit America; and **I** think that both *geomys* and *diplostoma* will eventually prove to be good genera: — the *Sand-rats* belonging to the former having cheek-pouches, which are filled from within the mouth, and the *gauffres* or *camas-rats* of the latter genus having their cheek-pouches exterior to the mouth, and entirely unconnected with its cavity. I have had no opportunity of examining the *geomys pinetis,* which is the type of the genus; but **Mr. Leadbetter**, with his wonted liberality, has permitted me to inspect an individual of a hitherto undescribed species from Cadadaguios; and **Mr. David Douglas** very kindly sent me a specimen of another species, which he captured on the banks of the Columbia, and which forms the subject of the following article. From these two the characters of the genus, given in the preceding pages, were drawn up, the description of the teeth, and the views of the scull, being made from the latter. With regard to the Canada pouched-rat, great doubt still exists as to whether it belongs properly to geomys or to diplostoma. It was first described by **Dr. Shaw**, and an engraving published in the Linnean Transactions, from a drawing by **Major Davies**, of a specimen sent to Governor Prescot, from the interior of Canada. Judging merely from that figure and description, **I** should have little doubt of the cheek-pouches opening into the mouth, and of their being precisely similar in form and functions to the cheek-pouches of the sand-rats; but I have been told, on good authority, that the identical specimen described by **Shaw** (and which, on the dispersion of **Mr. Bullock's** collection, passed into the hands of **M. Temminck**) is, in fact, similar to the gauffres, in having cheek-pouches that open exteriorly, and that consequently **Major Davies's** drawing represented them in an unnatural, inverted position. **Mr. Say**, under the generic name of pseudostoma, gives the characters of a Missouri gauffre, with cheek-pouches opening exteriorly, and he identifies his specimen with the *mus bursarius*. He alludes to the Georgia hamster, as belonging to the same genus, without giving any further account of its characters than merely quoting **Dr. Barton's** remark, of its being only half the size of the Missouri one. His account of the dentition of the Missouri gauffre corresponds, as far as it goes, pretty closely with that of the Columbia geomys. **Dr. Harlan** and **Dr. Godman** refer the Georgia, Canada, and Missouri animals, to one species.

[62.]   1. Geomys Douglasii.   (Rich.)   *Columbia Sand-Rat.*

G. (Douglasii), super fuligneus subter pedibusque pallidior, caudâ dimidium corporis superanti.
Columbia Sand-rat, of a dusky-brown colour above, paler beneath, and on the feet, with a tail exceeding half the body
      in length.

Plate XVIII C.   Fig. 1 to 6.

DESCRIPTION.

The *head* is large and depressed ; the *nose* obtuse, particularly when viewed sideways. The *nostrils* are small and round, situated at the extremity of the   nose, and separated by a furrowed septum about a line wide ; they have a small naked margin, and the narrow upper lip betwixt them and the roots of the upper incisors is covered with short hair. *Mouth* moderately large ; lips hairy. *Incisors*, strong, exserted, orange-coloured ; upper ones with a fine but distinct furrow on their anterior surface, close to their inner edges ; lower ones with a similar furrow on their sides, close to their outer edges. *Cheek-pouches* large, much resembling the thumb of a lady's glove, in form and size, and hanging down by the sides of the head ; they have a pale buff-colour, and are of a soft membranous texture, nearly bare outside, having merely some very short, soft, scarcely visible, white hairs, scattered over them, with a reticulation of darker nerves : within they appear glandular, and their openings into the mouth are sufficiently wide to admit the point of the little finger, being nearly equal to the diameter of the pouch itself.   The fore-side of the pouch is posterior to the eye, and the hind-side is opposite to the ear ; its tip, which must touch the ground when the animal walks, is very obtuse.   The *whiskers* are short and soft.

Body, shaped like that of a mole, and covered with short, soft, dense, velvety fur, of an uniform dusky-brown colour.   The fur on the belly and feet has a lighter hue.   *Tail*, more than half the length of the body, round, tapering, and obtuse ; covered with hair, particularly near its root.

Legs, short and thick.   *Fore-toes* short, but very flexible ; the three middle ones united at their bases by skin ; the outer one is smaller and further back, and the thumb is very small, but is armed with a claw similar in form to the others, though it is much smaller.   The *claws* are very sharp-pointed, compressed, curved, and about as long again as their respective toes. The *palm* is naked, and its posterior part is filled by a large, rounded, callous eminence. The hind-feet are a little more slender than the fore-ones, and they are armed with smaller claws, shaped like those on the hind-feet of the spermophiles.   The hind-soles are entirely naked, without any conspicuous tubercles ; the heel is naked and narrow.

DIMENSIONS.

	Inches.	Lines.		Inches.	Lines.
Length of head and body . . 6	6		Length of the middle fore-toe, excluding the		
,, head . . . 1	10		claw 0	4	
,, tail (vertebræ) . . 2	10		,, from the heel to end of middle claw,		
,, cheek-pouches . . 1	3		measured along the sole . . . 1	2	
Diameter of cheek-pouches, about . . 0	6		*Dimensions of the scull.*		
Distance from the end of the nose to the eye 0	11		Length from the extremity of the upper jaw		
,, ,, of the nose to the			to the occipital crest (by calipers) . 1	6	
auditory opening . . . 1	8		Breadth, including the zygomatic arches 1	0	
,, between the eyes . . 0	7		,, of frontal bone between the orbits . 0	3	
Length from wrist-joint to end of the middle-			Length of the lower jaw from the condyles to		
claw 1	0		its anterior extremity . . . 1	0	

PLATE XVIII C.

Fig. 1, 2, and 3. Views of the scull (nat. size.)
— 4. View of the lower jaw (nat. size.)

Fig. 5. View of the palate and upper teeth (magnified.
— 6. View of the first upper grinders (magnified.)

The specimen here described is a female, which was taken in her nest, with three young ones, by Mr. Douglas, near the mouth of the Columbia. When put into my hands, the fur had mostly fallen off, but the specimen was in other respects perfect, and what was wanting has been supplied in the description from Mr. Douglas's notes. The state of ossification of the scull shewed the animal to be an old one. Mr. Douglas acquaints me, that the outside of the pouches was cold to the touch, even when the animal was alive, and that on the inside they were lined with small, orbicular, indurated glands, more numerous near the opening into the mouth. When full, the pouches had an oblong form, and, when empty, they were corrugated or retracted to one-third of their length ; but they are never inverted so as to produce the hood-like form of the pouch of a *diplostoma*. When in the act of emptying its pouches, the animal sits on its hams like a marmot or squirrel, and squeezes his sacks against the breast with his chin and fore-paws.

These little sand-rats are numerous in the neighbourhood of Fort Vancouver, where they inhabit the declivities of low hills, and burrow in the sandy soil. They feed on acorns, nuts (*corylus rostrata*), and grass, and commit great havoc in the potatoe-fields adjoining to the Fort, not only by eating the potatoes on the spot, but by carrying off large quantities of them in their pouches. The specific name is a small tribute of respect for the zeal and intelligence of its active and diligent discoverer.

† 2. Geomys umbrinus. (Rich.) *Leadbeater's Sand-Rat.*

G. (umbrinus), super umbrinus subter griseus, gulâ pedibusque albidis, caudâ griseâ vestitâ longitudine capitis.
Leadbeater's Sand-Rat, of an umber-brown colour on the dorsal aspect, gray below, with white feet and throat, and a gray hairy tail, as long as the head.

DESCRIPTION.

The *head* is large; the nose, wide and obtuse, and, with the exception of the naked margins of the nostrils, covered with fur similar in colour and quality to that on the crown of the head. The *nostrils* are small round openings, half a line apart, with a furrowed septum, and having their superior margins naked and vaulted; a narrow, hairy, upper lip, not exceeding a line in width, separates the nostrils from the upper incisors. The *whiskers* are white, and are shorter than the head. The *incisors* are much exserted, and are without grooves on their anterior surfaces, which are slightly convex, and of a deep yellow colour. The lips unite behind the upper incisors, so as to form a naked furrow leading towards the mouth, which is rendered more complete by the stiffness of the hairs on each side of it. The *cheek-pouches* are of a soiled buff-colour, and are clothed throughout their exterior surface with very short, soft, whitish hairs, which do not lie so close as entirely to conceal the skin. The middle of the pouch is opposite to the ear, and its anterior margin extends forwards to between the eye and the angle of the mouth; its tip is rounded.

The *body*, in shape, resembles that of a mole. It is covered with a smooth coat of fur, of the length and quality of that of a meadow-mouse; but possessing more nearly the lustre and appearance of the fur of a musk-rat. For the greater part of its length from the roots upwards, it has a blackish-gray colour. On the upper and lateral parts of the head, and over the whole of the back, the tips of the fur are of a nearly pure umber-brown colour, deepest on the head, and slightly intermixed with chestnut-brown on the flanks. The belly, and fore and hind legs, are pale gray, with, in some parts, a tinge of brown. The sides of the mouth are dark-brown, with a few white hairs intermixed. The chin, throat, feet, and claws, are white. The *tail* is round and tapering, and is well covered with short grayish-white hairs; the hairs on the sides of the fore-feet are rather stiff, and curve a little over the naked palms; those on the hind-feet are shorter; the posterior extremities are situated far forward.

DIMENSIONS.

	Inches.	Lines.		Inches.	Lines.
Length of the head and body	7	0	Distance from the posterior angle of the orbit to the auditory opening	0	6
„ head	1	8	„ „ posterior part of the wrist		
„ tail	1	9	tubercle to the tip of the middle fore-claw	0	10
Distance from the end of the nose to the anterior angle of the orbit	0	9	Length of the middle fore-claw	0	4½
Diameter of the orbit, about	0	2	Distance from the heel to the tip of the middle hind-claw	1	0

Although this animal is not an inhabitant of the fur countries, the above description has been inserted with the view of rendering the account of the genus more complete. I received no information respecting its manners or food. The specimen came from Cadadaguios, a town in the south-western part of Louisiana.

[63.]   3 ? GEOMYS ? BURSARIUS.   *Canada Pouched-Rat.*

Mus bursarius. SHAW, *Linnean Trans.*, vol. v. p. 227, pl. 8.
Canada rat. SHAW, *Zool.*, vol. ii. part 1. p. 100.
Geomys cinereus. " RAFINESQUE-SMALTZ, *Amer. Month. Mag.*, 1817." DESMAREST, *Mamm. in notis ad pag.* 315.
Hamster du Canada. DESMAREST, *Mamm.*, p. 312.

This animal was not seen by us on the late expeditions, and, as has been mentioned in a preceding page (199), it is still a matter of doubt whether it ought to be included in this genus or in the following one. The specimen figured by Major Davies, in the Linnean Transactions, was of a pale gray colour, and nine inches and a half long from the nose to the root of the tail, which measured two inches and a half. The belly was paler than the back, and the cheek-pouches were covered with very short pale hairs. Its superior incisors were deeply grooved in the middle, and more faintly close to their inner margins.

The *tucan* of Fernandez has been considered by some as identical with the *mus bursarius* of Shaw, but without sufficient grounds. Fernandez describes it merely as a fat, thick, *gnawer*, a span long, clothed with tawny fur, having a *long murine nose*, short round ears, a short tail, very short legs, crooked claws ; and he adds that it is scarcely able to see in day-light, leads a subterranean life, and feeds on roots and seeds, which it hoards up in its burrows. He also mentions that there are several other kinds of moles in New Spain which cannot see at all.

[64.]   4 ? GEOMYS? TALPOIDES.   (Richardson.)
Mole-shaped Sand-Rat.

Cricetus ? talpoides. RICHARDSON, *Zool. Journ.*, No. 12, p. 518.
Ootaw-chee-gœshees. CREE INDIANS ?

G. ♀ (*talpoides*), *super subterque cinerascenti-niger, gulá caudáque brevi albis, pedibus posticis sub-tetradactylis.*
Mole-shaped Sand-rat ? of a grayish-black colour, with white chin, throat, and tail, and only four perfect toes on the hind-feet.

The specimen described in this article was presented to the Zoological Society by Mr. Leadbeater, who obtained it from Hudson's Bay, but it was not accompanied by any notice of its precise habitat or a description of its manners. I am inclined to identify it with a small animal inhabiting the banks of the Saskatchewan, which I know only from the accounts of the residents and the mounds it throws up in the form of mole-hills, but generally rather larger. It lives entirely under ground, and during the winter it must either sleep or confine itself to its old paths, as the soil is then too much frozen to permit it to make new roads. As soon as the snow disappears in the spring, and whilst the ground is as yet only partially thawed, little heaps of earth newly thrown up attest the activity of this animal. I could not, however, procure a specimen, the soil being, at the period I was residing on the banks of the Saskatchewan, still too much frozen to permit me to reach the animal by digging. The earth thrown up then I suppose to have been merely the clearings of the galleries which it had made during the preceding year. It inhabits only sandy banks, and its food probably consists principally of roots. It cannot, like the English mole, feed on earth worms, for none exist in those latitudes *.

As the teeth of the specimen could not be examined, the genus to which it belongs is uncertain ; but from its strong general resemblance to *G. Douglasii* and *G. umbrinus*, it is placed with them at present. Some uncertainty also exists as to the form of its cheek-pouches, which have been partially inverted in mounting, probably from an attempt of the artist to imitate the cheek-pouches of a *diplostoma;* but if so, he has been unable to give them the hood-like form of the pouches of the latter.

* I was told by a gentleman who has for forty years superintended the cultivation of considerable pieces of ground on the banks of the Saskatchewan, that during the whole of that period he never saw an earth-worm turned up.

DESCRIPTION.

Body shaped like that of the mole ; *head* rather small, but when the pouches are distended it must have considerable breadth. The obtuse nose is covered with short hairs. The *incisors* are very strong, and have flat, anterior surfaces ; the upper ones are short and straight, and are each marked with a single very fine groove, close to its inner edge ; the under ones are long, curved inwards, and not grooved. The *whiskers* are composed of fine hairs as long as the head. The eyes are small and far back. The auditory opening is capable of receiving the head of a pin, and is slightly margined. The *pouches* are covered on the outside with fur of the same colour with that on the back, but beneath and on their posterior margins their hairy covering is white. On the head and body the *fur* is of a grayish-black colour its whole length, with a faint brownish reflection in some lights, and it is as fine as that of the common mole, but not quite so close and velvety. The chin and throat are white. The *tail* is very short and cylindrical, and is covered by a close smooth coat of short white hairs.

The *extremities* are very short ; the fore-foot has four toes and the rudiment of a thumb. Of these the middle toe is the longest, and has the largest claw ; the first and third are equal to each other in length ; the outer one is shorter and far back, and the thumb is still farther back, and consists merely of a short claw. The *fore-claws* are long, compressed, slightly-curved and pointed. There are four short toes on the *hind-foot,* armed with compressed claws much shorter than the fore-ones, and the rudiment of a fifth toe, so small that it was discovered only after very minute inspection.

DIMENSIONS
Of the specimen in the Zoological Museum.

	Inches.	Lines.		Inches.	Lines.
Length of head and body . .	7	4	Length of the fur on the back . .	0	6
„ tail	1	10	„ from the tubercle at the posterior part of the palm to the end of the middle		
„ from end of nose to the eye .	0	9			
„ from ditto to the auditory opening	1	3	fore-claw	0	10½
„ from back part of the eye to the auditory opening . . .	0	6	„ of middle fore-claw . .	0	4
			„ from the heel to the tip of the middle hind-claw . .	0	11
Height of the back . . .	2	0			
Length of the lower incisors . .	0	5			

[65.] 1. Diplostoma? bulbivorum. (Rich.) *Camas-Rat.*

Genus. Diplostoma. Rafinesque-Smaltz? (Desmarest, *Mamm.*)

Plate xviii B.

There is a specimen of a quadruped in the Hudson's Bay Museum, which Mr. David Douglas informs me is the animal known on the banks of the Columbia by the name of the *Camas-rat,* because the bulbous root of the Quamash or Camas plant (*Scilla esculenta*) forms its favourite food. The scull is wanting, and the animal, therefore, cannot be with certainty referred to a genus, but the form of its exterior cheek-pouches leads me to think that it may belong to the *diplostoma* of M. Rafinesque-Smaltz. There is, however, a discrepancy in the number of its toes and in the presence of a tail, which, if M. Rafinesque's specimen was perfect, is decisive against this arrangement. The characters of the genus *diplostoma,* as quoted by M. Desmarest, are as follows :—

" *Diplostoma* (Am. Month. Mag. 1817.)—Mouth double ; the exterior one formed by two great pouches, which extend as far back as the shoulders, and meet before the incisors, all of which are furrowed ; four molar teeth of a side in each jaw ; body cylindrical ; neither tail nor ears ; eyes hid by the fur ; four toes on each foot. This genus is nearly allied to that of the mole-rat, but differs in its cheek-pouches, and in the number of its toes. Two species were discovered by Bradbury on the Missouri. They live beneath the surface of the earth and eat roots. The early French travellers named them *gauffres.*"

DESCRIPTION.

Form.—Body like that of a great mole, with a head that appears large and clumsy owing to the swelling out of the cheek-pouches. The nose being margined by a slight prolongation of the superior edge of the cheek-pouch appears flat and broad, but its tip and nostrils are comparatively small ; it does not project in the least beyond the plane of the incisors.

The *incisors* are entirely exserted, are stronger than those of the musk-rat, and have three convex sides. The anterior side is the broadest, is without grooves, and has a yellowish colour. The upper incisors have even cutting edges, and project forwards and downwards immediately from under the nostrils, instead of standing out from a cleft in the upper lip. The lower ones are linear, with round tips, and project nine lines above their sockets, being longer than the upper ones. The true *mouth* is a vertical slit, nearly an inch long, situated

PLATE 18 B

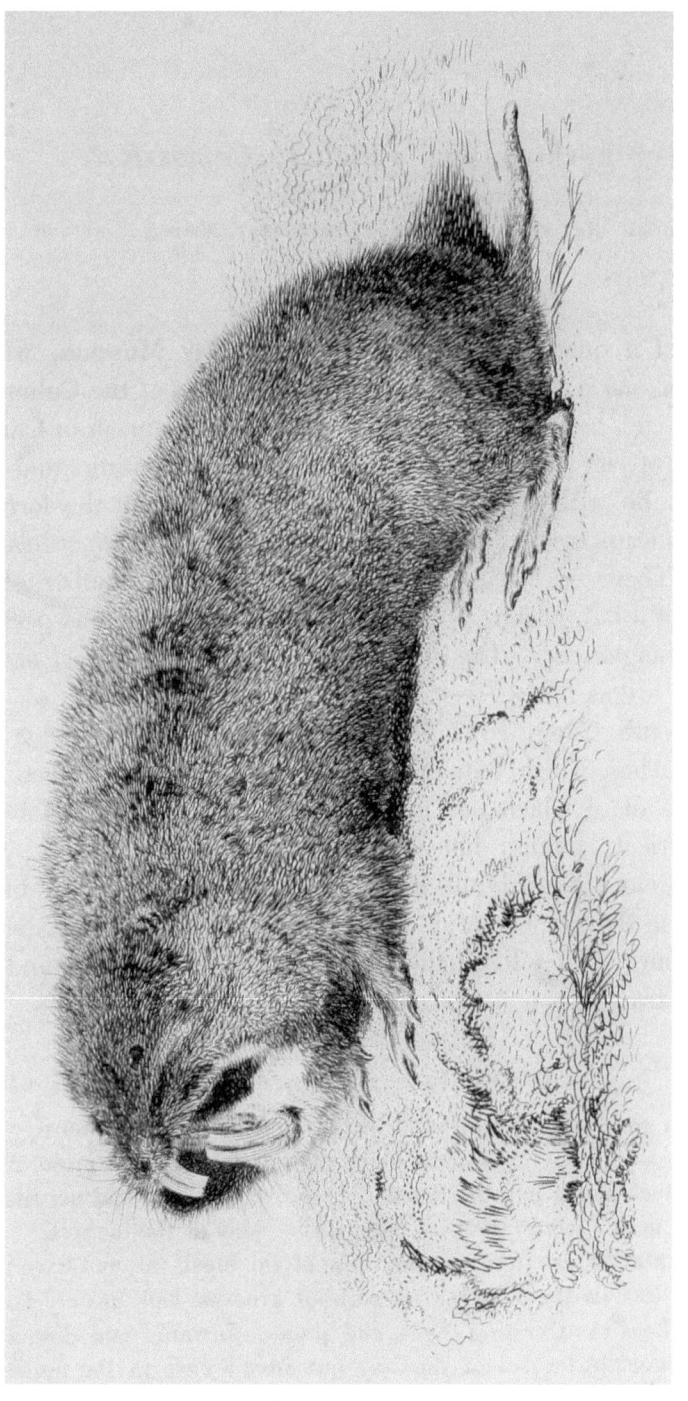

DIPLOSTOMA DOUGLASII.

Published by John Murray. January 1829.

behind the incisors and hidden by them. The lips, which in fact are right and left, and not upper and under, are covered with white hair. There is a hollow space of about half an inch in length between the upper incisors and the upper corner of the mouth, which is partially naked, but is protected by some coarse white hairs that incline over it. The inferior incisors are situated in the lower angle of the mouth, but the lips come into apposition behind them. No part of either upper or under incisors that projects beyond the sockets is covered by the lips. On each cheek there is a wide *pouch*, not communicating with the cavity of the mouth, but opening forwards, and with its fellow forming a kind of hood, in the middle of which are placed the mouth, incisors, and extremity of the lower jaw, the latter having an upward direction. The pouch is widest at its mouth ; its anterior margin commences on the side of the nose, about half an inch from its tip, and curving downwards is united with the lower jaw, a little more than an inch from the insertion of the incisors. The integument forming the outer parietes of the pouch is covered externally by fur of the same quality and colour with that on the head and body ; and when the animal is viewed in profile, the cheek appears merely a little puffed up, but exhibits no membranaceous or bag-like projection like the pouches in the genus *geomys*. Interiorly the pouches are clothed with a shorter and coarser hair, particularly the side forming the parietes of the mouth, which is well covered with short white hairs ; the opposite side of the lining of the pouch is furnished, however, merely with scattered patches of hair, and is in some places quite naked. Each pouch has a semi-cup-shaped cavity when distended :—the distance from the union of its upper margin with the nose to that of its lower margin with the chin, is about two inches, and its depth is nearly as much. The *whiskers* are very short. The *eyes* (which appear to have been small) are situated about an inch from the tip of the nose. The *auditory openings* are moderately large, but there are no external ears.

Fur.—The body and head are covered with short fur like that which clothes the meadow-mice. On the dorsal aspect it has a colour intermediate between chestnut and yellowish-brown, darker on the crown of the head than elsewhere. On the belly the brown is mixed with a considerable portion of gray. The lips, the lower jaw, the lining of the pouches, and a narrow space round the anus are covered with white-fur. Close to the upper part of each side of the mouth there is a rhomboidal mark, which is clothed with hair of a liver-brown colour. The fur on the back has that dark, shining, lead-gray colour, from the roots to near the tips, which is usually seen in the meadow-mice.

The *tail* is short, round, and tapering, with an obtuse tip, and is thinly clothed with hairs of a pale brown colour. The *extremities* are short, and are covered down to the wrist and ankle joints with fur similar to that on the body. There are five toes on each foot. The *fore-feet* are hairy above, with naked palms, which have a large callous tubercle at their posterior part, resembling a heel, as in the genus *geomys ;* behind this tubercle there is a tuft of strong white hairs. The toes are short; the middle one is the longest, the one on each side of it are a little shorter ; the fifth or exterior toe is much shorter and considerably further back, its extremity (without the claw) reaching only to the root of the third toe. The thumb is still shorter and further back than the fifth toe. The claws are long, strong, slightly curved, and much compressed. Their edges are in contact beneath at their insertions into the ends of the toes, and separate a little towards their points, being very similar in form to the claws of the spermophiles, but not so strong in proportion to the size of the animal. The middle claw is the longest; those belonging to the thumb and outer toe are much shorter and more conical than the rest, but they are in other respects similar, the thumb-claw differing in that respect from the obtuse, rounded flat thumb-claw of a spermophile.

The *hind-feet* are covered above with whitish hairs. The soles are naked and narrow, and the toes short. The first and fifth toe are so much smaller and further back than the others, that, at first sight, there appear to be only three hind-toes. Of these three the middle one is longer than the one on each side of it; and of the other two, the first toe is a little further back, but somewhat larger than the outer or fifth one. The hind nails are short, conical, obtuse, and more or less excavated underneath. The nail of the fourth toe is more spoon-shaped than the others.

DIMENSIONS.

	Inches.	Lines.		Inches.	Lines.
Length of the head and body . . .	11	0	Length of middle hind-toe . . .	0	4
„ „ head	3	0	„ middle hind-claw . . .	0	2
Breadth of the head behind the eyes, when			„ upper incisors (the exposed portion)	0	6
the pouches are distended . . .	3	6	„ lower incisors ditto .	0	9
Length of the tail	2	6	„ from the orbit to the tip of the nose	1	2
„ palm, middle fore-toe and claw .	1	0	„ of the orbit, about . .	0	3
„ middle fore-toe . . .	0	3	„ from the orbit to the auditory opening	0	6
„ middle fore-claw . . .	0	4½	Height of the back, about . . .	4	0
„ sole, middle hind-toe and claw	1	6	Length of the fur on the back . . .	0	6

M. Rafinesque has not detailed the characters of his *diplostoma fusca* from the Missouri sufficiently to enable us to judge how far it differs from the camas-rat; but the furrows on the upper and lower incisors of his species, and no mention being made of the white fur about the mouth, lead me to consider it as distinct. The want of a tail, and the smaller number of toes on his specimen, may have been owing to an injury the skin had sustained, as he had not an opportunity of examining the recent animal.

Mr. Schoolcraft gives a description of a "gopher" that he procured at the Falls of St. Anthony, on the Mississippi, which I shall transcribe, as it contains the fullest account of the habits of these animals which I have met with. "It is about ten inches long from the nose to the tail, with a body shaped very much like that of a large wharf-rat, which it also resembles in the colour of its hair and the length and nudity of its tail. Its legs are short, and each foot is furnished with *five* long and sharp claws. It has two large fore-teeth in each jaw, resembling those of the squirrel, but its most remarkable character is a pouch on each side of the jaw, formed by a duplicature of the skin of the cheek. These project inwardly, where they are accommodated by an unusual width and flattening out of the head.

As the animal lives wholly under ground like a mole, these pouches serve the purpose of bags for carrying the earth out of their holes. They are filled with the fore-claws, and emptied at the mouth of the hole by a power which the animal possesses of ejecting the pouches from each cheek in the manner that a cap or stocking is turned. In this way it works its path under ground, and ploughs up the prairies in many places in such a manner, that the white hunters of the Missouri and Arkansas frequently avail themselves of the labours of the gopher by planting corn upon the prairies which have been thus mellowed. It lives entirely upon the roots of plants, eating all with indiscriminate voracity, and has been found particularly destructive to beets, carrots, and other tap rooted plants in the military gardens at St. Peter's*."

Mr. Schoolcraft's account of the manners of the Mississippi gauffre, and the mode in which it uses its cheek-pouches, is evidently the testimony of an eye-witness, and may be compared with Mr. Douglas's equally clear and precise description of the habits of the Columbia sand-rat. A minute examination of the specimens in my possession induces me to place implicit reliance on both these accounts. The skin of the *geomys Douglasii*, even when thoroughly soaked, cannot be made to fold in, so as to produce the hood-like cheek-pouch of a gauffre, neither can the pouch of the *diplostoma bulbivorum* be everted, so as to become pendulous. Its bottom alone can be turned out, by which it is emptied of its contents in the manner mentioned by Mr. Schoolcraft ; but the lining of the exterior parietes of the pouch is firmly united to the external skin, and is incapable of being everted. The incisors in form and position, the form of the mouth, the ears, eyes, extremities, and tail of the sand-rats, bear, however, a very close resemblance to those of the gauffres, and they cannot be finally established as separate genera, until their dentition has been compared. The Camas-rats are very common on the plains of the Multnomah River, and may, as Mr. Douglas informs me, be easily snared in the summer.

* SCHOOLCRAFT, *Journ.*, p. 365.

2 E

APLODONTIA. (Richardson.) *Sewellel.*

Aplodontia. RICHARDSON, *Zool. Journ.* January, 1819.

CHARACTERS.

Dental formula ; incisors $\frac{2}{2}$, canines $\frac{0-0}{0-0}$, grinders $\frac{5-5}{4-4} = 22.$

Incisors, very strong, flatly-convex anteriorly without grooves; narrower behind. *Grinders* simple, remarkably even on the crowns. The first in the upper jaw, small, cylindrical, and pointed, is placed within the anterior corner of the second one, and exists in the adult. The rest of the grinders are perfectly simple in their structure, without roots, and have slightly concave crowns, which are merely bordered with enamel without any transverse ridges or eminences. On the exterior side of the four posterior pairs of upper grinders, and the inner side of all the lower ones, there is an acute vertical ridge extending the whole length of the tooth, formed by a sharp fold of enamel. When the grinders are *in situ*, there is a wide semicircular furrow between each pair of ridges, formed by the two adjoining teeth. The side of each tooth opposite the ridge is convexly semicircular. The second grinder in the upper jaw, and the first in the lower one, are a little larger than the more posterior ones, and the former has a projection of enamel at its anterior corner, producing a second though smaller vertical ridge, within which the first small grinder is situated and leans towards it. There is a slight furrow on the exterior sides of the lower grinders, most conspicuous in the first one.

Palate narrow, bounded by perfectly parallel and straight rows of grinders.

Head flat and broad, nose a little arched, thick and obtuse. *Lower jaw* thick and strong, with a large triangular process, concave behind, projecting at its posterior inferior angle further out than the zygomatic arch. The transverse diameter of the articulating surface of the condyle is greater than the longitudinal one. The jaw is altogether stronger than is usual in the *Rodentia*.

" *Cheek-pouches* none " *.

Eyes very small. *Ears* short and rounded, approaching in form to the human ear, and thickly clothed on both sides with short hair.

Body thick and short, clothed with fur like that of a musk-rat, but not so long or fine.

Limbs robust, short; feet moderately strong, with naked soles. Five toes on all the feet, rather short, but well separated. The thumb of the fore-feet is considerably shorter than the other toes. *Claws*, particularly the fore-ones, very long, strong, much compressed, and but little curved.

Tail very short, concealed by the fur of the hips. *Mammæ* six, the anterior pair situated between the fore-legs.

Habits.—Animals forming small societies, feeding on vegetable substances, and living in burrows.

* Mr. David Douglas.

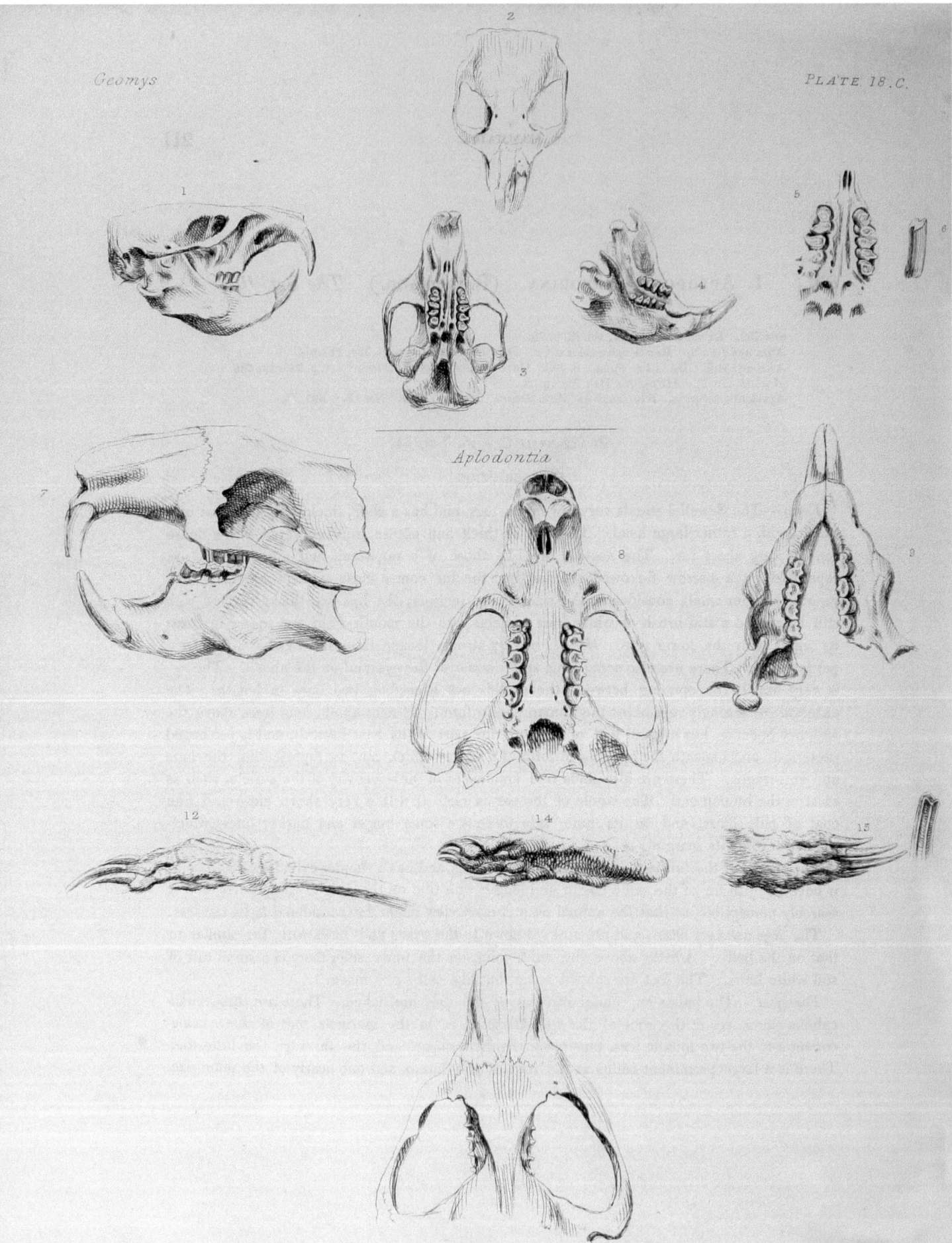

Aplodontia

[66.] 1. APLODONTIA LEPORINA. (Richardson.) *The Sewellel.*

Sewellel. LEWIS and CLARK, vol. iii. p. 39.
Anisonyx ? rufa. RAFINESQUE-SMALTZ. DESMAREST, *Mamm.*, p. 330, in notis.
Arctomys rufa. HARLAN, *Fauna*, p. 308. GRIFFITH's *Anim. Kingdom*, vol. v. p 245. sp. 636.
Marmot, No. 17. HUDSON's BAY MUSEUM.
Aplodontia leporina. RICHARDSON, *Zool. Journ.* January, 1829. No. 15. p. 335.

PLATE XVIII C. FIG. 7 TO 14.

DESCRIPTION.

Form.—The Sewellel stands very low on its legs, and has a short thick body like that of a rabbit, with a rather large head. The nose is thick and obtuse, and is covered with a dense coat of very short fur. The nostrils are like those of a rat, small and roundish, and are separated by a narrow furrowed septum, but the fur comes close to their margins. The mouth is rather small, considering the size of the incisors, the lips are thick, clothed with stiff hairs, and a stiff brush of white hair projects into the mouth from the upper lip, near its union with the lower one. *Whiskers* very strong, longer than the head, partly black, partly white. There are also some long stiff hairs over the eye and on the cheek. The *eye* is very small, the opening between the eyelids not exceeding two lines in length. The external *ear* strongly resembles the human one in form. It rises about four lines above the auditory opening, has a small fold of the anterior part of its base inwards, and is prolonged posteriorly and beneath the opening in form of a narrow thick margin representing the lobe and anti-tragus. There are also folds and eminences in the concavity of the auricle, such as exist in the human ear. The whole of the ear is clothed with a very short, close and fine coat of pale hairs, and on its inner side there are some longer and darker ones, which project beyond its margin.

The stump of the *tail* is scarcely half an inch long, and has a slender cylindrical form. It is covered with fur of the same colour and length with that on the neighbouring parts, and is scarcely perceptible, so that the animal on a cursory view might be considered to be tail-less.

The *legs* are very short, and are covered down to the wrists and heels with fur similar to that on the body. A little above the wrist joint, on the inner side, there is a small tuft of stiff white hairs. The feet are shaped somewhat like those of a marmot.

Fore-feet.—The palms and under surfaces of the toes are naked. There are three small callous eminences at the root of the toes, disposed as in the marmots, one of them being common to the two middle toes, one proper to the fore-toe, and the third to the little toe. There is a large prominent callus at the root of the thumb, and one nearly of the same size on the opposite side of the palm. The thumb is of sufficient length to be of use in grasping, and its upper phalanx is closely covered by a smooth rounded nail. As in the marmots, the second toe is the longest, the third is a little shorter, the first is about two lines shorter than the second, and the fourth or last is scarcely shorter than the first, being considerably longer than the thumb. The claws are large and very much compressed, so that their edges are in contact beneath nearly their whole length. They are slightly arched above and nearly straight below. The *hind-feet* are more slender than the fore ones, and the claws are about one-half smaller, rather more arched, and less compressed, their edges separating beneath so as to form a narrow oblong groove towards their points. The soles are longer than the palms, and are naked to the heel. They are furnished with four callous eminences situated at the roots of the toes, and two placed further back, all more conspicuous than those on the hind-feet of the American spermophiles. The innermost and outermost toe are nearly equal in length, and are shorter than the three others.

Fur.—The quality of the fur is very much like that of a rabbit when out of season. It consists of a close short fur, four or five lines long, mixed with longer hairs. The latter are most numerous about the sides of the neck and fore-part of the back. They are scattered over the posterior part of the back and belly, and are numerous on the shoulders and thighs. The sides and upper part of the nose are covered with short fur, and the fur on all the feet is short.

Colour.—Incisors yellow; claws white. The general hue of the *back* is intermediate between umber and chestnut browns, without any tendency towards a rufous hue, and it is rendered darker by most of the long scattered hairs on that part being black. The belly is grayish, or clove-brown, and many of the long hairs there, and on the sides, are tipped with white. The nose is clothed with short hairs, nearly of the colour of the back; the lips are whitish, and there is a pretty large spot of pure white on the throat. The position of the *mammæ* in the female is indicated by brown circular marks. The fur has no lustre on its surface, and little beauty; that on the back, when blown aside, exhibits a grayish-black shining colour, from the roots to the brownish tips.

DIMENSIONS

Of a full-grown specimen.

	Inches.	Lines.		Inches.	Lines.
Length of head and body . .	14	0	*Dimensions of the scull of same specimen.*		
„ tail	0	6	Breadth of the scull, measured from the outside of one zygomatic arch to the outside of		
„ from the elbow to the wrist joint	2	0	the other, posteriorly . .	2	4
„ „ the wrist joint to the end of the middle-claw	1	9	Length of the orbit . . .	1	0
„ of thumb . . .	0	3	Width of ditto, posteriorly .	1	0
„ middle toe . .	0	6	Smallest distance between the orbits, measured		
„ middle-claw . . .	0	6	across the frontal bone . .	0	3
„ from the knee-joint to the heel	2	2	Width of the palate . .	0	3
„ „ heel to the tip of the middle claw	2	0	Length of the nasal bones .	1	1
„ of middle hind toe and claw .	0	9	Greatest breadth of each nasal bone .	0	3

Fig. 7, 8, 10. Different views of the anterior half of the skull (nat. size.)

— 9. Lower jaw with one condyle broken off.

Fig. 11. Upper molar tooth.

— 12, 13. Views of the upper surface of the fore-feet.

— 14. Sole of the hind-feet.

Amongst Mr. Douglas's specimens, there is a young one, with more white hairs interspersed through its fur, and some differences in the form of its scull, which seem to point it out as a second species. The breadth of its frontal bone, between the orbits, where least, is six lines, being twice the breadth of the same bone in *A. leporina.* Its nasal bones are as broad as in the latter, but are three lines shorter. The dentition is perfectly the same in both, but in the young specimen there is a new set of grinders in the lower jaw, which have destroyed the greater part of the bodies of the old grinders, leaving merely a long process before, another behind, in each socket, resembling fangs. The specimen is not sufficiently perfect to enable me to give its characters as a distinct species, but I have little doubt of its being so.

Since the account of this genus was published in the Zoological Journal, Mr. Douglas has placed in my hands an Indian blanket or robe, formed by sewing the skins of the sewellel together. The robe contains twenty-seven skins, which have been selected when the fur was in prime order. In all of them the long hairs are so numerous as to hide the wool or down at their roots, and their points have a very high lustre. The general colour of the surface of the fur is between chestnut and umber browns, lighter, and with more lustre on the sides. Some of the skins, which are in the best order, have the longer hairs on the back of the head, and between the shoulders almost black. It is probable, however, that these are the skins of two species of sewellels, in the robe, and that one of them wants the white mark on the throat. The down of all the skins of the robe has a shining blackish-gray colour.

[67.]   1. Hystrix pilosus.   (Catesby.)   *Canada Porcupine.*

Genus Hystrix.   Linn.   Erethizon.   F. Cuvier.
Hystrix pilosus, Americanus.   Catesby, *Carol. App.*, p. 30.   An. 1741.
The Porcupine, from Hudson's Bay.   Edwards, *Birds*, p. 52.   Fig.
Cavia Hudsonis.   Klein, *Quadr.*, p. 51.   An. 1751.
Hystrix Hudsonius.   Brisson, *Regn. An.*, p. 148.   An. 1756.
Hystrix dorsata.   Linn., *Syst.*, p. 57.   An. 1757.
L'Urson.   Buffon, vol. xii. p. 426. t. 55.   An. 1776.
Canada Porcupine.   Forster, *Phil. Trans.*, vol. lxii. p. 374.   Pennant, *Quadr.*, vol. ii. p. 126.
        Arct. Zool., vol. i. p. 109.
The Porcupine.   Hearne, *Journal*, p. 381.   Hutchins, *MSS.*
Erethizon Dorsatum.   F. Cuvier, *Mem. de Mus.*, vol. ix. p. 413.
Hystrix dorsata.   Sabine, *Franklin's Journ.*, p. 664.   Harlan, *Fauna*, p. 190.
Canada Porcupine.   Godman, *Nat. Hist.*, vol. ii. p. 160.   The figure represents only four toes on the
        hind feet, instead of five.
Cawquaw.   Cree Indians.   Ooketook.   Esquimaux.

This sluggish and unsightly animal early attracted the notice of travellers to the northern parts of America.   Buffon invented for it the appellation " urson," by which he intended to recall the memory of Henry Hudson, the illustrious but unfortunate discoverer of the country, where the animal chiefly abounds, and also to denote its spiny armature, resembling that of the common hedgehog (*l'herisson*). Linnè gave it the specific name of *dorsata*, but I have preferred Catesby's prior epithet of *pilosus*, which seems to be equally, if not more, appropriate.

The Canada porcupine is found on the banks of the Mackenzie, as high as latitude 67°, and, according to American writers, it ranges as far south as latitude 37°.   It is said to be very rare in Virginia ; but to be numerous in some parts of Kentucky ; and it is reported to have multiplied greatly, of late years, near Oneida Lake, in the state of New York*.   In the fur countries, it is most numerous in sandy districts, covered with the *pinus Banksiana*, on the bark of which it delights to feed.   It also eats the bark of the larch and spruce firs, and the buds of various kinds of willow.   In the more southern districts, it is said to feed chiefly on the bark and leaves of the *pinus Canadensis* and *tilia glabra*, and to be fond of sweet apples and young maize, which it eats in a sitting posture, holding them to its mouth with the fore-paws.   It travels slowly, and Hearne remarks, that " the Indians, going with packets from fort to fort, often see them in the trees, but not having occasion for them at the time, leave them until their return, and should

---

* Cozzens, *Lyceum Nat. Hist., New York*, vol. i. p. 190.

their absence be a week or ten days, they are sure to find them within a mile of the place where they had seen them before." Mr. Hutchins observes, that, " in walking, the tail is drawn along the snow, making a deep track, which is often the means of betraying the animal ; but its haunts are most readily discovered by the barked trees on which it has fed, which, if done the same winter, is a sure sign that the porcupine is near the spot. They are usually found on the branches, and, on approaching them, they make a crying noise, like a child. The tree, being cut down, the animal is despatched by only striking it on the nose." It is readily attacked by the Indian dogs, and soon killed, but not without injury to its assailants, for its quills, which it erects when attacked, are rough, with minute teeth directed backwards, that have the effect of rendering this seemingly weak and flexible weapon a very dangerous one. Their points, which are pretty sharp, have no sooner insinuated themselves into the skin of an assailant than they gradually bury themselves, and travel onwards until they cause death, by wounding some vital organ. These spines, which are detached from the porcupine by the slightest touch, and probably by the will of the animal, soon fill the mouths of the dogs, which worry it, and unless the Indian women carefully pick them out, seldom fail to kill them. Wolves occasionally die from the same cause. The Canada porcupine makes its retreat amongst the roots of an old tree, and is said to pass much of its time in sleeping. When disturbed, it makes a whining or mewing noise. It pairs in the latter end of September, and brings forth two young ones in April or May. Its flesh, which tastes like flabby pork, is relished by the Indians, but is soon nauseated by Europeans. The bones are often deeply tinged with a greenish yellow colour. Like other animals, which feed on coarse vegetable substances, it is much infested by intestinal worms. The quills or spines are dyed of various bright colours by the native women, and worked into shot-pouches, belts, shoes, and other ornamental articles of dress.

DESCRIPTION.

Form.—Body thick and clumsy, back much arched in a regular curve from the nose to the buttocks, when it droops more rapidly to the tail, which is very low. *Legs* very short. *Tail* short, thick, rounded at the tip, and turned a little upwards. *Nose* flattish above, broad and abrupt. There is a narrow naked margin round the nostrils, but there is no smooth dividing line on the upper lip. *Eyes* lateral, very small, and round. *Ears* situated behind and above the auditory opening, covered as thickly with fur as the neighbouring parts, and entirely concealed by it. *Incisors* nearly as strong as those of the beaver. They curve forwards a little so as to project beyond the nose, are convex anteriorly, narrower behind, and are not much compressed. They have a yellow colour. The crowns of the grinders as they wear acquire an even surface.

Fur.—The upper-lip is covered with short hair of a dull yellowish-brown colour. The cheeks and forehead are clothed with liver-brown hair, moderately long, interspersed with a very few black and white hairs. The hair on the body, both above and below, is long, and of a dull liver-brown colour, intermixed on all the upper parts and on the hips with still longer hairs, some of which are entirely black, others entirely white, and a third set black at the roots and white at the tips. The white hairs are most numerous on the posterior part of the body. There are also many round spindle-shaped, sharp-pointed *spines* or quills fixed amongst the hair which covers the upper parts. The spines commence on the crown of the head, and are there short, thick, very sharp-pointed, and very numerous. There are a good many longer and more slender ones on the shoulders and fore-part of the back. There are also many on the sides and middle of the back, but these are still more slender and flexible as well as less conspicuous. The buttocks and thighs are thickly set with long, very strong, and sharp spines. Some of the spines are entirely white, others brown at the tips. The *throat* and *belly* are covered with brown hair, not so long as that on the back, lying more smoothly, and unmixed with either white hairs or spines. The *tail* is covered with brown hair above and below, and soiled white hair on its margin and tip. There are many small spines amongst the hair on its upper surface.

The *legs* are covered with brown hairs, mixed on their exterior surfaces with some white ones. The *palms* are nearly oval, or rather egg-shaped, being semicircular before and narrower behind. There are four very short toes on the fore-feet, which are armed with long, compressed, curved, blackish claws, grooved underneath their whole length. Their points are not acute. The middle or second fore-toe is rather the longest, the one on each side of it is scarcely inferior in length, and the outer one is a little smaller and somewhat further back. The *hind-soles* are oval, approaching to circular, larger than the palms, destitute of hair and covered with a rough skin like shagreen. There are five toes on the hind-foot, which do not differ much from each other in length, but their roots and consequently their extremities are arranged in a curved line, corresponding with that of the anterior margin of the soles. The hind-claws resemble the fore-ones. The hair which covers the upper surface of the feet curves down by the sides of the soles, and being worn even, as if clipped off, it forms a thick marginal brush, which considerably increases the diameter of the soles, and fits them for walking on the snow.

The Canada porcupines vary in the depth of their colours. Pennant informs us that Sir Ashton Lever had a white one.

DIMENSIONS.

	Inches.	Lines.		Inches.	Lines.
Length of the head and body	30	0	Diameter of the eye	0	2
„ tail	8	0	Breadth of the nose	1	0
Height of the centre of the back	14	0	Length of the longest claw	1	6
Length of hair on the body	3	9			

[68.]   1. Lepus Americanus.   (Erxlebein.)   *The American Hare.*

Genus. Lepus. Linn.
Lièvre, (Queutonmalisia.) Sagard-Theodat, *Canada,* p. 747. An. 1636.
Hare, Hedge-coney. Lawson, p. 122. Catesby, *App.,* xxviii.
Rabbit. Smith, *Voy.,* vol. i. p. 156. An. 1748.
American Hare. Kalm, *Travels,* vol. i. p. 105 ; vol. ii. p. 45.
Lepus Americanus. Erxlebein, *Syst.,* An. 1777.
Lepus Hudsonius. Pallas, *Glires,* p. 30. An. 1778.
American Hare. Forster, *Phil. Trans.,* vol. lxii. p. 376. Pennant, *Arct. Zool.,* vol. i. p. 90. Hearne,
          Journ., p. 384.
Lepus Americanus. Sabine, *Franklin's Journ.,* p. 664. Richardson, *Append. Parry's Second Voy.,* p. 324.
          Harlan, *Fauna,* p. 193.
The American Hare. Godman, *Nat. Hist.,* vol. ii. p. 157.
Wawpoos. Cree Indians. Kah. Chepewyans.
Rabbit. European Residents at Hudson's Bay.
Le lapin. French Canadians.

This is a common animal, in the woody districts of North America, from one extremity of the continent to the other.   It abounds in Mackenzie's River as high as the sixty-eighth parallel of latitude*; but on the barren grounds to the eastward of the Coppermine, and on the extensive plains or prairies through which the Missouri and Saskatchewan flow, it is replaced by other and larger species.

The American Hare does not burrow. In the northern districts it resides mostly in willow thickets, or in woods where willows or dwarf birch constitute much of the underwood. The bark of the willow forms a great part of its food in the winter, but in the summer it eats grass and other vegetables. It is reported to do much damage in cultivated districts, to fields of cabbage or turnips. In the fur countries, few are killed in the summer, because the natives can then procure abundance of water-fowl and game of various kinds. In the winter, however, they are more sought after, and in the Hare-Indian country, on the banks of the Mackenzie, where larger animals are scarce during that season, they constitute the chief food of the natives. They are principally taken in snares set in the paths that they make through the snow, and fixed to a pole which springs up when the noose is drawn, care being taken to obstruct their passage on one side of the noose by a small hedge of branches. To prevent them from cutting the snare

---

* From a clerical error in the appendix to Capt. Franklin's Narrative, it is stated that the American hare does not exist " further north than Carlton-house." It should have been " further north than Fort Enterprise."

instead of endeavouring to pass through it, it is occasionally rubbed with a little of their own dung.   The Hare-Indians, when they come to a place where the hare-tracks are numerous, begin their operations by beating a circular path in the snow, so as to enclose a pretty large clump of wood, knowing that the hares will not readily cross such a path.   They next bar the ways by little hedges, in the gaps of which they set snares, and then they enter the circle and beat amongst the bushes with their dogs to drive the hares into the nooses.   On the success of this operation the supper of a whole horde often depends, as, with the usual improvidence attendant on a hunter's life, these Indians seldom keep any stock of provision by them.   Unless when disturbed, the American hare rarely runs about during the day.   It has numerous enemies, such as wolves, foxes, wolverenes, martins, ermines, snowy owls, and various hawks; but the Canada lynx is the animal which perhaps most exclusively feeds upon it.   It has been remarked that lynxes are numerous only when there are plenty of hares in the neighbourhood. At some periods a sort of epidemic has destroyed vast numbers of hares in particular districts, and they have not recruited again until after the lapse of several years, during which the lynxes were likewise scarce.   In the spring and summer the hares are much infested by a large species of *cimex*.   In the fur countries this hare becomes white in the winter.   This change takes place in the northern districts in the month of October, and the animals retain their white coat until the end of April, when it begins to fall off, and is replaced by their shorter and coloured summer dress.   The white colour is less perfect in more southern districts, and to the southward of New England, according to Pennant, the brown dress endures all the year.   The same author says that the winter coat, in northern districts, consists of a multitude of long white hairs, twice the length of the summer fur, which still remains beneath.   After a careful examination, however, of many specimens in different states, I agree with the clerk of the California* in thinking that the change to the winter dress takes place by a lengthening and blanching of the summer fur; whilst the change in the beginning of summer consists in the winter coat falling off during the growth of new and coloured fur.

The winter skins of this animal are imported by the Hudson's Bay Company under the name of rabbit-skins; but from their small value the importation does not at present exceed eight or ten thousand in a year, as they will not cover the expenses of carriage from the interior.   Mr. Jeremie relates that in one season

* *Voyage in search of a North-West Passage.*

twenty-five thousand were taken at the post at which he resided in Hudson's Bay, and great numbers might still be obtained in some districts, were it an object to do so. In some parts of the fur countries the natives line their dresses with hare-skins, and the Hare-Indians sometimes tear the skins with the fur into strips, and plait them into a kind of cloth. They resort to this expedient, however, only from the scarcity of deer-skins and moose-leather, which form closer and better dresses.

DESCRIPTION.

The *form* of this animal is similar to that of the other species of the genus amongst which there is a great resemblance, and it is so like that of the common European rabbit, that it is universally called " the rabbit" by the English residents at Hudson's Bay. Its average weight is about four lbs.

Dental formula; incisors, $\frac{4}{2}$; canines $\frac{0-0}{0-0}$; grinders, $\frac{6-6}{5-5}$ = 28.

Incisors, white; superior ones linear, flattened anteriorly with a deep groove near their inner margins, rounded laterally, without a groove there; inferior ones quite flat and smooth anteriorly, and on the sides; somewhat narrower behind; with slightly oblique cutting edges.

In the *winter* this animal is covered with a thick coat of fine long fur, which, when lying smooth, appears every where of a pure white colour, except a narrow border on the posterior margins of the ears, and round their tips, and about one-third down their anterior margins, which are blackish-brown, on account of the dark roots of the hair being visible on these parts. The *whiskers,* which are three inches long, are some of them black throughout, whilst others are black only at the base. There are four or five long black hairs ovei the eyes, and a narrow margin of the eyelids is blackish-brown.

The *fur* on the back, when blown aside, shews a blackish-gray colour for more than one-third of its length from the roots upwards; then a clear yellowish or wood-brown for rather a shorter space; and, lastly, a pure snow-white to the tips. There are also interspersed many longer and rather stronger hairs, which are white their whole length. The fur on the throat is similar to that on the back, but on the belly it is almost entirely white, there being merely a slight tinge of gray at the roots. The fur on the upper aspect of the head is shorter than that on the body, and the brown colour beneath the white tips is much darker. The fur on the *ears* is blackish-brown from beneath the white tips to the roots. The *tail* appears entirely white in winter, but the fur is coloured towards its base like that on the back, though with less of the pale brown in its middle parts. The fur on the outer and anterior aspect of the *extremities* corresponds in colour with that of the back, whilst the fur on the inner aspects is white nearly its whole length, as on the belly. The *fore-nails* are narrow, nearly straight, and very sharp. The hind-ones are broader and longer. Their colour on both feet is nearly white.

In its *summer dress,* the fur on the upper parts is shining blackish-gray at the roots as in

2 F 2

winter, but towards the tips it is ringed with yellowish-brown and black.   On the back, the black is in large proportion, and the resulting colour of the surface is a dark umber-brown, mixed with yellowish-brown.   On the head there is more brown, and it has a brighter tint. There is more black on the crown of the head than on the cheeks; the sides of the muzzle are paler, and are sprinkled with white.   A white circle surrounds the eye, but the margins of the eyelids are black as in winter.   The under jaw is smoke-gray, the throat unmixed yellowish-brown.   The white colour commences between the fore-legs, extends over all the belly, and predominates on the extremities; but wherever the fur on the latter is ruffled, the brown colours of the roots of the fur are seen.   The sides present a dull, pale, yellowish-brown colour, with a few scattered black hairs.   The *ears* are nearly naked in the summer, but the fur generally remains on their margins of a mixed white and blackish-brown, the latter colour prevailing at their tips.   The *tail* is white underneath; on its upper surface the gray and brown colours appear through the white.   Whiskers as in winter.

DIMENSIONS

Of full-sized individuals.

	Inches.	Lines.		Inches.	Lines.
Length of head and body . .	19	0	Length of middle toe and claw . .	1	3
„ from nose to the tip of the middle claw of the hind leg, when stretched out .	27	6	„ the middle fore-claw alone	0	7
„ of head, measured with a line over the curvature of the forehead	4	3	*Hind Extremities.*		
			„ from knee joint to end of middle claw . .	9	6
„ of head, measured with callipers .	3	6	„ of tibia . .	4	6
„ ears, in their winter fur (posteriorly)	3	2	„ from heel to end of middle claw	5	6
„ ears, from rictus to apex .	2	9	„ heel to root of middle toe .	3	3
„ tail (vertebræ) .	1	6	„ of middle toe and claw .	2	4
„ tail, including fur . .	2	6	„ middle claw . .	0	6½
„ whiskers . .	3	9	„ the scull, from the insertion of the incisors to occipital spine (measured by calliper compasses) . . .	2	9
Fore Extremities.					
„ ulna . . .	3	6			
„ from carpal joint to point of middle claw . . .	2	9			

[69.]   2. LEPUS GLACIALIS. (Leach.) *Polar Hare.*

Varying Hare. PENNANT, *Arct. Zool.*, vol. i. p. 94. HEARNE, *Journey*, p. 382.
Lepus timidus. FABRICIUS, *Fauna Grœnl.*, p. 25.
Lepus glacialis. LEACH, *Ross's Voyage.* Capt. SABINE, *Suppl. Parry's First Voy.*, clxxxviii. SABINE (Mr.),
      FRANKLIN'S *Journ.*, p. 664. RICHARDSON, *Appendix Parry's Second Voy.*, p. 321. HARLAN,
      Fauna, p. 194.
The Polar Hare. GODMAN, *Nat. Hist.*, vol. ii. p. 162.
Kaw-choh. COPPER and HARE INDIANS.
Ookalik. ESQUIMAUX. Rekaleek. GREENLANDERS.

This animal was, down to the period of Captain Ross's voyage to Baffin's Bay, considered as the same with the varying hare; although Pennant had remarked that its size was greater than that of the latter animal. Dr. Leach first noted Captain Ross's specimens as belonging to a distinct species; and Captain Sabine enumerated its specific characters in the Appendix to Captain Parry's First Voyage. Many specimens, brought home by the late arctic voyagers, exist in various museums in Great Britain. The Polar Hare inhabits both sides of Baffin's Bay, and is common on the Barren Grounds, at the northern extremity of the American continent. Its most southerly known habitat is the neighbourhood of Fort Churchill, on Hudson's Bay, which is in the 58th parallel of latitude; but it may, perhaps, extend further to the southward on the elevated ridge of the Rocky Mountains, or on the eastern coast of Labrador. It is not found in wooded districts; hence, it does not come further south on the line of the Mackenzie and Slave Lake than latitude 64°. It was found in latitude 75° on the North Georgian Islands. Although it does not frequent thick woods, it is often seen near the small and thin clumps of spruce fir, which are scattered on the confines of the Barren Grounds. It seeks the sides of hills, where the wind prevents the snow from lodging deeply, and where, even in the winter, it can procure the berries of the alpine arbutus, the bark of some dwarf willows, or the evergreen leaves of the Labrador tea-plant (*ledum*) *. It does not dig burrows, but shelters itself amongst large stones, or in the crevices of rocks, and in the winter time its form is generally found in a wreath of snow, at the base of a cliff. The Polar Hare is not a very shy animal, and on the approach of a hunter it merely runs to a little distance, and sits down, repeating this manœuvre as often as its pursuer comes

* On the barren coast of Winter Island, the hares went out on the ice to the ships, to feed on the tea-leaves, thrown overboard by the sailors.—LYON's *Private Journal.*

nearly within gunshot, until it is thoroughly scared by his perseverance, when it makes off.   It is not difficult to get within bow-shot of it, by walking round it, and gradually contracting the circle—a method much practised by the Indians.   In the late boat-voyage along the northern coast, we landed on a rocky islet, off Cape Parry, which, though not above three hundred yards in diameter, was tenanted by a solitary alpine hare.   The whole party went in pursuit of this poor animal ; but it availed itself so skilfully of the shelter of the rocks, and retreated with so much cunning and activity from stone to stone, that none of us could obtain a shot at it, although it never was able to conceal itself from our search for more than a minute or two at a time.

The winter fur of the Polar Hare is of a snow-white colour to the roots, and is more dense, and of a finer quality than that of the American hare.   It bears a close resemblance to swan-down.   The fur is in prime order in latitude 65°, about the end of October, and begins towards the end of April to be replaced by the summer coat, which is more or less coloured.   I have killed individuals at the time they were losing their winter fur, and have seen others exhibiting dark colours later in the season, but have not been able to obtain a full-grown summer specimen.   Fabricius informs us that, in Greenland, the Polar Hares retain their white colour all the summer.   Captain Sabine states, that some full-grown specimens, killed on Melville Island (lat. 75°), in the height of summer, had the hair of the back and sides of a grayish-brown colour towards the points, but the mass of fur beneath still remained white ; the face and front of the ears were of a deeper gray.   The fur was interspersed with long solitary hairs, which, in many individuals, were banded with brown and white in the middle of summer.   The weight of a full-grown Polar Hare varies, according to its condition, from 7 to 14 lbs., and a similar variation, in the weight of the common British hare, is known to exist.   Its flesh is whitish, and well flavoured, being greatly superior to that of the American hare ; and also much more juicy than the alpine or varying hare of Scotland.

According to Indian information, the Polar Hare brings forth once in the year, and from two to four young at a time.   Fabricius says that, in Greenland, they produce eight young at a birth, in the month of June : they pair in April.

DESCRIPTION
Of a full-grown *winter* specimen, from Bear Lake.

Size.—Equal to that of the largest English hare, superior to that of the varying or Alpine hare of Scotland.   *Scull* one-third larger than that of the American hare, with a larger orbital cavity, and smaller space for containing the brain. The breadth of the frontal bone or distance between the orbits is not greater than in the smaller scull of the American hare. The margins of the orbits project considerably, so as to produce a well-marked depression in the anterior part of the frontal bone, included between them.

Dental formula, incisors $\frac{4}{2}$, canines $\frac{0-0}{0-0}$, *grinders* $\frac{6-6}{5-5}$, = 28.

Incisors white, four-sided. Upper ones with a conspicuous but rather shallow groove near their inner margins anteriorly, and another groove on their sides. The posterior or supplemental upper incisors have two grooves on their posterior faces, which give them a prismatical form. The cutting edges of the incisors are nearly even.

The *fur* is every where entirely white, except on the tips of the ears, which are brownish-black. The back and margins of the ears are covered with a close coat of hair, which is white to the roots. The hairs lining the interior of the ear are white and moderately long, but they are not so close as to prevent the dark skin from partially appearing. The hair on the tips of the ears is mostly brownish-black to its base, a little of it only, where it adjoins to the white fur, shewing a wood-brown colour near its roots. The whiskers in some specimens are entirely white, in others partially black. The fur on the back is remarkably close and fine, that on the belly is longer and not quite so close. The extremities are covered with a smooth coat of hair of a pure white colour to the roots. The brush on the soles has a soiled yellowish-white colour. The fore-toes are short, and their claws are of a dark-brownish horn colour, and are very long and considerably curved, but their general shape nearly resembles that of the claws of the common hare. They are more curved and blunter than the claws of the American hare, and project much further beyond the fur. The hind-claws are rather broader than the fore-ones, dark at the roots and pale at the tips. The tail is covered with pretty long fur, woolly at the roots, and of a pure white colour its whole length. The *irides* are of a honey-yellow colour. The skin of the polar-hare, when in full winter dress, is so tender, that it is difficult to take it off without tearing it.

DIMENSIONS.

	Inches.	Lines.		Inches.	Lines.
Length of the head and body	22	6	*Fore Extremities.*		
„ from the nose to the point of the middle claw, when the hind leg is stretched out	30	6	Length from wrist joint to point of the middle claw	3	4
„ of the head, from the occipital spine to the end of the nose, measured over the forehead, and pressing down the fur	5	6	„ of middle toe and claw	1	10
„ of the head, measured with a pair of calliper compasses	4	6	„ the middle fore-claw	1	0
„ of the ears, including the fur, measured posteriorly	4	6	*Hind Extremities.*		
„ the ears, from rictus to apex	3	6	„ from the heel to the base of the middle toe	4	0
„ „ black fur at the tip of the ear	0	6	„ of middle toe and claw	2	6
„ „ tail (vertebræ)	1	6	„ from the heel to the end of the middle claw	6	6
„ „ tail, including the fur	3	6	„ of the scull, from the insertion of the incisors to the occipital crest, measured by callipers	3	10
„ „ whiskers	3	6			

DESCRIPTION

Of a Polar hare, three months old, killed on the 12th of August at Repulse Bay.

The head and back are hoary, from an intermixture of hairs entirely black, with others which are black at the base and white at the tips.   When these hairs are blown aside, they permit a shorter yellowish-gray down to be seen.   On the breast, flanks, and thighs, the longer hairs have fewer white tips, and are more thinly scattered, allowing much of the down to become visible : the down on these parts has a bluish-gray colour.   The belly, feet, and tail are entirely white.   The hairs on the belly are very long.   The ears have a similar colour with the back, but the proportion of black hairs is rather greater.   Their margins are white, and there is a small brownish-black spot at their tips.

DIMENSIONS.

	Inches.	Lines.
Length of the head and body	17	6
„ „ ears	3	6

A nearly mature fœtal specimen was of a blackish-brown colour on all the upper parts and outsides of the extremities.

[70.]   3. Lepus Virginianus.   (Harlan.)   *Prairie Hare.*

> Varying Hare.   Lewis and Clark, *Journey,* &c., vol. ii. p. 178.
> The Varying Hare ?   Godman, *Nat. Hist.*, vol. ii. p. 163.
> Lepus Virginianus.   Harlan, *Fauna,* p. 312.
> Prairie Hare.   Fur Traders.

The servants of the North-west and Hudson's Bay Companies have long been acquainted with this animal, but it is still very imperfectly known to naturalists.   The best account of it is contained in the narrative of Lewis and Clark's interesting Journey to the Columbia ; but Dr. Harlan first named it as a species distinct from the *Lepus variabilis.*   It is a common animal on the plains through which the north and south branches of the Saskatchewan flow, and which extend as far eastward as the Winepegoosis and southern extremity of Winepeg lake, and to the southward, unite with the plains of the Missouri, where this hare is also found, as well as on the great plains of the Columbia.   I have not heard of its existing further north than latitude 55°.

It frequents the open plains, where it lives much after the solitary manner of the common European hare, without burrowing; it is also occasionally met with among the small clumps of poplars and willows, with which the plains are studded near their confines; but it does not resort to the thick woods, like the American hare. It possesses great speed. I was not successful in the attempts I made to obtain specimens of this hare, a mutilated hunter's skin, in the winter dress, being all I could procure. Mr. Drummond killed a full-grown individual on the banks of the Saskatchewan, in the month of September, and remarked that, as far as his recollection went, there was no difference betwixt it and the common English hare. Owing to a succession of wet weather, and want of convenience for drying specimens, the skin unfortunately became putrid, and was thrown away.

DESCRIPTION
Of a mutilated winter skin.

The *fur* is not quite so dense and fine as that of the Polar hare, but more so than that of the American hare. It is everywhere of a pure white colour on the surface, except on the borders of the ears. The whiskers and muzzle are white. There are no coloured rings round the eye, but when the fur there is blown aside, it is seen to be of a very pale wood-brown or fawn colour for about two-thirds of its length from the roots upwards. On the upper aspect of the head, the wood-brown colour of the concealed parts of the fur is deeper, and is mixed with a little bluish-gray. On the cheeks the fur is longer, and white to very near the roots, where it is bluish-gray. On the sides of the neck, the fur is bluish-gray for a short space at the roots, then of a buff colour intermediate between pale wood-brown and cream-yellow for two-thirds of its length; and, lastly, white at the tips. On the *back*, the fur is white for one-third of its length from the roots, then pale brownish-yellow or buff colour for less than a third; and, lastly, white to the tips. The fur on the *belly* and *legs* is white its whole length. The *ears* have a pretty broad wood-brown or fawn-coloured border along their anterior margin, and a narrower one towards the base of the posterior margin; the fur on these borders is blackish-brown towards its roots. The back of the ear between the fawn-coloured margins is covered with entirely white fur. The ear has a brownish-black tip about the same size with the black tip of the ear of the Polar hare.

DIMENSIONS.

	Inches.	Lines.		Inches.	Lines.
Length of the head and body . .	22	0	Length of the ears, measured posteriorly, including fur . . .	4	0
,, fur of back . .	1	4	,, ,, from rictus to apex, without fur	3	0
,, fur of belly . .	2	4	,, fur, at tip of the ear .	0	6
,, whiskers . .	3	6			

The DESCRIPTION of this species of hare, by Lewis and Clark, is as follows:—

" They weigh from seven to eleven pounds; the eye is large and prominent, the pupil of a deep sea-green, the iris of a bright yellow and silver colour; the head, neck, back, shoulders, and outer parts of the legs and thighs are of a lead colour; the sides, as they approach the belly, become gradually more white; the belly, breast, and inner parts of the legs and thighs are white, with a light shade of lead colour; the tail is covered with white, soft fur, not quite so long as on the other parts of the body; the body is covered with a deep, fine, soft, close fur. The animal assumes these colours from the middle of April to the middle of November; during the rest of the year it is of a pure white, except the black and reddish-brown of the ears, which never changes. In March, a few reddish spots are sometimes mixed with the white on the head and upper parts of the neck and shoulders. This animal can leap twenty-one feet. Its food is grass and herbs, and in winter it feeds much on the bark of aromatic shrubs, which grow on the plains. These hares are generally found separate, and never associate in greater numbers than two or three."

Dr. Godman has given the following account of a specimen belonging to the Prince of Musignano, which was killed on the Blue Mountains of Pennsylvania. This species is said to be common throughout the mountainous regions of the United States; but its identity with the *Lepus Virginianus* of Harlan has not been ascertained, and it may be observed, that nothing is said in Dr. Godman's description respecting the fawn-coloured margins of the ear, which distinguish the *Lepus Virginianus*, in its winter dress, from the Polar hare.

" In its *summer dress* the general colour of this hare is a light reddish-brown, which is lighter on the breast and head, becoming darker from the superior parts of the shoulders to the posterior parts of the body. The hairs are coloured in the following manner :—They are plumbeous at the base, then light yellow, then dusky, then reddish-brown, and finally, black at tip. The under jaw is white, and this colour extends backwards until opposite the bases of the ears. The belly and legs are white, faintly tinged with light reddish-brown; the tail is whitish, which colour is superiorly mingled with bluish or lead colour. The *ears* are externally bluish-white, and darker at the tip; internally they are of a faint reddish-white.

" In *winter dress* the general colour is pure white, the fur being long, soft, fine, and in greatest quantity upon the breast. The hairs are plumbeous at the base, then reddish, and at tip of a snowy whiteness. The ears are slightly tipped with dark lead colour, and edged within by brown and white hairs intermixed. The whiskers are entirely white or black at the base and white at tip. The feet are thickly clothed with hair, that conceals the slightly-curved nails, which are long and narrow at the base."

DIMENSIONS
Of a recent specimen.

	Feet.	Inches.			Feet.	Inches.
" Total length	2	7	Length of fore-arm		0	4
Height to top of fore-shoulder	0	10	,, fore-paw		0	2¾
,, top of thigh	1	2	,, the thigh		0	6
Length of head	0	4	,, hind-foot		0	6
,, the ears	0	4	,, the tail		0	1½ "
Distance from eyes to the end of the nose	0	1¾				

LEPUS (LAGOMYS) PRINCEPS.

Published by John Murray, January 1829.

[71.] 4. LEPUS (LAGOMYS) PRINCEPS. (Richardson.)
The Little-Chief Hare.

GENUS, Lepus. LINN. *Sub-genus*, Lagomys. CUVIER.
Lepus (Lagomys) princeps. RICHARDSON, *Zool. Journal.* No. 12, p. 520. March, 1828.

LEPUS LAGOMYS (*princeps*), *fuscus subter griseus, capite brevi, auriculis rotundatis.*
The Little-Chief Hare: tailless; colour blackish-brown, beneath gray; head short and thick; ears rounded.]

This highly interesting little animal inhabits the Rocky Mountains, from latitude 52° to 60°. Through the kindness of Mr. Macpherson, I obtained some specimens from the River of the Mountains, or south branch of the Mackenzie; and Mr. Drummond killed several near the sources of the Elk River. There is likewise a good specimen in the museum of the Hudson's Bay Company.

Mr. Drummond informs me, that the Little-Chief Hare frequents heaps of loose stones, through the interstices of which it makes its way with great facility. It is often seen at sun-set, mounted on a stone, and calling to its mates by a peculiar shrill whistle. On the approach of a man, it utters a feeble cry, like the squeak of a rabbit when hurt, and instantly disappears, to re-appear in a minute or two, at the distance of twenty or thirty yards, if the object of its apprehension remains stationary. On the least movement of the intruder, it instantly conceals itself again, repeating its cry of fear, which, when there are several of the animals in the same neighbourhood, is passed from one to the other. Mr. Drummond describes their cry as very deceptive, and as appearing to come from an animal at a great distance, whilst, in fact, the little creature is close at hand; and, if seated on a gray limestone rock, its colour is so similar, that it can scarcely be discerned. These animals feed on vegetables. Mr. Drummond never found their burrows, and he thinks that they do not make any, but that they construct their nests amongst the stones. He does not know whether they store up hay for the winter or not, but is certain that they do not come abroad during that season.

The trivial name which I have adopted for the species, is a translation of the Indian appellation, *buckathræ kah-yawzæ*. The Little-Chief Hare resembles the pika (*lagomys alpinus*) in its alpine habits and general form. It is, however, a smaller animal, the largest of our specimens falling short of seven inches, which is the length of the smallest pika seen by Pallas. I have not had an opportunity of comparing it with a specimen of the pika; but a scull of the latter, preserved in the museum of the College of Surgeons, is twice the size, and differs in form. The pika has not only a larger head, but its fur is described as coarse, and its colours as dissimilar to those of the Little-Chief Hare. The pika is said, by Pallas, to inhabit Kamskatcha; and, by Pennant, to have been discovered on the Aleutian Islands.

The Little-Chief Hare presents differences in its teeth from those of the true hares, which fully entitle it to rank in a distinct genus; and it is further entitled to that distinction from the naked tubercles at the end of its toes, and its very different habits.

DESCRIPTION.

Size somewhat less than the *Alpine Pika* of Siberia; length $6\frac{3}{4}$ inches.

On comparing the *scull* of this animal with those of the true hares, there appears a larger cavity in proportion to its size for the reception of the brain. The breadth of the scull too behind is increased by very large and spongy auditory processes. The bone anterior to the orbit is not cribriform as in the hares, although it is thin, and there is no depression of the frontal bone between the orbits.

Dental formula, incisors $\frac{2-2}{2}$, canines $\frac{0-0}{0-0}$, grinders $\frac{5-5}{5-5} = 26$.

Incisors white, anterior upper ones marked with a deep furrow nearer their interior margins, and having cutting edges, which present conjointly three well-marked points, the middle one of which is common to both teeth, and is shorter than the exterior one. These incisors are much thinner than the incisors of a hare, and are scooped out like a gouge behind. The small round posterior or accessary upper incisors, have flat summits. The *lower incisors* are thinner than those of the hares, and are chamfered away towards their summits more in form of a gouge than like the chisel-shaped edge of the incisors of a hare. *Grinders.*—The upper grinders are not very dissimilar to those of the hare on the crowns, but the transverse plates of enamel are more distinct. They differ in each tooth, having a very deep furrow on its inner side, which separates the folds of enamel. This furrow is nearly obsolete in the hares, whilst in this Lagomys it is as conspicuous as the separation betwixt the teeth. The small posterior grinder which exists in the upper jaw of the adult hares is entirely wanting in the different specimens of the little-chief hare which I have examined. The *lower grinders*, from the depth of their lateral grooves, have at first sight a greater resemblance to the grinders of animals belonging to the genus *arvicola* than to those of a hare; their crowns exhibit a single series of acute triangles with hollow areas. The first grinder has three not very deep grooves on a side, and is not so unlike the corresponding tooth of a hare as those which succeed it. The second, third and fourth, have each a groove in both sides, so deep as nearly to divide the tooth, and each of their crowns exhibits two triangular folds of enamel. The posterior grinder forms only one triangle.

Shape.—The body of the little-chief hare is moderately thick, and the head is short and broad with an arched forehead. The *whiskers* are longer than the head. The *ears* are large and nearly round, but do not, as far as I can judge from the prepared specimens, appear to have the incurvation of their anterior margins, which gives the funnel shape to the ears of the pika, as described by Pallas. An obtuse projection of the rump is the only vestige of a *tail.*

The *fur* is soft to the touch, and differs in quality from that of the hare, being less downy and having more the character of the fur of a meadow-mouse. It is of an uniform, shining grayish-black colour for three-fourths of its length from the roots upwards, then partly yellowish-brown, and partly white, and on the superior parts of the body, most of the hairs have short black tips. The black predominates on the posterior part of the back, but even there it is mixed with brown. Yellowish-brown prevails on the shoulders and sides; the under part of the protuberance which represents the tail is white, and all the under parts of the body are smoke-gray, tinged on the chest and some parts of the belly with brown. The fur on the back is about three-quarters of an inch long, that on the belly is somewhat shorter. In some specimens the principal colour of the head is yellowish-brown, in others there are many black hairs scattered over the crown. The ears have a narrow white border, and are pretty well clothed anteriorly with white hairs tipped with black. The hairs which cover them posteriorly are longer, and nearly black for their whole length.

The *extremities* are white with a brownish tinge. The soles of the feet are clothed with dusky-brown hair. The claws are black, short, arched, much compressed, but not very sharp, and are concealed by the fur of the toes. There is a large naked black tubercle at the root of each of the four fore-claws, and a fifth minute tubercle far back near the exterior margin of the rather broad and flat palm. The thumb is a little further back than the outer toe, but not so far as the last-mentioned tubercle. It is very short, and has no naked callus at its base, but its claw is as large as those of the toes. There are four toes on the hind-foot, each terminated by a naked callus and claw, similar to those on the fore-toes.

DIMENSIONS.

	Inches.	Lines.		Inches.	Lines.
Length of head and body . . .	6	9	Length of largest whiskers . .	2	9
„ head . . .	2	2	„ fore-foot, from wrist-joint to end		
„ from nose to auditory opening .	1	9	of middle claw	0	9
„ from nose to centre of pupil	0	9	Breadth of fore-palm at the thumb .	0	4
Height of the ear . . .	1	0	Length of the scull from incisors to occipital		
Breadth of the ear . . .	0	9	spine	1	6
Length from heel to the middle claw of the			Total breadth of the scull at the auditory		
hind-foot	1	1½	openings		9
„ fur on the back . .	0	10			

† 1. LIPURA HUDSONIA.   (Illiger.)   *Tail-less Marmot.*

GENUS, Lipura. ILLIGER.
Tail-less Marmot. PENNANT, *Arct. Zool.*, vol. i. p. 112.   *Hist. Quadr.*, vol. ii. p. 137.
              BEWICK, *Quadr.*, p. 374.
Daman de la baie d'Hudson. SCHREBER, t. 240. C.
Arctomys Hudsonius. TURTON, LINN., vol. i. p. 90.
Hyrax Hudsonius. SHAW, *Zool.*, vol. ii. p. 225.

This animal was first described by Pennant from a specimen preserved in the Leverian Museum, and said to have been brought from Hudson's Bay.   It has not been obtained from that quarter since Pennant's time, and there is much reason to doubt the habitat assigned to the animal, though there appears to be none to question the genuineness of the specimen.

The characters attributed to the genus *lipura*, by Illiger, are: " two superior incisors; four inferior ones, obliquely truncated; an interval between the incisors and the grinders, which are composed of folded layers of enamel; a pointed muzzle; body covered with coarse hair; no tail; feet, with four toes, armed with flat nails.

Pennant describes the animal as being of the *size* of the common marmot (*i. e.* head and body, 16 inches long) with short ears; head and body of a cinereous brown; the ends of the hairs white; two cutting teeth above, four below; no tail."

[72.]   1. Equus caballus. (Linn.)   *The Horse.*

The Horse. Warden, *United States*, p. 234.
Wild Horse. Long, *Journ.*, vol. ii. p. 313 ; vol. iii. p. 107.

Herds of wild horses, the offspring of those which have escaped from the Spanish possessions in Mexico, are not uncommon on the extensive prairies that lie to the west of the Mississipi.   They were once numerous on the Kootannie Lands, near the northern sources of the Columbia, on the eastern side of the Rocky Mountain ridge, but of late years they have been almost eradicated in that quarter.   They are not known to exist in a wild state to the northward of the fifty-second or fifty-third parallel of latitude.   The young stallions live in separate herds, being driven away by the old ones, and are easily ensnared by using domestic mares as a decoy.   The Kootannies are acquainted with the Spanish-American mode of taking them with the *lasso*.   Major Long mentions that " horses are an object of a particular hunt to the Osages.   For the purposes of obtaining these animals, which in their wild state preserve all their fleetness, they go in a large party to the country of the Red Canadian River, where they are to be found in considerable numbers.   When they discover a gang of the horses, they distribute themselves into three parties, two of which take their stations at different and proper distances on their route, which by previous experience they know the horses will most probably take when endeavouring to escape.   This arrangement being completed, the first party commences the pursuit in the direction of their colleagues, at whose position they at length arrive.   The second party then continues the chase with fresh horses, and pursues the fugitives to the third party, which generally succeeds in so far running them down as to noose and capture a considerable number of them."

The domestic horse is an object of great value to the nomadic tribes of Indians that frequent the extensive plains of the Saskatchewan and Missouri, for they are not only useful in transporting their tents and families from place to place, but one of the highest objects of the ambition of a young Indian is to possess a good horse for the chase of the buffalo, an exercise of which they are passionately fond.   To steal the horses of an adverse tribe is considered to be nearly as heroic an exploit as killing an enemy on the field of battle, and the distance to which they occasionally travel, and the privations they undergo on their horse-stealing excursions, are almost incredible. An Indian who owns a horse scarcely ever ventures to sleep after nightfall, but sits at his tent door with the halter in one hand and his gun in the other, the horse's fore-legs being at the same time tied together with thongs of leather. Notwithstanding all this care, however, it often happens that the hunter, suffering himself to be overpowered by sleep for only a few minutes, awakes from the noise made by the thief gallopping off with the animal.

The Spokans, who inhabit the country lying between the forks of the Columbia, as well as some other tribes of Indians, are fond of horse-flesh as an article of food ; and the residents at some of the Hudson's Bay Company's posts on that river, are under the necessity of making it their principal article of diet.

[73.]   1. CERVUS ALCES.   (Linn.)   *Moose Deer.*

GENUS, Cervus. LINN.
Ellan, stagg, or aptaptou. DE MONT's *Nova Francia,* p. 250. An. 1604.
Eslan ou orignat. SAGARD-THEODAT, *Canada,* p. 749. An. 1636.
Orignal. LA HONTAN, *Voy.,* p. 72. An. 1703.
Moose deer. DUDLEY, *Phil. Trans,* No. 368. p. 165. An. 1721.
Orignac. *Hist. de l'Amerique.* An. 1723.
Orignal. CHARLEVOIX, *Nouv. France,* vol. v. p. 185. An. 1744. DENYS, *Descr. de l'Amer.,* vol. i. p. 27. p. 163 ;
          vol. ii. pp. 321, 425. DU PRATZ, *Louis,* vol. i. p. 301.
Moose deer. PENNANT, *Arctic Zool.,* vol. i. p. 17. Cum fig. An. 1784.
Moose. UMFREVILLE, *Huds. Bay.* An. 1790. HERIOT's *Trav.* An. 1807. With a good figure.
Moose deer. WARDEN, *United States,* vol. i. p. 328. GODMAN, *Nat. Hist.,* vol. ii. p. 274.
Cervus alces. HARLAN, *Fauna,* p. 229. GRIFFITH, *An. King.,* vol. iv. p. 72. A good figure of the head.
" Orignac. BASQUE SETTLERS IN CANADA." (De Monts.)
Orignal. FRENCH CANADIANS of the present day.
Moosöä. CREE INDIANS, Denyai. CHEPEWYANS.
" Sondareinta. HURONS." (Theodat.)

The Moose Deer is said to derive its present English name from its Algonquin and Cree appellation of mongsoa or moosoa. It early attracted the attention of travellers in America, and various descriptions of it appear in their works, some of which are quoted above. Live specimens have occasionally been brought to

England, and one was sent to his late Majesty, from Churchill, in Hudson's Bay. Naturalists have generally considered the moose deer to be the same species with the elk of the northern parts of the old world*. The Anglo-Americans, however, having given the trivial name of elk to the Canada stag or red deer, some confusion has occasionally crept into the accounts published by travellers, of the size, manners, and geographical distribution of the moose; and it has also sometimes been confounded with the rein deer, from its possessing, in common with that animal, palmated horns. The fact, that few of the American quadrupeds have been found precisely similar to their European representatives, ought to excite doubts of the identity of the moose and Scandinavian elk, until it is established by satisfactory comparisons. This does not appear, however, to have been hitherto done, and some differences between them are hinted at by La Hontan. Major Smith also mentions, that the lower parts of the antlers of the American animal more often separate into branches than those of the European one.

Du Pratz informs us that, in his time, moose deer were found as far south as the Ohio; and Denys says, that they were once plentiful on the island of Cape Breton, though at the time that he wrote they had been extirpated. At present, according to Dr. Godman, they are not known in the state of Maine; but they exist in considerable numbers in the neighbourhood of the bay of Fundy. They frequent the woody tracts in the fur countries to their most northern limit. Several were seen on Captain Franklin's last expedition, at the mouth of the Mackenzie, feeding on the willows, which, owing to the rich alluvial deposits on that great river, extend to the shores of the Arctic sea, in lat. 69°. Further to the eastward, towards the Coppermine river, they are not found in a higher latitude than 65°, on account of the scarcity on the Barren Grounds of the aspen and willow, which constitute their food. I have not been able to ascertain whether they occupy the whole width of the continent or not. Mackenzie saw them high up on the eastern declivity of the Rocky Mountains, near the sources of the Elk river; but I suspect that they are rarely, if ever, found to the westward of the mountains. Authors mention that the moose generally form small herds in Canada. La Hontan, who travelled in that country in 1683, says, that whilst he accompanied the Indians they hunted the elk with dogs, when there was a crust on the snow; and that, after a chase of a few leagues, they generally found ten, fifteen, or twenty of them in a body: in three months his party killed fifty-six, and might have taken as many more. It is probable, however, that

* According to Buffon, the elk was unknown to the Greeks; and the word *alce* first occurs in the writings of Julius Cæsar, and was probably adopted by him from the Celtæ. Its Celtic name is elch; and Swedish, ælg.

2 H

La Hontan in this passage confounds the Canada stag and moose deer together. He mentions the animal being able to run, in the summer season, for three days and nights in succession, and the excellent flavour of its flesh,—facts which apply to the moose deer, but not to the Canada stag; on the other hand, the weight of the horns, which, he says, sometimes amounts to four hundred weight, is true only of the stag. In like manner, the accounts of the other early writers on Canada are liable to suspicion. In the more northern parts, the moose deer is quite a solitary animal, more than one being very seldom seen at a time, unless during the rutting season, or when a female is accompanied by her fawns. It has the sense of hearing in very great perfection, and is the most shy and wary of all the deer species; and on this account the art of moose-hunting is looked upon as the greatest of an Indian's acquirements, particularly by the Crees, who take to themselves the credit of being able to instruct the hunters of every other tribe. The skill of a moose-hunter is most tried in the early part of the winter; for during the summer the moose, as well as other animals, are so much tormented by musquitoes, that they become regardless of the approach of man. In the winter the hunter tracks the moose by its foot-marks in the snow, and it is necessary that he should keep constantly to leeward of the chase, and make his advances with the utmost caution, for the rustling of a withered leaf, or the cracking of a rotten twig, is sufficient to alarm the watchful beast. The difficulty of approach is increased by a habit which the moose deer has of making daily a sharp turn in its route, and choosing a place of repose so near some part of its path, that it can hear the least noise made by one that attempts to track it. To avoid this, the judicious hunter, instead of walking in the animal's footsteps, forms his judgment, from the appearance of the country, of the direction it is likely to have taken, and makes a circuit to leeward, until he again finds the track. This manœuvre is repeated, until he discovers, by the softness of the snow in the foot-marks and other signs, that he is very near the chase. He then disencumbers himself of every thing that might embarrass his motions, and makes his approach in the most cautious manner. If he gets close to the animal's lair, without being seen, it is usual for him to break a small twig, which, alarming the moose, it instantly starts up; but, not fully aware of the danger, squats on its hams, and voids its urine, preparatory to setting off. In this posture it presents the fairest mark, and the hunter's shot seldom fails to take effect in a mortal part. In the rutting season the bucks lay aside their timidity, and attack every animal that comes in their way, and even conquer their fear of man himself. The hunters then bring them within gun-shot, by scraping on the blade-bone of a deer, and by whistling, which, deceiving the male, he blindly hastens to the spot, to assail his supposed rival. If the hunter fails in giving it a mortal wound as it approaches, he shelters himself from its fury behind a tree ; and I have heard of several instances in which the enraged animal has completely stripped the bark from the trunk of a large tree, by striking with its fore-feet. In the spring time, when the snow is very deep, the hunters frequently run down the moose on snow-shoes. An instance is recorded in the narrative of Captain Franklin's second journey, where three hunters pursued a moose-deer for four successive days, until the footsteps of the chace were marked with blood, although they had not yet got a view of it. At this period of the pursuit the principal hunter had the misfortune to sprain his ankle, and the two others were tired out ; but one of them, having rested for twelve hours, set out again, and succeeded in killing the animal, after a further pursuit of two days' continuance. Notwithstanding the lengthened chase which the moose can sustain, when pursued on the snow, Hearne remarks that it is both tender-footed and short-winded ; and that, were it found in a country free from underwood, and dry under foot, it would become an easy prey to horsemen and dogs. The same author informs us, that in the summer moose-deer are often killed in the water by the Indians, who have the fortune to surprise them while they are crossing rivers or lakes, and that at such times they are the most inoffensive of animals, never making any resistance. " The young ones, in particular," says he, " are so simple, that I remember to have seen an Indian paddle his canoe up to one of them, and take it by the poll, without experiencing the least opposition ; the poor, harmless animal seeming, at the same time, as contented alongside the canoe, as if swimming by the side of its dam, and looking up in our faces with the same fearless innocence that a house-lamb would, making use of its fore-foot almost every instant, to clear its eyes of mosquitoes, which at that time were remarkably numerous. The moose is the easiest to tame and domesticate of any of the deer kind."

With respect to the food of the moose, the same traveller says, " Their legs are so long, and their necks so short, that they cannot graze on the level ground like other animals, but are obliged to browze on the tops of large plants and the leaves of trees in the summer, and in winter they always feed on the tops of willows and the small branches of the birch tree, on which account they are never found during that season but in such places as can afford them a plentiful supply of their favourite food ; and although they have no fore-teeth in the upper jaw, yet I have often seen willows and small birch trees cropped by them in the same manner as if they had been cut by a gardener's shears, though some of them were not smaller than a common pipe-stem*; they seem particularly partial to red willows" (*cornus alba.*)   To the eastward of the Rocky Mountains the evergreen leaves of the *gualtheria shallon* form, according to Lewis and Clark, a favourite part of the food of the moose-deer.

The flesh of the moose is more relished by the Indians and residents in the fur countries than that of any other animal, and principally, I believe, on account of its soft fat.   It bears a greater resemblance in its flavour to beef than to venison. "The flesh of the moose," says Hearne, "is very good, though the grain is but coarse, and it is much tougher than any other kind of venison.   The nose is most excellent, as is also the tongue, though by no means so fat and delicate as that of the common deer (rein-deer.)   The fat of the intestines is hard like suet ; but all the external fat is soft like that of a breast of mutton, and, when put into a bladder, is as fine as marrow.   In this they differ from all the other species of deer, of which the external fat is as hard as that of the kidnies."

The moose acquires a large size, particularly the males, which, I have been informed, occasionally attain a weight of eleven or twelve hundred pounds.   Moose dung is in form of oval, brown pellets.   Their skins, when properly dressed, make a soft, thick, pliable leather, excellently adapted for moccasins, or other articles of winter clothing.   The Dog-ribs excel in the art of dressing the skins, which is done in the following manner.   They are first scraped to an equal thickness throughout, and the hair taken off by a scraper, made of the shin-bone of a deer, split longitudinally ; they are then repeatedly moistened and rubbed, after being smeared with the brains of the animal, until they acquire a soft, spongy feel ; and lastly, they are suspended over a fire, made of rotten wood, until they are well impregnated with the smoke.   This last-mentioned process imparts a peculiar odour to the leather, and has the effect of preventing it from becoming so hard, after being wet, as it would otherwise do.

The DESCRIPTION of the moose, by Major Smith, being the fullest and most correct I have met with, I have quoted almost the whole of it.

" This animal is the largest of the genus, being higher at the shoulders than the horse ; its horns weigh sometimes near fifty pounds † : accordingly, to bear this heavy weight, its neck is short and strong, taking away much of the elegance of proportion so generally predominant in the deer ; but when it is asserted that the elk wants beauty or majesty, the opinion can

---

* The wooden pipe-stems used in Hudson's Bay are about the thickness of the little finger.

† Hearne says, that the horns of the moose sometimes exceed 60lbs., and have a harder texture than any other deer-horns to be found in the fur countries.

be entertained by those who have seen the female only, the young, or the mere stuffed specimen: for us who have had the opportunity of viewing the animal in all the glory of his full grown horns, amid the scenery of his own wilderness, no animal could appear more majestic or more imposing. It is, however, the aggregate of his appearance which produces this effect; for when the proportions of its structure are considered in detail, they certainly will seem destitute of that harmony of parts which in the imagination produces the feeling of beauty. The head, measuring above two feet in length, is narrow and clumsily shaped, by the swelling upon the upper part of the nose and nostrils; the eye is proportionably small and sunk; the ears long, hairy, and asinine; the neck and withers are surmounted by a heavy mane, and the throat furnished with long coarse hair, and in younger specimens encumbered with a pendulous gland: these give altogether an uncouth character to this part of the animal. Its body, however, is round, compact, and short; the tail not more than four inches long, and the legs, though very long, are remarkably clean and firm; this length of limbs, and the overhanging lips, have caused the ancients to fancy that it grazed walking backwards. The hair of the animal is coarse and angular, breaking if bent.

" Its movements are rather heavy, and the shoulders being higher than the croup, it does not gallop, but shuffles or ambles along, its joints cracking at every step, with a sound heard to some distance. Increasing its speed, the hind-feet straddle to avoid treading on its fore-heels, tossing the head and shoulders like a horse about to break from a trot to a gallop. It does not leap, but steps without effort over a fallen tree, a gate, or a split fence. During its progress it holds the nose up, so as to lay the horns horizontally back. This attitude prevents it seeing the ground distinctly; and as the weight is carried very high upon its elevated legs, it is said sometimes to trip by treading on its fore-heels, or otherwise, and occasionally to give itself a heavy fall. It is probably owing to this occurrence that the elk was believed by the ancients to have frequent attacks of epilepsy, and to be obliged to smell its hoof before it could recover; hence the Teutonic name of Elend (miserable), and the reputation, especially of the fore-hoofs, as a specific against the disease."

[73.]   2. CERVUS TARANDUS. (Linn.)   *The Rein-Deer,*
                                          or Caribou.

GENUS Cervus. LINN. *Sectio,* Rangiferini.
Caribou ou Asne sauvage. SAGARD-THEODAT, *Canada,* p. 751. An. 1636. LA HONTAN, t. i. p. 77. An. 1703.
    CHARLEVOIX, *Nouv. France,* t. v. p. 190.
Rein-deer, or Rain deer. DRAGE, *Voy.,* vol. i. p. 25. DOBBS, *Huds. Bay,* pp. 19, 22. PENNANT, *Arctic Zool.,*
    vol. i. p. 22. CARTWRIGHT, *Labrador,* pp. 91, 112, 133. FRANKLIN, *First Journey,* &c., pp. 240, 245.
    GODMAN, *Nat. Hist.,* vol. ii. p. 283.
Common deer. HEARNE, *Journ.,* p. 195, 200. PARRY's and LYON's *Narratives,* passim.
Cervus tarandus. SABINE, *Suppl. Parry's First Voy.,* p. cxc. RICHARDSON, *App. Parry's Second Voy.,* p. 326.
    ROSS, *Parry's Third Voy.* HARLAN, *Fauna,* p. 232.
Carrè-bœuf, or Caribou. FRENCH CANADIANS.
Attehk. CREE INDIANS. Etthin. CHEPEWYANS and other NORTHERN INDIANS.
Tooktoo. ESQUIMAUX. Tukta. GREENLANDERS (Fabricius.)

The rein-deer inhabits the arctic islands of Spitzbergen and the northern
extremity of the old continent; its range, according to Baron Cuvier, never
having extended to the southward of the Baltic. It has long been domesticated,
and its manners are well ascertained, and have been carefully described by able
naturalists. It varies in size according to the district in which it is fed; the
breed which the Laplanders train to the sledge being of small stature when compared with the large kind reared in the north of Asia by the Tungusians, who ride
upon them. The rein-deer or caribou of North America are much less perfectly
known. They have indeed so great a general resemblance in appearance and
manners to the Lapland deer, that they have been always considered to be the
same species, without the fact having ever been completely established. Pennant
states that the rein-deer are most numerous in the countries surrounding
Hudson's Bay, and that their most southern residence is the northern parts of
Canada *. They exist in Greenland, Labrador, and Newfoundland, but are not
known in Iceland. They extend most probably completely across the northern
parts of the American continent, and are mentioned both by Pennant and
Langsdorff, as inhabitants of the coast opposite to the Fox or Aleutian islands.
They do not appear, however, to extend so far to the southward on the Pacific
coast as they do in Labrador, and on the shores of Hudson's Bay. Some parts

* Dr. Harlan informs us, that the rein-deer extend as far south as the district of Maine, but without quoting his
authority. Charlevoix says, that it *is* so unusual for them to come so far south as Quebec, that he knew of only one
individual having wandered thither.—The one he alludes to, on being chased, precipitated itself from Cape Diamond,
and, swimming across the St. Lawrence, was killed by some Indians, who were encamped on Point Levi.

of New Caledonia seem to be altogether destitute of them. According to Pennant, they are not found on the islands that lie between Asia and America, but are numerous in Kamtschatka. The Koreki, a nation bordering on the latter country, are said to keep immense herds of rein-deer, some rich individuals possessing to the enormous extent of ten or twenty thousand. The limits assigned by writers to the rein-deer in America are liable to some uncertainty, because the term of caribou, by which they are generally known, has, particularly in Canada, been applied to very distinct species of deer *. Be this, however, as it may, there are two well-marked and permanent varieties of caribou that inhabit the fur-countries, one of them confined to the woody and more southern districts, and the other retiring to the woods only in the winter, but passing the summer on the coast of the arctic sea, or on the Barren Grounds, so often mentioned in this work. The early French writers on Canada, and Jeremie, Ellis, Dobbs, Umfreville, and others, who have given an account of that part of the Hudson Bay Company's possessions which lie to the southward of Churchill River, treat of the woodland variety only. Hearne's descriptions of the rein-deer, on the other hand, relate principally to the Barren Ground kind, with which he was thoroughly acquainted; and it is of this variety that specimens have been brought home by the late arctic expeditions. Neither variety has as yet been properly compared with the European or Asiatic races of rein-deer, and the distinguishing characters, if any exist, are still unknown. Major Smith, indeed, observes that " a probable distinction, by which some, if not all the varieties of caribou may be distinguished from the rein-deer of the old continent, is, that their horns are always shorter, less concave, more robust, the palm narrower, and with fewer processes than those of the former." I have had but little opportunity of ascertaining how far these remarks apply to the woodland variety of caribou, but I can with confidence say, after having seen many thousands of the Barren Ground kind, that the horns of the old males are as much if not more palmated than any antlers of the European rein-deer to be found in the British museums. The annexed cuts were made from drawings by Captain Back, of the antlers of two old buck caribou, killed on the Barren Grounds in the neighbourhood of Fort Enterprise. It is to be recollected, however, that the antlers of the rein-deer assume an almost infinite number of forms, no two individuals having them alike.

* Thus, Mr. Henry, when he mentions Caribou that weigh 400lbs., must have some other species of deer in view.

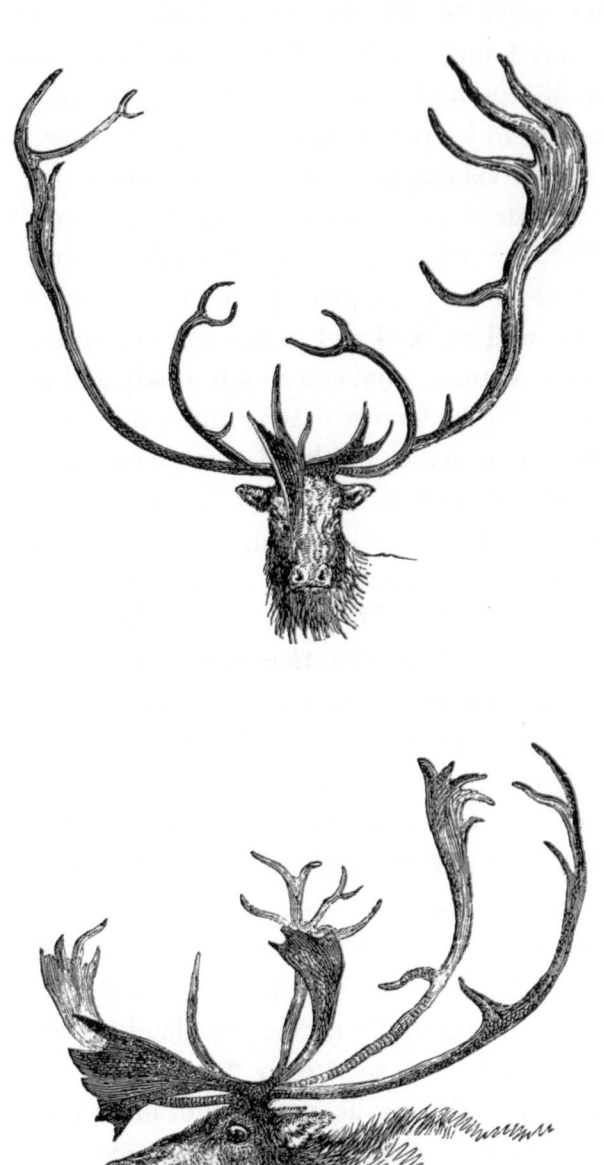

Cervus tarandus, var. α., arctica. *Barren Ground Caribou.*

Common Deer. Hearne, *Journey,* p. 195–200.
Bedsee-awseh. Copper Indians and Dog-ribs.
Bedsee-choh. (*Male*) Iidem. Tsootai. (*Female*) Iidem.
Tampeh. (*Female, with a fawn*) Iidem.
Took-too. Esquimaux (took-took, *dual;* took-toot, *plural.*)
Tukta. Greenlanders. (Pangnek, *male;* kollowak, *female;* norak, *young.*—Fabricius.)

This variety of rein-deer is of small stature, and weighs so little, that I have seen a Canadian voyager throw a full-grown doe on his shoulders, and carry it as an English butcher would a sheep. The bucks are of larger size, and weigh, exclusive of the offal, when in good condition, from 90 to 130 lbs. The old males have in general the largest and most palmated horns, while the young ones and the females have them less branched and more cylindrical and pointed; but this is not uniformly the case, and the variety of forms assumed by the horns of the Caribou is indeed so great, that it is difficult to comprehend them all in a general description. Some have the branches and extremities broadly palmated, and set round with finger-like points; others have them cylindrical, and even tapering, without any palmated portion whatever. The majority of adult males have a brow antler, in form of a broad vertical plate running down betwixt the eyes and hanging over the nose. In some, this plate springs from the right horn, in others from the left; in many there is a plate from each side, and in a considerable number it is altogether wanting: the plate is in general widest at its extremity, and is set with four or five points which are sometimes recurved. The main stem of the horns also exhibits an endless variety in its thickness, altitude, and curvature. During the growth of the horns they are covered with a hairy skin, which is soft and velvety to the touch, and in an early stage their interior consists of a substance which has the flavour of marrow, and resembles it much in appearance, but has a finer consistence, and is furnished with more conspicuous blood-vessels. The horns become indurated as they increase in size, and when they have attained their full growth, their velvety covering shrivels and peels off in ragged filaments. This takes place in the males in September, previous to the commencement of the rutting season, and by the end of November most of the old bucks have shed their horns. The young males retain theirs much longer, and the females do not

2 I

lose their horns until they are about to drop their young in the month of **May**. Hearne observes, that the Barren-Ground Caribou bears horns twice the size of those of the woodland variety, notwithstanding that the latter is a much larger animal.

In the month of July the Caribou sheds its winter covering, and acquires a short, smooth coat of hair, of a colour composed of clove-brown, mingled with deep reddish and yellowish browns ; the under surface of the neck, the belly, and the inner sides of the extremities, remaining white in all seasons. The hair at first is fine and flexible, but as it lengthens it increases gradually in diameter at its roots, becoming at the same time white, soft, compressible, and brittle, like the hair of the moose-deer. In the course of the winter the thickness of the hairs at their roots becomes so great that they are exceedingly close, and no longer lie down smoothly, but stand erect, and they are then so soft and tender below, that the flexible, coloured points are easily rubbed off, and the fur appears white, especially on the flanks. This occurs in a smaller degree on the back ; and on the under parts the hair, although it acquires length, remains more flexible and slender at its roots, and is, consequently, not so subject to break. Towards the spring, when the deer are tormented by the larvæ of the gad-fly making their way through the skin, they rub themselves against stones and rocks, until all the coloured tops of the hair are worn off, and their fur appears to be entirely of a soiled white colour.

The closeness of the hair of the Caribou, and the lightness of its skin, when properly dressed, renders it the most appropriate article for winter clothing in the high latitudes. The skins of the young deer make the best dresses, and they should be killed for that purpose in the months of August or September, as after the latter date the hair becomes too long and brittle. The prime parts of eight or ten deer-skins make a complete suit of clothing for a grown person, which is so impervious to the cold, that, with the addition of a blanket of the same material, any one, so clothed, may bivouack on the snow with safety, and even with comfort, in the most intense cold of an Arctic winter's night. The hoofs of this variety of rein-deer are very large, and spread greatly ; and the posterior or accessory ones make a loud clattering noise when the animal runs. The forms of the latter are almost always visible in its foot-marks, unless the ground be so hard that even the principal hoofs make little impression.

The Barren-Ground Caribou, which resort to the coast of the Arctic Sea, in summer, retire in winter to the woods lying between the sixty-third and the sixty-sixth degree of latitude, where they feed on the *usneæ, alectoriæ,* and other lichens, which hang from the trees, and on the long grass of the swamps. About the end of April, when the partial melting of the snow has softened the *cetrariæ, cornicularice,* and *cenomyces,* which clothe the barren-grounds like a carpet, they make short excursions from the woods, but return to them when the weather is frosty. In May the females proceed towards the sea-coast, and towards the end of June the males are in full march in the same direction. At that period the power of the sun has dried up the lichens on the barren-grounds, and the Caribou frequent the moist pastures which cover the bottoms of the narrow vallies on the coasts and islands of the Arctic Sea, where they graze on the sprouting carices, and on the withered grass or hay of the preceding year, which is at that period still standing, and retaining part of its sap. Their spring journey is performed partly on the snow, and partly, after the snow has disappeared, on the ice covering the rivers and lakes, which have, in general, a northerly direction. Soon after their arrival on the coast the females drop their young ; they commence their return to the south in September, and reach the vicinity of the woods towards the end of October, where they are joined by the males. This journey takes place after the snow has fallen, and they scrape it away with their feet to procure the lichens, which are then tender and pulpy, being preserved moist and unfrozen by the heat still remaining in the earth. Except in the rutting season, the bulk of the males and females live separately : the former retire deeper into the woods in the winter, whilst herds of the pregnant does stay on the skirts of the Barren Grounds, and proceed to the coast very early in spring. Captain Parry saw deer on Melville Peninsula as late as the 23d of September, and the females, with their fawns, made their first appearance on the 22d of April. The males in general do not go so far north as the females. On the coast of Hudson's Bay the Barren-Ground Caribou migrate further south than those on the Coppermine or Mackenzie Rivers ; but none of them go to the southward of Churchill.

The lichens, on which the Caribou principally feed whilst on the Barren-Grounds, are the *cornicularia tristis, divergens,* and *ochrileuca,* the *cetraria nivalis, cucullata,* and *islandica,* and the *cenomyce rangiferina.* When in condition, there is a layer of fat deposited on the back and rump of the males to the depth of two or three inches or more, immediately under the skin, which is termed *depouillè* by the Canadian voyagers ; and as an article of Indian trade, it is often of more value than all the remainder of the carcass. The *depouillè* is thickest at the commencement of the rutting season ; it then becomes of a red colour, and acquires a high flavour, and soon afterwards disappears. The females at that period are lean ; but in the course of the winter they acquire a small *depouillè,* which is exhausted soon after they drop their young. The flesh of the caribou is very tender, and its flavour when in season is, in my opinion, superior to that of the finest English venison ; but when the animal is lean it is very insipid, the difference being greater between well-fed and lean caribou than any one can conceive who has not had an opportunity of judging. The lean meat fills the stomach but never satisfies the appetite, and scarcely serves to recruit the strength when exhausted by labour. The flesh of the moose-deer and buffalo, on the other hand, is tough when lean, but is never so utterly tasteless and devoid of nourishment as that of a caribou in poor condition. The Chepewyans, the Copper Indians, the Dog ribs and Hare Indians of Great Bear Lake, would be totally unable to inhabit their barren lands were it not for the immense herds of this deer that exist there. Of the caribou horns they form their fish-spears and hooks; and previous to the introduction of European iron, ice-chisels and various other utensils were likewise made of them. The hide dressed with the fur is, as has been already mentioned, excellent for winter clothing, and supplies the place of both blanket and feather-bed to the inhabitants of the Arctic wilds. When subjected to the process described in the article on the moose-deer, it forms a soft and pliable leather, adapted for mocassins and summer clothing, or when sixty or seventy skins are sewed together, they make a tent sufficient for the residence of a large family. The shin-bone of the deer, split so as to present a sharp edge, is the knife that is used to remove the hair in the process of making the leather. The undressed hide, after the hair is taken off, is cut into thongs of various thickness, which are twisted into deer-snares, bow-strings, net-lines, and in fact supply all the purposes of rope. The finer thongs are used in the manufacture of fishing nets or in working snow-shoes; while the tendons of the dorsal muscles are split into fine and excellent sewing thread.

Besides these and many other uses to which the Indians appropriate different parts of the caribou in their domestic economy, the animal is no less useful in the way of food. The hunter breaks the leg bones of a recently-slaughtered deer, and while the marrow is still warm devours it with much relish. The kidneys and part of the intestines, particularly the thin folds of the third stomach or many-plies, are likewise occasionally eaten when raw, and the summits of the antlers, as long as they are soft, are also delicacies in a raw state. The colon or large gut is inverted, so as to preserve its fatty appendages, and is, when either roasted or boiled, one of the richest and most savoury morsels the country affords, either to the native or white resident. The remainder of the intestines, after being cleaned, are hung in the smoke for a few days and then broiled. The stomach and its contents, termed by the Esquimaux *nerrooks,* and by the Greenlanders *nerrokak* or *nerriookak,* are also eaten, and it would appear that the lichens and other vegetable matters on which the caribou feeds are more easily digested by the human stomach when they have are mixed with the salivary and gastric juices of a ruminating animal. Many of the Indians and Canadian voyagers prefer this savoury mixture after it has undergone a degree of fermentation, or lain to season, as they term it, for a few days. The blood, if mixed in proper proportion with a strong decoction of fat meat, forms, after some nicety in the cooking, a rich soup, which is very palatable and highly nutritious, but very difficult of digestion. When all the soft parts of the animal are consumed, the bones are pounded small, and a large quantity of marrow is extracted from them by boiling. This is used in making the better kinds of the mixture of dried meat and fat, which is named *pemmican,* and it is also preserved by the young men and females for anointing the hair and greasing the face on dress occasions. The tongue roasted, when fresh or when half dried, is a delicious morsel. When it is necessary to preserve the caribou meat for use at a future period, it is cut into thin slices and dried over the smoke of a slow fire, and then pounded betwixt two stones. This pounded meat is very dry and husky if eaten alone, but when a quantity of the back fat or *depouille* of the deer is added to it, is one of the greatest treats that can be offered to a resident in the fur countries. *Pemmican* is formed by pouring one-third part of melted fat over the pounded meat and incorporating them well together. If kept dry it may be preserved sound for three or four years, and from the quantity of nourishment it contains in small bulk, it is perhaps the best kind of food for those who travel through desert lands. *Thueehawgan* is a mixture of pounded deer's meat and dried fish or fish-roe, which is eaten raw, or when made into soup, by throwing a handful of it into boiling water.

The caribou travel in herds, varying in number from eight or ten to two or three hundred, and their daily excursions are generally towards the quarter from whence the wind blows. The Indians kill them with the bow and arrow or gun, take them in snares, or spear them in crossing rivers or lakes. The Esquimaux also take them in traps ingeniously formed of ice or snow. Of all the deer of North America, they are the most easy of approach, and are slaughtered in the greatest numbers. A single family of Indians will sometimes destroy two or three hundred in a few weeks, and in many cases they are killed for the sake of their tongues alone.

The following extract from Captain Lyon's interesting Journal, details some of the Esquimaux methods of killing them. " The rein-deer," says he*,

* *Private Journal,* p. 336.

" visits the polar regions at the latter end of May or the early part of June, and remains until late in September.   On his first arrival, he is thin, and his flesh is tasteless, but the short summer is sufficient to fatten him to two or three inches on the haunches.   When feeding on the level ground, an Esquimaux makes no attempt to approach him, but should a few rocks be near, the wary hunter feels secure of his prey.   Behind one of these he cautiously creeps, and having laid himself very close, with his bow and arrow before him, imitates the bellow of the deer when calling to each other.   Sometimes, for more complete deception, the hunter wears his deer-skin coat and hood so drawn over his head, as to resemble, in a great measure, the unsuspecting animals he is enticing.   Though the bellow proves a considerable attraction, yet if a man has great patience he may do without it, and may be equally certain that his prey will ultimately come to examine him ; the rein-deer being an inquisitive animal, and at the same time so silly, that if he sees any suspicious object which is not actually chasing him, he will gradually, and after many caperings, and forming repeated circles, approach nearer and nearer to it.   The Esquimaux rarely shoot until the creature is within twelve paces, and I have frequently been told of their being killed at a much shorter distance.   It is to be observed that the hunters never appear openly, but employ stratagem for their purpose ; thus, by patience and ingenuity, rendering their rudely-formed bows, and still worse arrows, as effective as the rifles of Europeans.   When two men hunt in company, they sometimes purposely shew themselves to the deer, and when his attention is fully engaged, walk slowly away from him, one before the other.   The deer follows, and when the hunters arrive near a stone, the foremost drops behind it and prepares his bow, while his companion continues walking steadily forward.   This latter, the deer still follows unsuspectingly, and thus passes near the concealed man, who takes a deliberate aim and kills the animal.   When the deer assemble in herds, there are particular passes which they invariably take, and on being driven to them are killed by arrows by the men, while the women, with shouts, drive them to the water. Here they swim with the ease and activity of water-dogs, the people in kayaks chasing and easily spearing them ; the carcasses float, and the hunter then presses forward and kills as many as he finds in his track.   No springs or traps are used in the capture of these animals, as is practised to the southward, in consequence of the total absence of standing wood."   The caribou entirely quit the districts which Captain Lyon visited, in the winter ; but the Esquimaux who inhabit the coast of the Welcome, to the southward of Chesterfield inlet, have an opportunity, by the animals continuing in their country, of shewing their ingenuity in the construction of deer-traps, of their   convenient and elegant building material, compact snow. The sides of the trap are built of slabs of that substance, cut as if for a snow-house ; an inclined plane of snow leads to the entrance of the pit, which is about five feet deep, and of sufficient dimensions to contain two or three large deer. The pit is covered with a large, thin slab of snow, which the animal is enticed to tread upon by a quantity of the lichens on which it feeds being placed conspicuously on an eminence beyond the opening. The exterior of the trap is banked up with snow so as to resemble a natural hillock, and care is taken to render it so steep on all sides but one, that the deer must pass over the mouth of the trap before it can reach the bait. The slab is sufficiently strong to bear the weight of a deer until it has passed its middle, when it revolves on two short axles of wood, precipitates the deer into the trap, and returns to its place again in consequence of the lower end being heavier than the other. Throughout the whole line of coast frequented by the Esquimaux, it is customary to see long lines of stones set on an end, or of turfs piled up at intervals of about twenty yards, for the purpose of leading the caribou to stations where they can be more easily approached. The natives find by experience that the animals, in feeding, imperceptibly take the line of direction of the objects thus placed before them, and the hunter can approach a herd that he sees from a distance, by gradually crawling from stone to stone, and remaining motionless when he sees any of the animals looking towards him. The whole of the Barren-Grounds are intersected by caribou paths, like sheep tracks, which are of service to travellers at times in leading them to convenient crossing places of lakes or rivers.

Hearne gives the following account of the deer pound in use amongst the Chepewyans :—

" When the Indians design to impound deer, they look out for one of the paths in which a number of them have trod, and which is observed to be still frequented by them. When these paths cross a lake, a wide river, or a barren plain, they are found to be much the best for the purpose ; and if the path run through a cluster of woods, capable of affording materials for building the pound, it adds considerably to the commodiousness of the situation. The pound is built by making a strong fence with brushy trees, without observing any degree of regularity, and the work is continued to any extent, according to the pleasure of the builders. I have seen some that were not less than a mile round, and am informed that there are others still more extensive. The door or entrance of the pound is not larger than a common gate, and the inside is so crowded with small counter-hedges as very much to resemble a maze, in every opening of which they set a snare, made with thongs of parchment deer-skins well twisted together, which are amazingly strong. One end of the snare is usually made fast to a growing pole; but if no one of a sufficient size can be found near the place where the snare is set, a loose pole is substituted in its room, which is always of such size and length that a deer cannot drag it far before it gets entangled among the other woods, which are all left standing, except what is found necessary for making the fence, hedges, &c. The pound being thus prepared, a row of small brush-wood is stuck up in the snow on each side of the door or entrance, and these hedge rows are continued along the open part of the lake, river, or plain, where neither stick nor stump besides is to be seen, which makes them the more distinctly observed. These poles or brush-wood are generally placed at the distance of fifteen or twenty yards from each other, and ranged in such a manner as to form two sides of a long acute angle, growing gradually wider in proportion to the distance they extend from the pound, which sometimes is not less than two or three miles, while the deer's path is exactly along the middle, between the two rows of brush-wood.

" Indians employed on this service always pitch their tents on or near to an eminence that affords a commanding prospect of the path leading to the pound; and when they see any deer going that way, men, women, and children walk along the lake or river side under cover of the woods, till they get behind them, then step forth to open view, and proceed towards the pound in form of a crescent. The poor timorous deer finding themselves pursued, and at the same time taking the two rows of brushy poles to be two ranks of people stationed to prevent their passing on either side, run straight forward in the path till they get into the pound. The Indians then close in, and block up the entrance with some brushy trees that have been cut down and lie at hand for that purpose. The deer being thus enclosed, the women and children walk round the pound to prevent them from jumping over or breaking through the fence, while the men are employed spearing such as are entangled in the snares, and shooting with bows and arrows those which remain loose in the pound. This method of hunting, if it deserve the name, is sometimes so successful, that many families subsist by it without having occasion to move their tents above once or twice during the course of a whole winter; and when the spring advances, both the deer and Indians draw out to the eastward, on the ground which is entirely barren, or at least what is called so in these parts, as it neither produces trees nor shrubs of any kind, so that moss and some little grass is all the herbage which is to be found on it."

Captain Franklin observes that " the rein-deer has a quick eye, but the hunter by keeping to leeward of them, and using a little caution, may approach very near ; their apprehensions being much more easily roused by the smell than the sight of any unusual object. Indeed their curiosity often causes them to come close up to and wheel round the hunter, thus affording him a good opportunity of singling out the fattest of the herd, and upon these occasions they become so confused by the shouts and gestures of their enemy, that they run backwards and forwards with great rapidity, but without the power of making their escape. The Copper Indians find by experience that a white dress attracts them most readily, and they often succeed in bringing them within shot, by kneeling and vibrating the gun from side to side, in imitation of the motion of a deer's horns when he is in the act of rubbing his head against a stone. The Dog-rib Indians have a mode of killing these animals, which, though simple, is very successful. It was thus described by Mr. Wentzel, who resided long amongst that people. The hunters go in pairs, the foremost man carrying in one hand the horns and part of the skin of the head of a deer, and in the other a small bundle of twigs, against which he, from time to time, rubs the horns, imitating the gestures peculiar to the animal. His comrade follows treading exactly in his footsteps, and holding the guns of both in a horizontal position, so that the muzzles project under the arms of him who carries the head. Both hunters have a fillet of white skin round their fore-heads, and the foremost has a strip of the same round his wrists. They approach the herd by degrees, raising their legs very slowly but setting them down some-what suddenly, after the manner of a deer, and always taking care to lift their right or left feet simultaneously. If any of the herd leave off feeding to gaze upon this extraordinary phenomenon, it instantly stops, and the head begins to play its part by licking its shoulders and performing other necessary movements. In this way the hunters attain the very centre of the herd without exciting sus-picion, and have leisure to single out the fattest. The hindmost man then pushes forward his comrade's gun, the head is dropt, and they both fire nearly at the same instant. The deer scamper off, the hunters trot after them ; in a short time the poor animals halt to ascertain the cause of their terror, their foes stop at the same moment, and having loaded as they ran, greet the gazers with a second fatal discharge. The consternation of the deer increases, they run to and fro in the utmost confusion, and sometimes a great part of the herd is destroyed within the space of a few hundred yards."

CERVUS TARANDUS, var. β, SYLVESTRIS.   *Woodland Caribou.*

Caribou. THEODAT, LA HONTAN, CHARLEVOIX, &c.
Rein-deer. DRAGE, DOBBS, &c.
Attehk. CREE INDIANS. Tantseeah. COPPER INDIANS.

Of the form of this variety I know little, having seen few of them alive or in an entire state when killed.   It is much larger than the Barren-Ground Caribou, has smaller horns, and even when in good condition is vastly inferior as an article of food.   The proper country of this deer is a stripe of low primitive rocks, well clothed with wood, about one hundred miles wide, and extending at the distance of eighty or a hundred miles from the shores of Hudson's Bay, from Athapescow Lake to Lake Superior.   Contrary to the practice of the Barren-Ground Caribou, the Woodland variety travels to the southward in the spring.   They cross the Nelson and Severn Rivers in immense herds in the month of May, pass the summer on the low, marshy shores of James' Bay, and return to the northward, and at the same time retire more inland in the month of September.   From November to April it is rare to meet with one within ninety or a hundred miles of the coast.   A few deer of this kind frequent the swamps near Cumberland-house in the winter, but it is extremely rare indeed for a stray individual to wander on that parallel so far to the westward as Carlton-house.   Mr. Hutchins mentions that he has seen eighty carcasses of this kind of deer brought into York Factory in one day, and many others were refused, for want of salt to preserve them.   These were killed when in the act of crossing Hayes River, and the natives continued to destroy them, for the sake of the skins, long after they had stored up more meat than they required.   I have been informed by several of the residents at York Factory that the herds are sometimes so large as to require several hours to cross the river in a crowded phalanx.   The rut takes place in the beginning of October, and the doe drops her young in June.   Mr. Hutchins said that several of the fawns have been brought up at the factories, and have become as tame as a pet lamb, but that they all died in the chops of the Channel when attempts were made to carry them to England.   The same gentleman mentions that the buck has a peculiar bag or cist in the lower part of the neck about the bigness of a crown-piece, and filled with fine flaxen hair neatly coiled round to the thickness of an inch.   There is an opening through the skin, near the head, leading to the cist, but Mr. Hutchins does not offer a conjecture as to its uses in the economy of the animal. Camper found a membranous cist in the rein-deer above the thyroid cartilage and opening into the larynx, but I have met with no account of a cist with a duct opening externally like that described by Mr. Hutchins, and, unfortunately, I was not aware of his remarks until the means of ascertaining whether such a sac exists in the Barren-Ground caribou were beyond my reach. Both the Barren-Ground and Woodland caribous are infested by the gadfly.

[75.] 3. CERVUS STRONGYLOCEROS. (Schreber.) *The Wapiti.*

Stag. PENNANT, *Arctic Zool.*, vol. i. p. 27.
Wewaskiss. HEARNE, *Journ.*, p. 360.
Waskeesews, or Red-deer. HUTCHINS, *MSS.*
Red-deer. UMFREVILLE, *Hudson's Bay*, p. 163. An. 1790.
The Elk. LEWIS and CLARK, *Voy.*, vol. ii. p. 167. An. 1816.
American Elk. BEWICK, *Quadr.*, p. 112.
Wapiti. "BARTON, *Med. and Phys. Journ.*, vol. iii. p. 36." WARDEN, *United States*, vol. i. p. 241.
Le Wapiti. F. CUVIER, *Hist. Nat. des Mamm.*, *Livr.* 20 and 28. An. 1820.
The Wapiti (C. Strongyloceros). SMITH, *Griffith's An. Kingd.*, vol. iv. p. 96.
Red-deer. HUDSON'S BAY TRADERS. La Biche. CANADIAN VOYAGERS.
Wawaskeeshoo; also, awaskees and moostoosh. CREES.

This animal does not extend its range further to the north than the 56th or 57th parallel of latitude, nor is it found to the eastward of a line drawn from the south end of Lake Winipeg to the Saskatchewan, in the 103d degree of longitude, and from thence till it strikes the Elk River in the 111th degree. To the South of Lake Winipeg it may perhaps exist further to the eastward. They are pretty numerous amongst the clumps of wood that skirt the plains of the Saskatchewan, where they live in small families of six or seven individuals. They feed on grass, on the young shoots of willows and poplars, and are very fond of the hips of the *rosa blanda*, which forms much of the underwood in the districts which they frequent. Hearne remarks that they are "the most stupid of all the deer kind, and frequently make a shrill whistling and quivering noise, not very unlike the braying of an ass." Mr. Drummond, who saw many of these deer in his journies through the plains of the Saskatchewan, informs me that the *wawaskeesh* does not *bell* like the English red-deer; and M. F. Cuvier describes the cry of the male as being prolonged and acute, and consisting of the successive sounds of the vowels *a, o, u,* (french,) uttered with so much strength as to offend the ear. The cry of the European stag, when compared to it, is dull and base, though not deficient in strength. The velvety covering shrivels and is rubbed off the horns of the Wapiti in the month of October, at the commencement of the rutting season, but the horns themselves do not fall until the months of March or April. Two male Wapitis were found near Edmonton-house, lying dead, with their horns locked into each other, and the moose and rein deer are reported to have occasionally died under similar circumstances. The flesh of the Wapiti is coarse, and is little prized by the natives, principally on account of its fat being hard like suet. It seemed to me to want the juiciness of venison, and to resemble dry but small grained beef. Its hide, when made into leather, after the Indian fashion, is said not to turn hard in drying after being wet, and in that respect to excel moose or rein-deer leather.

The wawaskeesh of the Saskatchewan River was long considered by the Fur Traders as the same with the red-deer or stag of Europe, and its re-semblance to that animal is, indeed, so great, that, as M. F. Cuvier states, their specific differences become apparent only when an opportunity occurs of comparing them with each other, and of attentively studying their manners when they are placed under similar circumstances. Pennant, without having seen a specimen of the true *wawaskeesh,* or, as he writes it, *waskesse,* applies that name to the moose, probably misled by the appellation of " grey moose," which was given to the *wapiti* in contradistinction to the name of " black moose," which was appro-priated to the *Cervus alces.* Its trivial name of " wapiti" has been only recently adopted in scientific works, but is preferable to the appellations either of elk, grey-moose, or red-deer, which have already been the means of confounding it with other species. A number of live specimens were brought from the Missouri to Europe some years ago, and were by several authors described as a new species, and introduced into the catalogues under the name of *Cervus Wapiti.* Several Hudson's-Bay Traders, however, well acquainted with the wawaskeesh, recognised it at once in the wapiti shewn in England ; and my recollections of two recent specimens of the wawaskeesh, which I had an opportunity of examining on the Saskatchewan, induce me to conform without hesitation to their opinion. It is also without doubt the Canada stag of various authors, but, as M. F. Cuvier has observed, the want of a pale mark on the rump in Perrault's figure is sufficient to excite a doubt of its being the *Cervus Canadensis* [*] of that author. Indeed, I do

---

[*] PERRAULT, *Mem. sur les An.,* vol. ii. p. 45.

not think it at all improbable that his figure is that of the *Cervus macrotis*, which may hereafter prove to be an inhabitant of Upper Canada *.

The height of the wapiti at the shoulders is 4½ feet, whilst that of the European stag is more than a foot less. They agree with each other in the form and proportions of their heads and limbs, but they differ in their respective tints of colour, that of the common stag being an uniform blackish-brown, whilst the wapiti has all its superior parts and the lower jaw of a pretty lively yellowish-brown, and a black mark extends from the angle of the mouth along the side of the lower jaw. There is a whitish circle round the eye of the European animal, but in the American one this circle is brown.

The common stag has generally the first antlers turned upwards at their points, whilst in the wapiti these antlers are depressed in the direction of the facial line, and this character appears to be constant. The neck in both species has a deeper tint of colour than the sides of the body; it is blackish-brown in the European stag, and mixed red and black in the American one, with coarse black hairs depending from it like a dewlap; and this colour, which changes to a brown mixed with white from the shoulders to the hips in the former, becomes a clear French-gray in the same parts of the latter. In both, the limbs have a deeper brown colour anteriorly than posteriorly; and both also have a very pale yellowish spot on the buttocks, bounded on the thighs by a black line, and the tail is likewise of this yellowish colour, but it is nearly seven inches long in the European stag, whilst it is scarcely two and a half in the Canadian one. The colours here mentioned are those which exist at the commencement of the autumn.

The hair of the wapiti is of mean length on the shoulders, the back, the flanks, the thighs, and the under part of the head: the sides and the limbs are clothed with shorter hairs; but they are very long on the sides of the head posteriorly and on the neck, particularly beneath, where they form, as has been mentioned above, a kind of dewlap; and there is on the posterior and outer aspect of the hind-leg a brush of tawny hair which surrounds a narrow, long, horny substance. The ears are white interiorly, and clothed with tufted hairs; exteriorly their colour is the same with that of the neighbouring parts. There is a naked triangular space round the lachrymal opening near the inner angle of the orbit. The hoofs of the wapiti are small.

The wapiti, like the common stag, has very large lachrymal or suborbital openings †, a muzzle, upper canine teeth, a soft tongue, coarse brittle hair, with a short wool beneath it, &c.

* The following passage occurs in the History of Canada by Theodat. " Les cerfs qu'ils appellent Sconoton, sont plus communs dans le pays des Neutres, qu'en toutes les autres contrées Huronnes, mais ils sont un peu plus petits que les nostres de deça, et tres legers de pied." The stag that he here speaks of as being smaller than the common one, cannot be the Wapiti. He mentions the Elk and Rein-deer, under their names of Eslan and Caribou, and he refers probably to the *Cervus Virginianus*, by the appellation of Le Dain, which he says is an animal that he knew merely by report as an inhabitant of North America. The country of the Neutres seems to have been on the northern shores of Lake Huron near the River Nattawasaga.

† The Crees probably on this account term the wapiti "stinking head."

[76.] 4. Cervus macrotis. (Say.) *The Black-tailed Deer.*

Jumping deer. Umfreville, *Hudson's Bay*, p. 164.
Black-tailed or Mule deer. Gass, *Journ.*, p. 55.
Black-tailed deer, Mule-deer. Lewis and Clark, vol. i. pp. 91, 92, 106, 152, 239, 264, 328; vol. ii. p. 152 ; vol. iii, p. 27, 125.
Mule-deer. Warden, *United States*, vol. i. p. 245.
" Cervus auritus. Idem, (Ed. Gall.,) t. v. p. 640."
Cerf mulet. Desmarest, *Mamm. in notis*, p. 443.
Le Dain fauve a queue noire. Idem, *loco citato.*
Black-tailed or Mule deer. James, *Long's Exped.*, vol. ii. p. 276.
Cervus macrotis. Say, *Long's Expedit.*, vol. ii. p. 254. (American ed., vol. ii. p. 88.) Sabine, *Franklin's Journ.*, p. 667. Harlan, *Fauna*, p. 243.
Black-tail deer. Godman, *Nat. Hist.*, vol. ii. p. 305.
Great-eared deer (Cervus macrotis). Griffith, *An. Kingd.*, vol. iv. p. 133.
Cervus macrotis. Idem, vol. v. No. 794.

Plate xx.

Lewis and Clark, in various parts of the narrative of their interesting journey, speak of a black-tailed deer, and of a mule-deer, which, on referring to Serjeant Gass' Journal, are found to be the same animal.   Mr. Say thinks that his *Cervus macrotis* is also the same species, and this opinion is confirmed by the observations of Mr. David Douglas.   I have seen no authenticated specimens of *Cervus macrotis,* but the skins of male and female deer killed in the vicinity of the Rocky Mountains, and presented to the Zoological Society, have all the characters ascribed by Mr. Say to his species.   The plains of the Saskatchewan are frequented by only four *Cervi,* two of which, the moose and wapiti, are well ascertained.   The other two have long been termed indiscriminately by the Canadian voyagers " *chevreuil,*" and by the Hudson's-Bay traders " jumping deer."   The Cree Indians call them both in their language *apeesee-mongsoos* (little moose), but when they wish to be more precise they distinguish one as the *atheeneetoo apeesee-mongsoos* (real little moose), and the other as the *kinwaithoo-wayoo apeesee-mongsoos* (long-tailed little moose), or simply as *kinwaithoos* (long-tail).   I used every endeavour whilst residing in that quarter to procure specimens of both kinds, and sent out Indian hunters for the purpose with the promise of a good reward if they succeeded ; but it happened to be a period of scarcity, and although some were killed, the appetites of the hunters proved superior to their love of gain, and they devoured them all even to

CERVUS MACROTIS.

Published by John Murray, January 1829.

the skins. I collected, however, information respecting them, which induces me to refer one to the *Cervus macrotis* of Say, and the other to the *Cervus leucurus* of Douglas, which is the *long-tailed common fallow deer* of Lewis and Clark. They are both said to be of a grey colour, but differ in size, in the length of their tails, and in their gait. The larger one (*C. macrotis*) when roused, makes off by uninterrupted bounds, raising all its feet from the ground at once, and vibrating its black-tipped tail from side to side; while the small one trots a few steps, makes a great bound, and trots again, much like the *Cervus Virginianus*.

If these indications be correct, the *Cervus macrotis* may be said to inhabit the whole extent of the plains of the Missouri, Saskatchewan, and Columbia; and according to Lewis and Clark, it is the only species to be found on the Mountains in the vicinity of the first falls of the Columbia. They are numerous on the Quamash Flats, which border on the Kooskooskee River. Their most northern range is the banks of the Saskatchewan, in about latitude 54°, and they do not come to the eastward of longitude 105° in that parallel. This deer being an inhabitant on the east side of the mountains of a district frequented by immense herds of buffalo, and also by the large moose deer and wapiti, is of small esteem amongst the Indians in that quarter, and has attracted but little attention from the traders: hence, with the exception of a brief notice by Umfreville, it was almost unknown to naturalists until Lewis and Clark's expedition gave some information respecting it.

DESCRIPTION

Of a male specimen, killed in January, 1827, noted as full-grown, but not old, now in the Zoological Museum.

Size.—Height about that of the Woodland-caribou; weight, said to be about two hundred-weight. *Horns*, cylindrical, twice forked, the first fork situated ten inches from the base, the second one six inches from the first. The stem has a direction upwards, outwards, and a little backwards, with a gentle curvature. It is two inches apart from its fellow at its union with the scull, and its lower part is rough, and somewhat knobbed. One of the anterior knobs in this specimen seems to be the rudiment of a snag or branchlet. At the first bifurcation one branch projects directly forwards, the other is nearly erect. Each fork subdivides into two tapering branchlets, nine or ten inches long; one branchlet of each pair curves forwards, and has its point turned a little inwards, the other is rather taller, and more upright. The horns are compressed for a short space at the forks: their total height is twenty inches, their summits, or the most erect of the posterior pair of points, are fifteen inches apart; and the superior points of the anterior pair are twenty inches apart. The horns want the small basal process which exists in Mr. Say's specimen. The lachrymal openings * are large, and situated close beneath the eye. The ears reach to the bifurcation of

* *Larmiers.* FRENCH AUTHORS. *Crumens* (th. *crumena*, a bag.) FLEMING.

the horns. *Colour.*—The nose and face are grayish-white, and a brown mark, originating between the nostrils, is continued behind their naked margins downwards to the lower jaw, to unite with a dark patch that is situated behind the chin; the chin and throat are white; the forehead is of a dull or soiled dark brown, mixed with umber. The neck, back, sides, and hips, are brownish-gray, the hairs clothing those parts being dark brown from their roots to near their tips, where they exhibit a pale yellowish-brown ring, surmounted by a black tip. The black tips are more conspicuous on the spinous processes of the neck, and form there a dark line, which is continued, though less distinctly, down the middle of the back. The colour of the chest is blackish-brown, and a dark line is continued from it down the centre of the belly; the fur on the posterior part of the belly is long and hairy; the anterior part of the belly is fawn-coloured; the posterior part is white, as are likewise the interior surfaces of the thighs. The *tail*, at its junction with the back, has a dark brown mark; the greater part of it is, however, white, with a tinge of brown, and its tip is black. The *legs* are of a mixed yellowish-brown and black colour, anteriorly, and of a very pale brownish-white, posteriorly.

DIMENSIONS.

	Feet.	Inches.		Feet.	Inches.
Length from the tip of the nose to the brow between the horns . .	1	0	Length of the tail (vertebræ) . .	0	6
			„ „ with the hair .	0	9
„ from the brow to the tail . .	4	4	Height at the fore-shoulder . .	2	6

DESCRIPTION
Of a full-grown female, killed April, 1827.

Has no antlers; the colours of the fur are more distinct; the upper parts are gray, with minute specklings; colour of the chest, between the fore-legs, dark hair-brown; between the hind-legs the fur is nearly white. The forehead is of a paler gray than the forehead of the male, and there is a large white patch surrounding the origin of the tail.

DIMENSIONS.

	Feet.	Inches.		Feet.	Inches.
Length of back, from the neck to the tail .	3	0	Length of head and body . . .	5	1
„ neck . . .	1	0	„ tail, with the hair, about .	0	8½
„ head, from nose to between the ears	1	1	„ ears . . .	0	8
			Height, about	2	5

CERVUS MACROTIS, var. β. Columbiana.

Black-tailed fallow deer. LEWIS and CLARK, vol. iii. pp. 26, 125.
Long-tailed deer. GRIFFITH, *Anim. Kingd.*, vol. iv. p. 134.
Cervus macrourus. IDEM, vol. v. No. 795.

Whether this be distinct or not from the *Cervus macrotis,* I am unable to say, having seen no specimens, and knowing it only from the short account by Lewis and Clark. Mr. Griffith, following Warden, has confounded this with the "long-tailed fallow deer" of these travellers, termed also by them "common red-deer with a long tail," to which the name of *macrourus* would have been appropriate, but which does not apply to this, which is said to have "a tail of the same length with that of the common deer."

Lewis and Clark's account of the variety or species is as follows. "The black-tailed fallow deer are peculiar to this coast (mouth of the Columbia), and are a distinct species, partaking equally of the qualities of the mule and the common deer. Their ears are longer than those of the common deer (*C. virginianus* or *leucurus*). The receptacle of the eye more conspicuous, their legs shorter, their bodies thicker and larger. The tail is of the same length with that of the common deer, the hair on the under side white, and on its sides and top of a deep jetty black; the hams resembling in form and colour those of the mule deer, which it likewise resembles in its gait. The black-tailed deer never runs at full speed, but bounds with every foot from the ground at the same time, like the mule-deer. He sometimes inhabits the woodlands, but more often the prairies and open grounds. It may be generally said that he is of a size larger than the common deer, and less than the mule deer. The flesh is seldom fat, and in flavour is far inferior to any other of the species."

[77.]   5. CERVUS LEUCURUS.   (Douglas.)   *Long-tailed Deer.*

Roe-buck.   DOBBS, *Hudson's Bay*, p. 41.   An. 1744.
Fallow or Virginian deer.   COOK, *Third Voy.*, vol. ii. p. 292.   An. 1778.
Long-tailed jumping deer.   UMFREVILLE, *Hudson's Bay*, p. 190.   An. 1790.
Deer, with small horns and a long tail.   GASS, *Journ.*, p. 55.   An. 1808.
Long-tailed red deer.   LEWIS and CLARK, vol. ii. p. 41.
Small deer of the Pacific.   IIDEM, vol. ii. p. 342.
Common red deer.   IIDEM, vol. iii. p. 26.
Common fallow deer, with long tails.   IIDEM, vol. iii. p. 85.
Apeesee-mongsoos.   CREE INDIANS.   Jumping deer.   HUDSON'S BAY TRADERS.
Chevreuil.   CANADIAN VOYAGERS.   Mowitch.   INDIANS west of the ROCKY MOUNTAINS.

This animal, from the general resemblance it has in size, form, and habits, to the *Cervus capreolus* of Europe, has obtained the name of *Chevreuil* from the French Canadians, and of Roebuck from the Scottish highlanders employed by the Hudson's Bay Company. These names occur in the works of several authors who have written on the fur countries, and Umfreville gives a brief, but, as far as it goes, a correct description of it. Lewis and Clark allude to it, as far as I can judge from their short notices under the different appellations quoted above, all of which indicate that they considered it to be a variety of the *Cervus virginianus*, which is named *red* or *fallow-deer* in different parts of the United States. The specific name of *Cervus macrourus* seems to have been intended to designate this deer; but the characters authors * have assigned to it appertain to var. β of the *Cervus macrotis*, having been compiled from Lewis and Clark's short account of their black-tailed fallow-deer. The black tip of the tail of the *Cervus macrotis* or mule-deer, renders it a more conspicuous object than that of the long-tailed deer, and the former is often termed *kinwaithoos* or *long-tail* by the Cree hunters, although the epithet is more appropriate to the latter. I could not, whilst residing on the banks of the Saskatchewan, procure a specimen of this animal, as has been mentioned in the preceding article, but have lately had an opportunity of examining the skin of a female one presented to the Zoological Museum by the Hudson's Bay Company. Mr. David Douglas has given an account of the habits of the species in the Zoological Journal, and I have adopted his specific name of *leucurus*, which is preferable to *macrourus*, because the original descriptions given under the latter name are not applicable to this species.

* WARDEN, *United States*, vol. i. p. 245. GRIFFITH'S *An. Kingd.*, vol. iv. p. 134.; vol. v. p. 316. No. 25.

This, like the preceding species, does not, on the east side of the Rocky Mountains, range further north than latitude 54°, nor is it found in that parallel to the eastward of the 105th degree of longitude. Mr. Douglas informs us that it is " the most common deer of any in the districts adjoining the river Columbia, more especially in the fertile prairies of the Cowalidske and Multnomah Rivers, within one hundred miles of the Pacific Ocean. It is also occasionally met with near the base of the Rocky Mountains on the same side of that ridge. Its favourite haunts are the coppices, composed of *Corylus, Rubus, Rosa,* and *Amelanchier,* on the declivities of the low hills or dry undulating grounds. Its gait is two ambling steps and a bound, exceeding double the distance of the steps, which mode it does not depart from even when closely pursued. In running the tail is erect, wagging from side to side, and from its unusual length is the most remarkable feature about the animal. The voice of the male calling the female, is like the sound produced by blowing in the muzzle of a gun or in a hollow cane. The voice of the female calling the young is *mœ mœ,* pronounced shortly. This is well imitated by the native tribes, with a stem of *Heracleum lanatum,* cut at a joint, leaving six inches of a tube. With this, aided by a head and horns of a full grown buck, which the hunter carries with him as a decoy, and which he moves backwards and forwards among the long grass, alternately feigning the voice with the tube, the unsuspecting animal is attracted within a few yards in the hope of finding its partner, when instantly springing up the hunter plants an arrow in his object. The flesh is excellent when in good order, and remarkably tender and well flavoured." " They go in herds from November to April and May, when the female secretes herself to bring forth. The young are spotted with white until the middle of the first winter, when they change to the same colour as the most aged."

Lewis and Clark say of it—" The common red deer inhabit the Rocky Mountains, in the neighbourhood of the Chopunnish, and about the Columbia, and down the river as low as where the tide-water commences. They do not differ essentially from those of the United States, being the same in shape, size, and appearance. The tail is, however, different, which is of unusual length, far exceeding that of the common deer. Captain Lewis measured one, and found it to be seventeen inches long." In another passage they remark, " the common fallow deer with long tails, though very poor, are better than the black-tailed fallow deer of the coast, from which they differ materially." As these intelligent travellers have remarked, this deer approaches very near to the *Cervus virginianus* in all its characters, and may eventually prove to be only a variety.

DESCRIPTION

Of a female specimen, killed February, 1827, and presented by the Hudson's Bay Company to the Zoological Museum.—Noted as full-grown.

Form, elegant ; limbs, very slender. *Lachrymal opening* apparently only a small fold in the skin close to the eye. Head and back fawn-colour, mixed with black; sides and cheeks paler; ears edged with dusky-brown; chin and throat white; the tail is fawn-coloured, inclining to rusty above, and pure white underneath and at the tip; hoofs, small and neat.

DIMENSIONS.

	Feet.	Inches.			Feet.	Inches.
Length of back	3	0	Length of tail		0	9
,, neck	1	1	,, ,, with fur		1	1
,, head	0	11	Height of the ears		0	5
,, ,, and body	5	0				

This is the smallest deer known in the fur countries, its weight falling short of that of the Barren-ground caribou.

Mr. Douglas brought home the horns of a full grown *male*. They have a close resemblance in form to the horns of the *Cervus virginianus*. The main stem rises at right angles to the facial line, and gives out near its base an erect, thick, conical snag; above this the horn makes a regular curve, nearly in a horizontal direction forwards, outwards, and at the extreme tip, a little inwards towards its fellow; two tapering erect antlers spring at right angles from the horizontal part of the main stem. The distance from the base of the horn to the tip of the snag or first antler is four inches; from the same place to the tip of the second antler, which springs from where the horn takes a horizontal direction, is ten inches, being the whole height of the horns; from the base to the tip of the third antler is also ten inches; and from the same place to the extreme tip of the horn, the distance, owing to the curvature of the main beam, is only eight inches. The lengths of the second and third antlers are respectively four and three inches. The distance from the tip of the first antler to that of its fellow on the other horn, is five inches; between the tips of the second antlers, thirteen inches; between the tips of the third antlers, sixteen inches; and between the extreme tips of each horn, twelve inches. Mr. Douglas describes the colour of the upper parts of the animal in summer as reddish-brown, which changes to a light gray in winter.

I received from Penetanguishene, on Lake Huron, the skin of a very young deer, which is of a dark yellowish-brown colour on the upper parts, interspersed with white round spots from the size of a pea to that of a small marble. On each side of the spine the spots are arranged in a pretty close even row, and on the

PLATE 21.

ANTILOPE FURCIFER.

Published by John Murray, January 1829.

neck there is a continuous white line prolonged from each row of dorsal spots. The spots on the sides of the animal are distributed without order. The belly is white, and between it and the middle of the back the hair is pale yellowish-brown, and this is also the colour of the chest. I cannot ascertain whether it is the young of the present species, or of the *Cervus virginianus*, the tail being mutilated; nor have I been able to discover the true *Cervus virginianus* within the district to which this work refers. It may probably exist, however, on the borders of Lakes Huron and Superior.

[78.] 1. ANTILOPE FURCIFER. (Smith.) *Prong-horned Antilope.*

GENUS. Antilope. PALLAS.
Teuthlalmaçame. HERNANDEZ, *Nov. Hisp.*, p. 324, 325. Fig. 324 ? An. 1651.
Le Squenoton. *Hist. de l'Amerique*, p. 175. An. 1723 ?
Squinaton. DOBBS, *Hudson's Bay*, p. 24. An. 1744 ?
Wild goats, or Matheh-tuckwuck. HUTCHINS, *MSS.*
Apistochickoshish. UMFREVILLE, *Hudson's Bay*, p. 165. An. 1790.
Antilope, cabre, or goat. GASS, *Journ.*, &c., pp. 49, 111.
Antilope. LEWIS and CLARK, *Journ.*, &c., vol. i. pp. 75, 208, 369 ; vol. ii. p. 169.
Antilope Americana. " ORD, *Guthrie's Geogr.* (Philad. ed.,) 1815."
Cervus hamatus. BLAINVILLE, *Nouv. Bull. Societ. Phil.*, 1816, p. 80.
Antilocapra Americana. ORD, *Journ. de Phys.*, 1818.
Cervus bifurcatus. RAFINESQUE.
Antilope furcifer. C. HAMILTON SMITH, *Lin. Trans.*, vol. xiii. Pl. 2. An. 1823. DESMAREST, *Mamm.*, p. 479.
     SABINE, *Franklin's First Journ.*, p. 667. GRIFFITH, *An. Kingd.*, vol. v. p. 323.
Antilope palmata. SMITH, *Opere citato.* DESMAREST, *Mamm.*, p. 478.
The prong-horned antilope. GODMAN, *Nat. Hist.*, vol. ii. p. 321. Cum figura. SMITH, *Griffith's An. Kingd.*,
     vol. iv. p. 170. Cum figura.
Antilocapra Americana. HARLAN, *Fauna*, p. 250.
Apeestat-chœkoos, also, My-attehk (*plur.* My-attekwuck.) CREE INDIANS.
Cabree. CANADIAN VOYAGERS. Goat. FUR TRADERS.

PLATE XXI.

An animal described and figured by Hernandez under the appellation of *Teuthlalmaçame*, has been referred by Major Smith to the same group of antilopes with the subject of this article ; and Messrs. Ord, Harlan, and others, are even of opinion that the figure represents this very species. The figure, however, is so rude, and the description so imperfect*, that the matter must remain doubtful,

* Hernandez's words, as reported by Recchus, are :—" *Teuthtlalmaçame, Temamaçame* ego potius computaverim inter Capreos." " *Teuthtlalmaçame,* que caprarum mediocrium, paulove majori constant magnitudine. Pilo teguntur cano, et qui facilè avellatur, fulvoque sed lateribus, et ventre candentibus, unde Berendos indigeni Hispani vocare solent. Cornua gestant juxta exortum lata, ac in *paucos,* parvosque teretes ac præacutos ramos divisa, et sub eis oculos quorum imaginem exhibemus."

and the inquiry can be interesting only in as far as it regards the change which time may have produced in the geographical distribution of the species, for there can be no great difficulty in ascertaining whether the animal is at present an inhabitant of California or not; and if the credit of having first noticed it be conceded to Hernandez, still it is to Lewis and Clark that naturalists owe their present knowledge of this animal. These intelligent travellers passed through the country where it chiefly abounds, and as it was often an object of chase with them, they had an opportunity of observing its manners; hence to the facts which they have recorded little of importance has been added by subsequent writers. It is a common animal on the plains lying betwixt the Saskatchewan and Missouri rivers; yet, although the English Fur Traders established themselves above the forks of the former river, in the year 1770, and the French Canadians had been in the habit of collecting furs along both rivers many years previous to that date, and must have been well acquainted with this animal, from the circumstances probably of its skin being of no value in trade, and its flesh little prized as food, nothing beyond its name was known to the civilized world until the year 1790, when Umfreville gave a short account of it in his " Present State of Hudson's Bay." The anonymous author of the *Histoire de l'Amerique Septentrionale* has a passage respecting an animal that he terms *Squenoton*, and which has been copied by Dobbs. His description of it is too short to be of any use in determining the species *, but from the habitat which he assigns to it, of the plains to the south west of Lake Winipeg, we know that he must allude either to the *Antilope furcifer,* or to the *Cervus macrotis*. The *Cervus mexicanus,* to which Pennant refers the Squenoton, does not inhabit those plains, and the author mentions the *Cervus leucurus* under the name of Chevreuil. The appellation of Squenoton has not descended to the present day, and this antilope has for a considerable time past been known among the Canadian voyagers by the name of *Cabrèe,* which is probably a Basque corruption of the Spanish word *Cabra* (goat), as it resembles the common goat in colour, and in the erect position of the hair covering the spine of the neck, and forming a kind of mane. The English Fur Traders still call it " the goat."

Lewis and Clark brought a specimen from the Missouri, which was deposited in Peale's Museum at Philadelphia, and described (according to Dr. Godman) by Mr. Ord, under the name of *Antilope Americana,* in an American edition of

---

* His words are :—" Le Squenoton ressemble au Chevreuil (*Cervus virginianus* or *leucurus ;*) il est plus haut, la jambe plus finè, et la tête plus longue et plus pointuè.' The word squenoton is perhaps derived from the Huron language. Theodat uses a nearly similar term (sconoton) to designate the stags which frequent the borders of Lake Huron.—*Vide* p. 753.

Guthrie's Geography, published in 1815 ; and in 1818, the same naturalist published in the *Journal de Physique* an account of a new genus founded upon it, which he termed *Antilocapra*. M. Blainville having inspected a pair of horns of this antilope in the Museum of the College of Surgeons, where they were attached to a board with their points in a wrong direction, published a notice of them in 1816, wherein he named the animal to which they belonged *Cervus hamatus ;* and an account and figure of Lewis and Clark's specimen, taken by Major Smith in 1817, appeared, in 1823, in the 13th volume of the Linnean Transactions. Major Smith considers the horns mentioned by M. Blainville to belong to a distinct species, whose name he has altered to *Antilope palmata*. From this detail it is evident that the specific name of *Americana* is the prior one; but the term *furcifer* having been generally adopted by naturalists, I have retained it here, including under it also the *Antilope palmata,* as I conceive the greater breadth of the horns to be merely the effect of age. The term *Americana* is objectionable as a specific name, where more than one species of the same genus exists in that country, and in reference to the present instance, the animal which will be afterwards described and figured as the *Capra Americana,* is by several eminent naturalists considered to be an antilope ; and if it is to be permanently placed in that genus, a change of name either of it or of the species at present under consideration would be indispensably necessary. The *Antilope furcifer* differs from the true antilopes, in having a snag or branch on its horns, and wanting the crumens or lachrymal openings, and also in being destitute of the posterior or accessory hoofs, there being only two on each foot.

The most northerly range of the prong-horned antilope is latitude 53°, on the banks of the north branch of the Saskatchewan. Some of them remain the whole year on the south branch of that river, but they are merely summer visitors to the north branch. They come every year to the neighbourhood of Carlton-house, when the snow has mostly gone ; soon after their arrival the females drop their young, and they retire to the southwards again in the autumn as soon as the snow begins to fall. Almost every year a small herd linger on a piece of rising ground not far from Carlton-house, until the snow has become too deep on the plains to permit them to travel over them. Few or none of that herd, however, survive until the spring, as they are persecuted by the wolves during the whole winter. They are found in the summer season in the fifty-third parallel of latitude, from longitude 106° to the foot of the Rocky Mountains. According to Lewis and Clark, they also abound on the plains of the Columbia to the west of the mountains, where they form the chief game of the Shoshonees. They frequent open prairies and low hills, interspersed with clumps of wood, but are not met with in the continuously wooded country. Major Smith has fallen into an unaccountable mistake in supposing that the palmated antilope inhabits " the bleak regions near the frozen ocean," and that specimens have been brought from Baffin's Bay. No specimens whatever of this antilope were obtained on any of the expeditions to Baffin's Bay, nor is there any mention made of the animal either in the narrative or Zoological appendices of Captain Ross's or of any of Captain Parry's voyages. If an imaginary line be drawn from the mouth of the Mackenzie, in longitude 135° to the intersection of the 100th degree of longitude, with the 53d parallel of latitude, it cuts off to the eastward a very large portion of the continent, which I am certain is not inhabited by either goat, sheep, or antilope. The only ruminating animals of that rocky but well-watered tract, which, to the south of latitude 60°, is in general woody, and to the north barren, are the moose, caribou, and musk-ox. The last is confined to the northern parts, the moose to the woody districts, and the caribou migrates from one to the other according to the season. The bison is found on the confines of the above-mentioned line, but I believe does not wander far to the eastward of the Slave and Churchill or Missinippi rivers.

The head and horns of a young male, and the entire skin of a very young fawn of this antilope were obtained at Carlton on Captain Franklin's first expedition, and deposited, the former at the College of Surgeons, and the latter at the British Museum. On his last expedition, heads of the adult male and female, together with the entire skin of a male two years old, were brought from the same place. The latter is now in the Zoological Museum, and the accompanying etching by Landseer was made from it, but the horns were added from the adult male head. Very lately the institution just mentioned has received several good specimens from the Hudson's Bay Company.

The prong-horned antilope appears on the banks of the Saskatchewan sometimes a solitary animal, sometimes assembled in herds of ten or twelve. Its sight and sense of smell are acute, and its speed is greater than that of any other inhabitant of the plains, although I have been informed by Mr. Prudens, who has resided forty years in that quarter, that when there is a little snow on the ground it may, with some little management, be run down by a high bred horse. The Indian hunters have no difficulty in bringing an antilope within gun-shot, by various stratagems, such as lying down on their backs and kicking their heels in the air, holding up a white rag, or clothing themselves in a white shirt, and shewing themselves only at intervals *.   By these and similar manœuvres, the curiosity of a herd of antilopes is so much roused that they wheel round the object of their attention, and at length approach near enough to enable the hunter to make sure of his mark.   From this disposition of the prong-horned antilopes, they are more easily killed than any of the deer of the district which they inhabit.   They are, however, objects of little interest to the Indians, who eat their flesh only when the bison, moose or wapiti are not to be procured, and their skins are of no value as an article of trade.   The Mandans on the Missouri are said to capture them in pounds.   The antilopes feed on the grass of the plains during the summer, and, according to Lewis and Clark, they migrate towards the mountains at the commencement of winter, and subsist there during that season on leaves and shrubs. They bring forth one, or more rarely two, young early in June.

DESCRIPTION

Of a male, killed at Carlton, in June 1827.—This individual must have attained a considerable age, as the sagittal and some other sutures of the scull were obliterated.

Dental formula, incisors $\frac{0}{8}$, canines $\frac{0-0}{0-0}$, grinders $\frac{6-6}{6-6} = 32$.

Incisors white ; the two exterior incisors are much smaller than the others, and they are all disposed with their edges tiled slightly over each other, and their points inclining outwards in the segment of a circle, adapted to the reception of the callous pad, which terminates the upper jaw.   The *upper grinders* gradually increase in size from the first to the fourth, which is considerably larger ; the fifth is of equal or, perhaps, greater size than the fourth ; and the sixth is somewhat smaller.   The three posterior ones have each a deep furrow on their inner sides, corresponding to a fold or ridge of enamel on the outer side, so, at first sight, each appears like two teeth.   The furrow is shallow in the third tooth, and does not exist in the two first.   In the *lower jaw*, the posterior grinder is the largest, and is divided into three portions by two deep furrows on its exterior side.   The fourth and fifth have, each, one deep furrow, and the third has two shallow furrows ; these three are nearly equal to each other in size, and are a little smaller than the sixth one.   The first and second lower grinders are much smaller than the others.

In the *scull* there is a considerable depression in the frontal bone, between the anterior parts of the orbits ; the orbits project considerably, and the solid osseous nucleus of the horn is seated on the projecting plate.   The supra-orbitar foramen is situated close to the inner

* " This same curiosity enables the wolves to make them a prey: for sometimes one of them will leave his companions to go and look at the wolves, which, should the antelope be frightened, at first crouch down, repeating the manœuvre, sometimes relieving each other, until they succeed in decoying it within their power, when it is pulled down and devoured."—GODMAN, *Nat. Hist.*, vol. ii. p. 323.

2 M

side of this bony projection. The nasal bones are very slightly arched, and there is no external suborbitar opening, but in the bone there is an oblong and somewhat irregular foramen, an inch and a quarter long, which, in the recent subject, is entirely closed by a tense membrane. The scull is smaller than that of the English domestic sheep; the jaws more slender, and the cavity for containing the brain smaller in proportion.

The *horns* are black, rise directly upwards and outwards, without any inclination either forwards or backwards, and curve sharply in towards each other at their tips. At the base the distance between them is 3½ inches; within 2 inches of the tip, where they begin to curve inwards, the distance between them is 10½ inches, and the tips are 7 inches apart. The horns are much compressed, in a lateral direction, to about half their height, where they give out a thin, triangular, or bracket-shaped process, which projects directly forwards for more than an inch. The surface of the lower half of the horns is striated, and is rough, with small warts and knobs, two or three of which project from a quarter to half an inch. The situation of these larger knobs varies in different specimens. The horns above the flat snag have a shining, striated surface, are nearly round, and taper considerably.

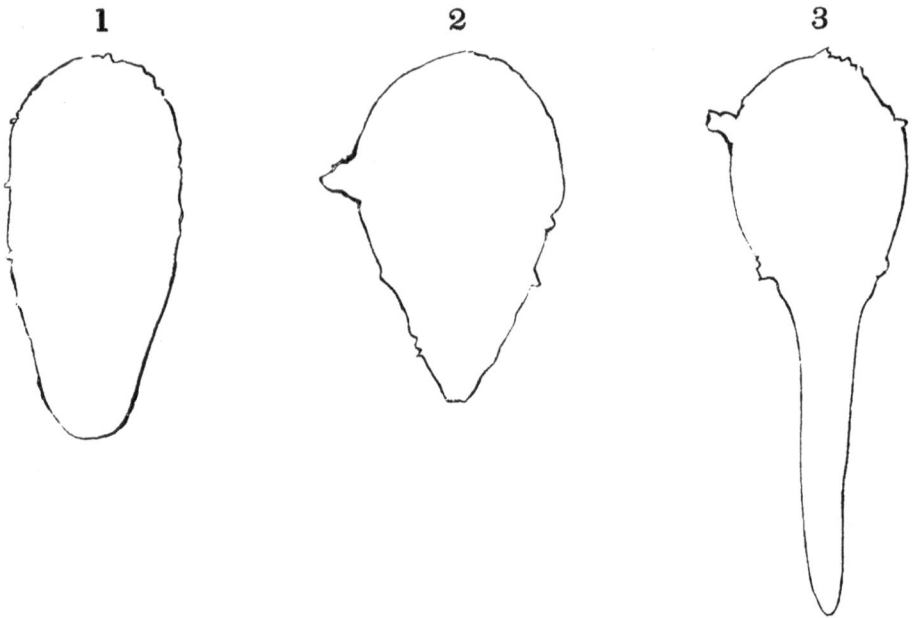

No. 1 is a section of the right horn, at its base; No. 2 a similar section at the commencement of the snag; and No. 3 is a section of the horn and snag;— natural size.

This animal has a graceful form, a slender head, with large eyes, and long and delicate limbs. The *nostrils* turn obliquely upwards from the raphè of the upper lip, and are separated by a small, tumid, triangular, naked space. The naked margins of the *lips* are blackish, but the lips and chin are covered with white hair. The *nose* is nearly straight, or very slightly arched, narrow, and is clothed towards its tip with short hair of a liver-brown colour, which gradually mingles towards the forehead with yellowish-brown hair. The *orbits* have a narrow, blackish-brown margin, and the eye-lashes, composed of a row of stiff, erect hairs, are black. The *cheeks* are covered with short hair, mostly of a wood-brown colour, and the forehead is clothed with longer bushy hair, and presents two white marks, one extending from ear to ear, the other a little anterior to it; the latter mark is slightly tinged with brown. The *ears* are upwards of six inches high, narrow, and have the inner side curving in for half their height; from thence to their acute tips they are flat. They are covered posteriorly by a smooth coat of short hair, of a yellowish-brown colour, mixed with dark umber, the latter colour prevailing near the tip. They are lined interiorly with longer hair of a grayish-white colour. There is a dark blackish-brown spot at the angle of each jaw, which exhales a strong hircine odour, and between this spot and the ear the hair is pale, or nearly white. There are no external indications whatever of a crumen or lachrymal opening. The upper parts of the body are of a clear, yellowish-brown colour, deepening on the ridge of the back into blackish-gray. The hairs are much longer between the ears, and on the back of the neck, where they form an erect mane, of a blackish brown colour on its tips. The sides and thighs are paler than the back, and approach in colour to a clear wood-brown. The under jaw has a very pale yellowish-brown colour, fading to white. The hair is bushy about the angle of the lower jaw, and has a wood-brown colour. This colour forms three belts across the throat, which differ from each other in breadth, and are separated by two patches of pure white. The chest, belly, insides of the thighs, and legs, the tail, and a large patch round it, which includes the rump, and upper part of the buttocks, are pure white. There is a pale yellowish mark at the root of the tail. The tail is 4½ inches long. The *legs* are slender, with long shank-bones; the fur, covering their anterior surfaces, is yellowish-brown. It has only two hoofs, there being no vestige of the posterior supplementary ones.

The hair, which clothes the body, resembles that of a moose or rein-deer in its structure. It is long, round, tapering from the root to the point, waved, and of a soft and brittle texture, particularly towards the root, where it is easily compressed, and does not regain its round form again. Its interior is white and spongy, like the pith of rush. When the hair makes its first appearance in the summer, it forms a smooth coat, and has the ordinary flexibility and appearance of hair; but as it lengthens it acquires the brittle, spongy texture, at its roots, and, increasing at the same time in diameter, it becomes erect, and forms a very close coat. As the spring approaches the fine and flexible points are rubbed off, particularly on the sides, where the hair appears as if it had been clipped. The mane on the hind-head and neck retains its darker points, even when the winter coat is dropping off. The nose, cheeks, part of the lower jaw, ears, and legs, are clothed at all times with short, flexible hairs, which lie smoothly.

DIMENSIONS.

	Feet.	Inches.		Feet.	Inches.
Length from the nose to the root of the tail	4	4	Girth behind the fore-legs	3	0
Height at the fore-shoulder	3	0	„ before the hind-legs	2	10
„ „ haunches	3	0	Length of the tail, with the hair	0	4½

2 M 2

The *females* are stated, by some American writers, to have horns like the males, although smaller; but in gravid, and, therefore, at least nearly full-grown individuals, which I have examined, there was merely a short, obtuse process, of the frontal bone, scarcely to be felt through the fur, and not covered with horn.

The *young*, at birth, are covered on the upper parts with short hair, of a clove-brown colour, more or less hoary. The situation of the mane is marked by a dark line. The tail is yellowish-brown, and the buttocks are pure white. The dark mark on the nose, the one behind the angle of the jaw, and the bands across the throat, exist as in the adult. The legs are of a pure wood-brown colour.

[79.]   1. CAPRA AMERICANA.   *Rocky-Mountain Goat.*

GENUS. Capra. LINN.
Antilope Americana et Rupicapra Americana. BLAINVILLE, *Bull. Soc. Phil.* An. 1816, p. 80.
Ovis montana.   ORD, *Journ. Phil. Acad.*, vol. i. pt. i. p. 8.   An. 1817.
" Mazama sericea. RAFINESQUE-SMALTZ, *Am. Month. Mag.* An. 1817, p. 44."
Rocky-Mountain sheep. JAMESON, *Wernerian Trans.*, vol. iii. p. 306. An. 1821 (read An. 1819.)
Antilope lanigera. SMITH, *Linn. Trans.*, vol. xiii. p. 38. t. 4. An. 1822.

PLATE XXII.

The Rocky Mountain Goat has been supposed to be an inhabitant of California, where it is said to have been discovered by Fathers Piccolo and De Salvatierra, as will be noticed in the article on the Rocky Mountain sheep. Vancouver brought home a mutilated skin which he obtained on the North-west coast of America; and Lieutenant-General Davies presented a specimen to the Linnean Society, of which an account was published by M. de Blainville, in 1816. Mr. Ord, in 1817, described a skin brought home by Lewis and Clark under a new specific name, and a detailed description by Major C. H. Smith, drawn up in the same year from General Davies' specimen, was published in the Linnean Transactions for 1821, under a third name * The animal has been known to the members of the North-west and Hudson's Bay Companies from the first establishment of their trading posts on the banks of the Columbia River, and in New Caledonia, and they have sent several specimens to Europe. One of these being

* *Griffith's An. Kingd.*, vol. iv. p. 286.

CAPRA AMERICANA.

Published by John Murray, January 1829.

presented to the Wernerian Society of Edinburgh, was submitted to a competent judge, who reported that " the wool, which forms the chief covering of the skin, is fully an inch and a half long, and is of the very finest quality. It is unlike the fleece of the common sheep, which contains a variety of different kinds of wool suitable to the fabrication of articles very dissimilar in their nature, and requires much care to distribute them in their proper order. The fleece under consideration is wholly fine. That on the forepart of the skin has all the apparent qualities of wool : on the back part it very much resembles cotton. The whole fleece is much mixed with hairs, and on those parts where the hairs are long and pendant, there is almost no wool." In consequence of this report, a suggestion was made to the Highland Society of the advantage likely to accrue from the introduction into Scotland of an animal bearing so valuable a fleece. The Hudson's Bay Company alone possess the power of effecting this patriotic design. Very lately that Company presented a perfect specimen of the goat to the Zoological Society, from which the accompanying etching was executed by Landseer. From the circumstance of this animal bearing wool, it has been occasionally termed a sheep by the voyagers and even by naturalists; and as it has often gone by the same name with the Rocky Mountain sheep of the following article, some little confusion has crept into the accounts of their habits which have been published from the reports of the traders.

The Rocky Mountain goat inhabits the most lofty peaks of the range, from whence it derives its English appellation, seldom descending so near the low country as the Rocky Mountain sheep does. Mr. Drummond saw no goats on the eastern declivity of the mountains near the sources of the Elk River, where the sheep are numerous, but he learnt from the Indians that they frequent the steepest precipices, and are much more difficult to procure than the sheep. Their manners are said to resemble greatly those of the domestic goat. The exact limits of the range of this animal have not been ascertained, but it probably extends from the 40th to the 64th or 65th degree of latitude. It is common on the elevated part of the Rocky Mountain range that gives origin to four great tributaries to as many different seas, viz. the Mackenzie, the Columbia, the Nelson, and the Missouri Rivers.

The fine wool which the animal produces grows principally on the back and hips, and is intermixed with long coarse hair. Its flesh is hard and dry, and is little esteemed*. The Indians make caps and saddles of its skin. I have followed Dr. Harlan in ranking this quadruped in the genus Capra, and have

---

* Mr. Donald M'Kenzie says, its flesh has a musky flavour.—HARLAN, *Fauna*, p. 258.

adopted M. de Blainville's specific name, though somewhat objectionable, because it is due to the first describer of an animal to retain if possible the appellation he bestows on it, in preference even to a better though later name *.

DESCRIPTION

Of the specimen in the Zoological Museum.

Size of the domestic sheep, and a resemblance exists to the merino breed, in the mode in which the fleece hangs down on the sides. The form of the body and neck is robust, like that of the common goat. *Nose* nearly straight. *Ears* pointed; lined with long hair. The *horns* are awl shaped, sharp pointed, and nearly erect, having but a slight curvature, and inclination backwards. They are marked at the base with rings, which disappear above half way up, and towards the tips they are remarkably even, smooth, and polished; their surface throughout is black and shining.

Colour.—The animal is totally white except the horns, hoofs, lips, and margins of the nostrils.

Fleece.—The body is covered with long straight hair, considerably coarser than the wool of sheep, but softer than that of the common goat. This long hair is abundant on the shoulders, neck, back, and thighs; a considerable tuft of it attached to the chin forms a beard, and there is likewise much of it on the chest and lower part of the throat. The *tail* is short, and though clothed with long hair is almost concealed by that which covers the rump. Under the hair of the body there is a close coat of fine white wool. The hair on the face and legs is short without wool. The *legs* are thick and short; the fetlocks are short, and with the hoofs are perpendicular. The latter are of a black colour, and are deeply grooved on the soles. They resemble those of the common goat. The small posterior hoofs do not touch the ground.

* Mr. Ord's specific name of *montana*, if hereafter adopted, may be inconvenient should this animal be arranged with the antelopes, because an antelope discovered on the banks of the White Nile by M. Rüppell, has been figured under the name of *Antelope montana*.

PLATE 23.

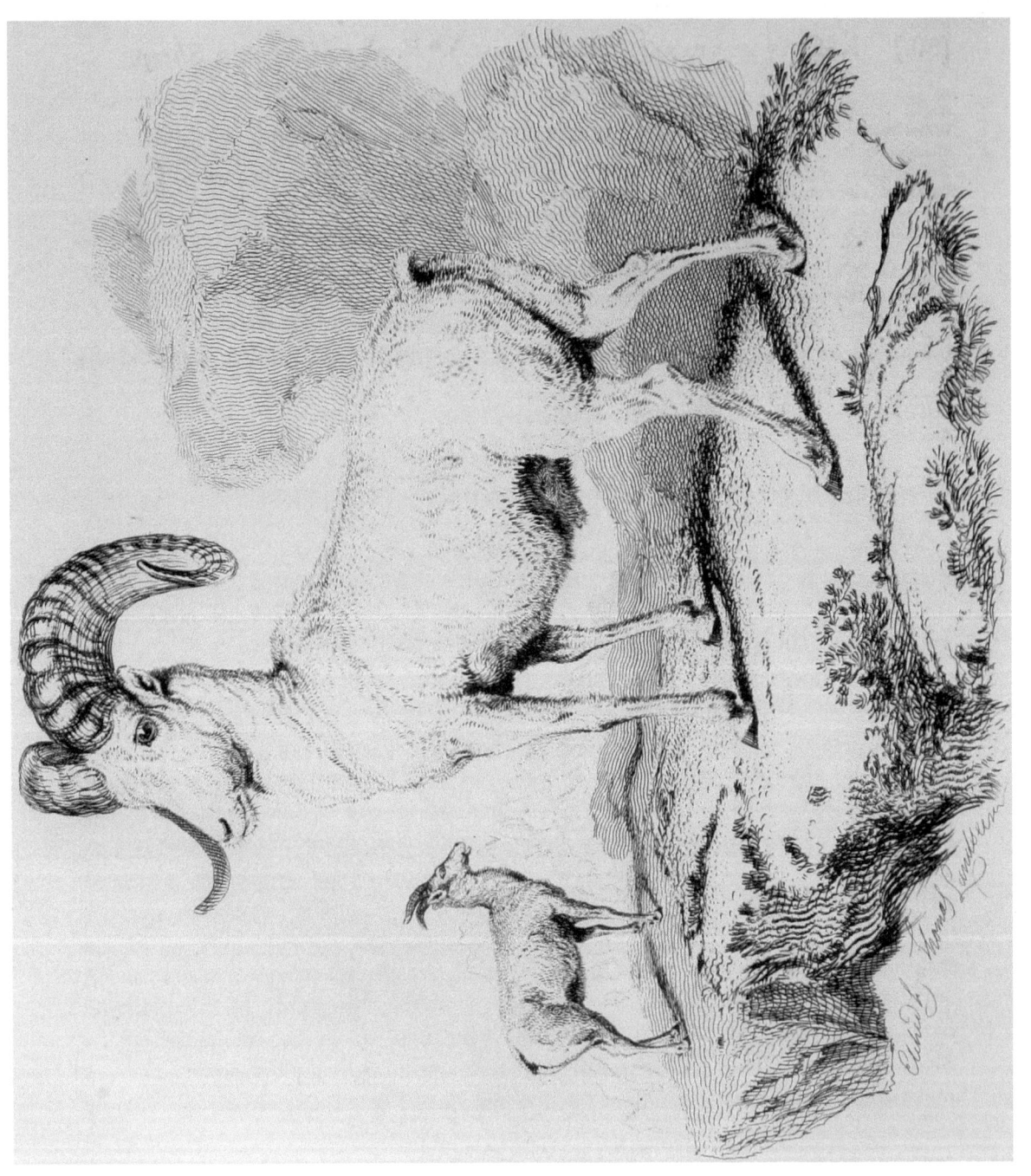

OVIS MONTANA.

Published by John Murray, January 1829.

[80.]   1. Ovis montana.   (Desmarest.)   *Rocky-Mountain Sheep.*

Argali. Cook, *Third Voy.* An. 1778.
White buffalo. Mackenzie, *Voy.*, p. 76. An. 1789. (The horn is mentioned page 208.)
Mountain goat. Umfreville, *Hudson's Bay*, p. 164.
Mountain ram. M'Gillivray, *New York Med. Reposit.*, vol. vi. p. 238, with a figure. An. 1803.
Big-horn. Lewis and Clark, vol. i. p. 144.
Belier sauvage d'Amerique. Geoffroy, *Ann. du Museum*, t. ii. pl. lx.
Ovis montana. " *Encyclop.*, pl.   , suppl. 14. Fig. 4. Schreber, pl. ccxiv D." Richardson, *Werner. Trans*
        vol. iv.   Part 1. p. 22.
Rocky-Mountain sheep. Warden, *United St.*, vol. i. p. 217.
Mouflon d'Amerique. Desmarest, *Mamm.*, p. 487.
Ovis Ammon. Harlan, *Fauna*, p. 259.
The Argali. Godman, *Nat. Hist.*, vol. ii. p. 329.
Ovis Ammon, var.? Pygargus. Griffith, *An. King.*, vol. iv. p. 318, with a figure; and vol. v. p. 359. No. 873.
Cul-blanc et grosse corne. Canadian Voyagers.
My-attehk. Cree Indians.
Ema-kee-kawnow. Pegans, Blood Indians, and Black-feet.
Ahsahta. Mandans.

Plate XXIII.

When Fathers Piccolo and de Salvatierra, in the year 1697, established the first mission in California nearly two centuries after the first discovery of that country, they found, says the former, " two sorts of *deer* that we know nothing of: we call them sheep, because they somewhat resemble ours in make.   The first sort is as large as a calf of one or two years old ; its head is much like that of a stag, and its horns, which are very large, are like those of a ram ; its tail and hair are speckled, and shorter than a stag's, but its hoof is large, round, and cleft as an ox's.   I have eaten of these beasts ; their flesh is very tender and delicious. The other sort of sheep, some of which are white and others black, differ less from ours.   They are larger, and have a great deal more wool, which is very good, and easy to be spun and wrought *."   Hernandez, Clavighiero, and other writers on California, likewise mention these animals, and Vanegas has given a figure of the first-mentioned one, which has, though evidently on insufficient grounds, been considered to be the same with the Siberian Argali, and with the subject of this article ; while the one noticed in the latter part of the quotation has been referred to the species already described under the name of Rocky Mountain goat.   The speckled hair does not agree with any descriptions I have met with of the Rocky Mountain sheep, nor have I heard that black individuals

* *Philos. Trans.* No. 318. p. 232 ; and Jones's *Abridg.*, vol. v. p. 194.

are ever met with in the herds of the Rocky Mountain goat; it is therefore probable that the Californian animals are different from the allied ones, which inhabit the more northern part of the Rocky Mountain ridge. Mr. David Douglas describes Piccolo's sheep under the name of *Ovis Californica* *. Pennant, who considers the Asiatic argali (*O. ammon*) and the Corsican mufro (*O. musimon*) to be varieties of the same species, says, in Arctic Zoology, that "the argali is suspected to be found in California, but not on the best authorities."

On Cook's third voyage he obtained the spoils of an animal on the North-west coast of America, which the editor of the journal takes for granted were those of the argali. They were, doubtless, skins of the Rocky Mountain sheep. Sir Alexander Mackenzie, in his voyage down the great river which bears his name, received an account of the mountain sheep, called by the natives of that district " white buffaloes," and in his subsequent journey across the Rocky Mountains at the sources of the Elk River, he saw some utensils made of their horns, which he not unaptly compares with the horns of the musk-ox. When, in consequence of the important discoveries of that adventurous and intelligent traveller, the English North-west Fur Company were led in the spirit of commercial enterprise to cross the Rocky Mountains twice every year, they became well acquainted with the mountain sheep, and sent several skins of it to Europe, which do not however appear to have fallen into scientific hands, as no account of them was published. The attention of naturalists was drawn to the animal in 1803, by a paper published in the Medical Repository of New York, by Mr. M'Gillivray, who also presented to the New York Museum a specimen procured by him three years previously on the mountains from whence the Elk River takes its origin. This specimen being afterwards sent to M. Geoffroy, he published a description of it with a figure in the *Annales du Museum*. Some years afterwards Lewis and Clark brought male and female specimens to Philadelphia, which have lately been figured by Griffith and Godman. Mr. Drummond shot many in the same district in which Mr. M'Gillivray procured his one; and two specimens obtained on the mountains which skirt the south branch of the Mackenzie, were presented to me by Mr. Macpherson, and are now in the Museum of the Zoological Society. They are male and female, and are the subjects from which the accompanying spirited and very accurate etchings were made by Landseer.

The Rocky Mountain sheep inhabit the lofty chain of mountains from whence they derive their name from its northern termination in latitude 68° to about latitude 40°, and most likely still further south. They also frequent the elevated

* *Zool. Journ.* April, 1829.

and craggy ridges with which the country between the great mountain range and the Pacific is intersected; but they do not appear to have advanced further to the eastward than the declivity of the Rocky Mountains, nor are they found in any of the hilly tracts nearer to Hudson's Bay. They collect in flocks consisting of from three to thirty, the young rams and the females herding together during the winter and spring, while the old rams form separate flocks, except during the month of December, which is their rutting season. The ewes bring forth in June or July, and then retire with their lambs to the most inaccessible heights. Mr. Drummond informs me that in the retired parts of the mountains, where the hunters had seldom penetrated, he found no difficulty in approaching the Rocky Mountain sheep, which there exhibited the simplicity of character so remarkable in the domestic species; but that where they had been often fired at, they were exceedingly wild, alarmed their companions on the approach of danger by a hissing noise, and scaled the rocks with a speed and agility that baffled pursuit. He lost several that he had mortally wounded, by their retiring to die amongst the secluded precipices. Their favourite feeding places are grassy knolls, skirted by craggy rocks, to which they can retreat when pursued by dogs or wolves. They are accustomed to pay daily visits to certain caves in the mountains, that are encrusted with a saline efflorescence, of which they are fond. These caves are situated in slaty rocks, and it was in them alone that Mr. Drummond found the *Weissia macrocarpa* growing. The same gentleman mentions that the horns of the old rams attain a size so enormous, and curve so much forwards and downwards, that they effectually prevent the animal from feeding on level ground. The flesh of the Rocky Mountain sheep is stated by Mr. Drummond and others, who have fed on it, to be quite delicious when it is in season, far superior to that of any of the deer species which frequent the same quarter, and even exceeding in flavour the finest English mutton. The Kamschatdales, in like manner, esteem the flesh of the argali as food fit for the gods.

The missionaries who first discovered the Rocky Mountain sheep, or the nearly allied species of California, described it correctly as possessing the hair of the stag and the horns of the ram; and M. Geoffroy has also briefly characterised it as having the head of a sheep with the body of a deer. Several naturalists of eminence have considered it as forming but one species with the argali; and Baron Cuvier supposes that it may have crossed Behring's Straits on the ice. It resembles the argali, indeed, perfectly in its manners, in the form of its body, and in the nature and colours of its hairy coat; but it seems to be a larger animal, and to present a constant difference in the form of curvature of its horns. Whether it

2 N

may eventually prove to be a distinct species, or merely a permanent variety, no inconvenience can result from describing it, for the present, under a name already appropriated to it. In the Museum of the Linnean Society there is a good specimen of a sheep from the mountains of Nepaul, which does not appear to differ from the Siberian argali, but seems very distinct from the American one.

DESCRIPTION.

Size, much greater than the largest-sized varieties of the domestic sheep. It is bigger than the argali.

The *horns* of the *male* are very large, arise a short way above the eyes, and occupy almost the whole space between the ears, but do not touch each other at their bases. They curve first backwards, then downwards, forwards and upwards, until they form a complete turn, during the whole course of which they recede from the side of the head in a spiral manner. They diminish in size rapidly towards their points, which are turned upwards. At their bases, and for a considerable portion of their length, they are three-sided, the anterior or upper side being, as it were, thickened, and projecting obtusely at its union with the two others. This side is marked by transverse furrows, which are less deep the further they are from the scull; and towards the tips the horns are rounded, and but obscurely wrinkled. The furrows extend to the other two sides of the horn, but are there less distinct. The intervals of the furrows swell out, or are rounded.

The *horns* of the *female* are much smaller, and nearly erect, having but a slight curvature, and an inclination backwards and outwards.

The *ears* are of a moderate size; the facial line straight, and the general *form* of the animal rather elegant, being intermediate betwixt that of the sheep and the stag. *Tail* very short. The *hair* is like that of the rein-deer, being, on its first growth in the autumn, short, fine, and flexible; but, as the winter advances, becoming much coarser, dry, and brittle, though at the same time it feels soft to the touch. In the latter season the hair is so close at its roots, that it is necessarily erect. The legs are covered with shorter hairs.

The head, buttocks, and posterior part of the belly, are white; the rest of the body and the neck are of a pale umber or dusky wood-brown colour. A deeper and more shining brown prevails on the anterior aspect of the legs. The tail is dark-brown, and a narrow brown line, extending from its base, runs up betwixt the white buttocks, to unite with the brown colour of the back. The colours reside in the ends of the hair, and as these are rubbed off during the progress of the winter, the tints become paler. The old rams are almost totally white in the spring. This is the case with the male specimen in our plate. The female, in the back ground, presented the colours mentioned above.

DIMENSIONS

Of an Old Rocky-Mountain Ram, killed on the south branch of the Mackenzie, and now in the Zoological Museum.

	Feet.	Inches.		Feet.	Inches.
Length of head and body . . .	6	0	Circumference of a horn at its base .	1	1
Height at fore shoulder . .	3	5	Distance from the tip of one horn to the tip		
Length of tail	0	2	of the other	2	3
„ one horn, measured along its curvature . . .	2	10			

[81.]   1. Ovibos moschatus.   (Blainville.)   *Musk-Ox.*

Genus. Ovibos. Blainville.
Le bœuf musquè. M. Jeremie, *Voy. au Nord*, t. iii. p. 314. Charlevoix, *Nouv. France*, t. v. p. 194.
Musk-Ox. Drage, *Voy.*, vol. ii. p. 260. Dobbs, *Hudson's Bay*, pp. 19, 25. Ellis, *Voy.*, p. 232. Pennant,
    Quadr., vol. i. p. 31. *Arctic Zool.*, vol. i. p. 9. Hearne, *Journey*, p. 137. Parry, *First Voy.*, p. 257,
    with a plate. *Second Voy*, pp. 497, 503, 512. British Museum. *Specimen on the stair.*
Bos moschatus. Gmelin, *Syst.* Sabine (Capt.), *Parry's First Voy., Suppl.*, p. clxxxix. Sabine (Mr.), *Frank-*
    lin's Journ., p. 668. Richardson, *Parry's Second Voy., Appendix*, p. 331.
Matheh-moostoos (ugly bison.) Cree Indians.
Adgiddah-yawseh (little bison.) Chepewyans and Copper Indians.
Oomingmak. Esquimaux.

We are indebted for the first notice of this animal to M. Jeremie, who brought some of its wool to France, and had some stockings made of it, which were said to have been more beautiful than silk. The earlier English voyagers also give us some information respecting it, but Pennant has the merit of being the first who systematically arranged and described it, from the skin of a specimen sent home by Hearne, the celebrated traveller. From its want of a naked muzzle and some other peculiarities, M. Blainville has placed it in a genus, intermediate, as its name denotes, between the sheep and the ox; but it is remarkable amongst the American animals for never having had more than one specific appellation, whilst other animals, of much less interest, have been honoured with a long list of synonyms.

The musk-ox inhabits the Barren-lands of America lying to the northward of the sixtieth parallel of latitude*. Hearne mentions that he once saw tracks of one within a few miles of Fort Churchill, in latitude 59°; and in his first journey to the north, he saw many in about latitude 61°. I have been informed that they do not now come so far to the southward even on the Hudson's Bay shore; and further to the westward they are rarely seen in any numbers lower than latitude 67°, although from portions of their sculls and horns, which are occasionally found near the northern borders of Great Slave Lake, it is probable that they ranged at no very distant period over the whole country lying betwixt that great sheet of water and the Polar sea. I have not heard of their having been seen on the banks of Mackenzie's River to the southward of Great Bear Lake, nor do they

* Pennant says, that they are found in the lands of the *Cris* or *Cristinaux*, and *Assinibouls;* this is, however, a mistake. The lands he alludes to, are the plains which extend from the Red River of Lake Winipeg to the Saskatchewan, and are inhabited by the *Crees* or *Natheh-wye-withinyoo*, and the *Asseeneepools* or *Asseeneepoytuck ;* but it is the bison that frequents that district, and not the musk-ox. He is correct, however, in saying that they are hunted by the *Attimospiquay*, who are the *Dog-ribs* of Great Bear Lake.

come to the southwestern end of that lake, although they exist in numbers on its north-eastern arm. They range over the islands which lie to the north of the American continent as far as Melville island, in latitude 75°, but they do not, like the rein-deer, extend to Greenland, Spitzbergen, or Lapland. From Indian information, we learn that to the westward of the Rocky Mountains, which skirt the Mackenzie, there is an extensive tract of barren country, which is also inhabited by the musk-ox and rein-deer. It is to the Russian traders that we must look for information on this head; but it is probable that, owing to the greater mildness of the climate to the westward of the Rocky Mountains, the musk-ox, which affects a cold barren district, where grass is replaced by lichens, does not range so far to the southward on the Pacific coast as it does on the shores of Hudson's Bay. It is not known in New Caledonia, nor on the banks of the Columbia, nor is it found on the Rocky Mountain ridge at the usual crossing-places near the sources of the Peace, Elk, and Saskatchewan rivers. It is, therefore, fair to conclude that the animal described by Fathers Marco de Niça and Gomara, as an inhabitant of New Mexico, and which Pennant refers to the musk-ox, is of a different species*. The musk-ox has not crossed over to the Asiatic shore, and does not exist in Siberia, although fossil sculls have been found there of a species nearly allied, which has been enumerated in the systematic works under the name of *Ovibos Pallantis*. The appearance of musk-oxen on Melville Island, in the month of May, as ascertained on Captain Parry's first voyage, is interesting, not merely as a part of their natural history, but as giving us reason to infer that a chain of islands lies between Melville Island and Cape Lyon, or that Wollaston and Banks' Lands form one large island, over which the migrations of the animals must have been performed.

The districts inhabited by the musk-ox are the proper lands of the Esquimaux†; and neither the Northern Indians nor the Crees have an original name for it, both terming it bison, with an additional epithet. The country frequented by the musk-ox is mostly rocky and destitute of wood, except on the banks of the larger rivers, which are generally more or less thickly clothed with spruce trees. Their food is similar to that of the Caribou, grass at one season and lichens at another; and the contents of its paunch are eaten by the natives with the same relish that they devour the " nerrooks " of the caribou. The dung of the musk-ox takes the form of round pellets, differing from those of the caribou only in their greater

---

* The Mexican animal is said to be a sheep, as large as a horse, with long hair, short tails, and enormous horns. The only horse which the musk-ox can be said to resemble in size, is a Shetland pony.

† The northern Indian appellation for an Esquimaux, is " Inhabitant of the Barren Land."

size. When this animal is fat, its flesh is well tasted, and resembles that of the caribou, but has a coarser grain. The flesh of the bulls is high flavoured, and both bulls and cows, when lean, smell strongly of musk, their flesh at the same time being very dark and tough, and certainly far inferior to that of any other ruminating animal existing in North America. The carcase of a musk-ox weighs, exclusive of the offal, about three hundred weight, or nearly three times as much as a Barren-ground caribou, and twice as much as one of the Woodland caribou.

Notwithstanding the shortness of the legs of the musk-ox, it runs fast, and it climbs hills and rocks with great ease. One, pursued on the banks of the Copper-mine, scaled a lofty sand cliff, having so great a declivity that we were obliged to crawl on hands and knees to follow it. Its foot-marks are very similar to those of the caribou, but are rather longer and narrower. These oxen assemble in herds of from twenty to thirty, rut about the end of August and beginning of September, and bring forth one calf about the latter end of May or beginning of June. Hearne, from the circumstance of few bulls being seen, supposes that they kill each other in their contests for the cows. If the hunters keep themselves concealed when they fire upon a herd of musk-oxen, the poor animals mistake the noise for thunder, and, forming themselves into a group, crowd nearer and nearer together as their companions fall around them; but should they discover their enemies by sight or by their sense of smell, which is very acute, the whole herd seek for safety by instant flight. The bulls, however, are very irascible, and, particularly when wounded, will often attack the hunter and endanger his life, unless he possesses both activity and presence of mind. The Esquimaux, who are well accustomed to the pursuit of this animal, sometimes turn its irritable disposition to good account; for an expert hunter having provoked a bull to attack him, wheels round it more quickly than it can turn, and by repeated stabs in the belly, puts an end to its life. The wool of the musk-ox resembles that of the bison, but is perhaps finer, and would no doubt be highly useful in the arts if it could be procured in sufficient quantity.

DESCRIPTION.

In *size*, the full-grown musk-ox nearly equals the small breed of cattle peculiar to the highland districts of Scotland. The *horns* are very broad at their origin, covering the brow and whole crown of the head, and touching each other for their entire breadth from before backwards. As each horn rises from its flatly convex base, it becomes round and tapering, and curves directly downwards between the eye and the ear, until it reaches the angle of the mouth, when it turns upwards in the segment of a circle to above the level of the eye. The horn, for half its length, is dull white, and is rough, with small longitudinal splinters of unequal length ; beyond that, it is smooth and shining, and near the point it becomes black. The *head* is large and broad, and the nose is very obtuse. The *nostrils* are oblong openings that incline towards each other from above downwards. Their inner margins, for the breadth of three lines, are naked, and they are united for half an inch at their base. There is no other vestige of a muzzle ; the rest of the end of the nose, the middle part of the upper lip, and the greater part of the lower lip and chin, are covered with a close coat of short, white hairs. There is no furrow on the upper lip. The remainder of the head, anterior to the horns, is covered with very dark umber-brown hair, which is long and bushy towards the root of the nose, giving an arched appearance to the facial line, which does not exist in the scull. There is also a quantity of long, straight hair, on the lateral margins of the mouth, and sides of the lower jaw. The *eyes* are moderately large, and the hair immediately round them is shorter and paler than on the rest of the cheeks. The *ears* are short, and being similar in colour to the long hair on the hind head, are not very conspicuous.

The general colour of the hair of the body is brown. On the neck and between the shoulders it is long, matted, and somewhat curled, and has more or less of a grizzled hue, being of a dull brown colour, fading on the tips into brownish-white. The bushy appearance of the hair on these parts causes the animal to seem humped. The hair on the back and hips is also long, but lies smoothly ; and on the shoulders, sides, and thighs, it is so long as to hang down below the middle of the leg. On the centre of the back it has a soiled brownish-white colour, forming a mark, which is aptly termed by Captain Parry the saddle ; the colour of the hips is a darker brown ; and on the thighs, sides, and belly, its surface is nearly blackish-brown. The hair on the throat and chest is very straight and long, and, together with the long hair on the lower jaw, hangs down like a beard and dewlap. The *tail* is so short, as to be concealed by the fur of the hips. There is a large quantity of fine brownish-ash-coloured wool or down among the hair covering the body.

The *legs* are short and thick, and are clothed with short, dull, brownish-white hair, unmixed with wool. The hoofs are narrower, and not longer than those of the caribou, but are so similar in form, that it requires the experience of a practised hunter to know the difference of the impressions they leave in the soil.

The *cow* differs from the bull, in having smaller horns, whose bases, instead of touching, are separated by a hairy space, and in the hair on the throat and chest being shorter. It is also considerably smaller than the bull.

[82.]   1. Bos Americanus.   (Gmelin.)   *American Bison.*

Genus. Bos.
Taurus Mexicanus. Hernandez, *Mex.*, p. 587. Fig. (malè.) An. 1651.
Taureau sauvage. Hennepin, *Nouv. Decouv.*, vol. i., p. 186. Fig. (malè.) An. 1699.
The buffalo. Lawson, *Carol.*, p. 115. Fig. Catesby, *Carol.*, Append. xxxvii. tab. 20. Harmon, *Journey,* p. 415.
        Franklin, *First Journey*, p. 113 ; with a plate of a buffalo-pound, p. 110.
Bison. Ray, *Synop. Quadr.*, p. 71. Pennant, *Arctic Zool.*, vol. i. p. 1. Long, *Exped.*, vol. iii. p. 68.
Bos Americanns. Gmelin, *Syst.* Sabine, *Franklin's Journey*, p. 668.
American wild ox, or bison. Warden, *United States*, vol. i. p. 248.
Peecheek. Algonquins. (Nochena peecheek. Bison cow.)
Moostoosh. Crees. Adgiddah. Chepewyans.
Buffalo. Hudson's Bay Traders. Le bœuf. Canadian Voyagers.
It is unknown to the Esquimaux of the Polar Sea.

At the period when Europeans began to form settlements in North America, this animal was occasionally met with on the Atlantic coast ; but even then it appears to have been rare to the eastward of the Apalachian mountains, for Lawson has thought it to be a fact worth recording, that two were killed in one season on Cape Fear River.   As early as the first discovery of Canada, it was unknown in that country, and no mention of it whatever occurs in the *Voyages du Sieur de Champlain Xaintongeois*, nor in the *Nova Francia* of De Monts, who obtained the first monopoly of the fur trade.   Theodat, whose history of Canada was published in 1636, merely says that he was informed that bulls existed in the remote western countries *.   Warden mentions that at no very distant date herds of them existed in the western parts of Pennsylvania ; and that as late as the year 1766, they were pretty numerous in Kentucky ; but they have gradually retired before the white population, and are now, he says, rarely seen to the south of the Ohio, or on the east side of the Mississippi.   They still exist, however, in vast numbers in Louisiana, roaming in countless herds over the prairies that are watered by the Arkansa, Platte, Missouri, and upper branches of the Saskatchewan and Peace rivers.   Great Slave Lake, in latitude 60°, was at one time the northern boundary of their range ; but of late years, according to the testimony of the natives, they have taken possession of the flat limestone district of Slave Point, on the north side of that lake, and have wandered to the vicinity of Great Marten Lake, in latitude 63° or 64°.   As far as I have been able to ascertain, the

* His words are,—"On tient qu'il y a des dains en quelques contrées ; mais pour des *buffles*, le P. Joseph m'a asseuré en avoir veu des peaux entieres entres les mains d'un sauvage de pays fort esloigné ; je n'en ay point veu, mais je croy ce bon Pere."—Sagard-Theodat, *Histoire du Canada*, p. 756.

limestone and sandstone formations, lying between the great Rocky Mountain
ridge and the lower eastern chain of primitive rocks, are the only districts in
the fur countries that are frequented by the bison. In these comparatively level
tracts there is much prairie land, on which they find good grass in the summer;
and also many marshes overgrown with bulrushes and carices, which supply them
with winter food. Salt springs and lakes also abound on the confines of the
limestone, and there are several well known salt-licks where bison are sure to be
found at all seasons of the year. They do not frequent any of the districts formed
of primitive rocks, and the limits of their range to the eastward within the
Hudson Bay Company's territories may be nearly correctly marked on the map by
a line commencing in longitude 97° on the Red River which flows into the south-end
of Lake Winipeg, crossing the Saskatchewan to the westward of Basquiau hill,
and running from thence by the Athapescow to the east end of Great Slave Lake.
Their migrations to the westward were formerly limited by the Rocky Mountain
range, and they are still unknown in New Caledonia and on the shores of the
Pacific to the north of the Columbia river; but of late years they have found out
a passage across the mountains near the sources of the Saskatchewan, and their
numbers to the westward are said to be annually increasing. In 1806, when
Lewis and Clark crossed the mountains at the head of the Missouri, bison skins
were an important article of traffic between the inhabitants on the east side and
the natives to the westward. Further to the southward, in New Mexico and California, the bison appears to be numerous on both sides of the Rocky Mountain
chain. One of the earliest accounts we have of the animal is by Hernandez; and
Recchus' edition of his observations, or rather commentary upon them, is illustrated
by an engraving which seems to have been made from a rude sketch of the bison,
altered by the European artist to a closer resemblance with the European ox.
Hennepin, in the narrative of his discovery of Louisiana, and his travels through
that country between the years 1669 and 1682, gives a very good description of the
bison, together with a figure, which is apparently a copy of that of Recchus. It
does not appear to have excited much attention in Europe until lately, when several
specimens, having been imported into England, were exhibited under the attractive
title of *Bonasus*, which, though described by the ancients, was asserted to have been
lost to the moderns until recognised in the American animal. The American
bison has in fact much resemblance to the *aurochs* of the Germans (*Bos urus*,
Boddært) identified by Cuvier with the *bonassus* of Aristotle, the *bison* of
Pausanias and Pliny, and the *urus* of Cæsar, and which, down to the reign of

Charlemagne, was not rare in Germany, but is now nearly confined to the hilly country lying between the Caspian and Black Sea.

The bison wander constantly from place to place, either from being disturbed by hunters or in quest of food. They are much attracted by the soft tender grass, which springs up after a fire has spread over the prairie. In winter they scrape away the snow with their feet to reach the grass. The bulls and cows live in separate herds for the greatest part of the year, but at all seasons one or two old bulls generally accompany a large herd of cows. In the rutting season, the males fight against each other with great fury, and at that period it is very dangerous to approach them. The bison is, however, in general, a shy animal, and takes to flight instantly on winding an enemy, which the acuteness of its sense of smell enables it to do from a great distance. They are less wary when they are assembled together in numbers, and will then often blindly follow their leaders, regardless of, or trampling down the hunters posted in their way. It is dangerous for the hunter to shew himself after having wounded one, for it will pursue him, and although its gait may appear heavy and awkward, it will have no great difficulty in overtaking the fleetest runner. While I resided at Carlton-house, an accident of this kind occurred. Mr. Finnan M'Donald, one of the Hudson's Bay Company's clerks, was descending the Saskatchewan in a boat, and one evening having pitched his tent for the night, he went out in the dusk to look for game. It had become nearly dark when he fired at a bison-bull, which was galloping over a small eminence, and as he was hastening forward to see if his shot had taken effect, the wounded beast made a rush at him. He had the presence of mind to seize the animal by the long hair on its forehead as it struck him on the side with its horn, and being a remarkably tall and powerful man, a struggle ensued, which continued until his wrist was severely sprained, and his arm was rendered powerless; he then fell, and after receiving two or three blows became senseless. Shortly afterwards he was found by his companions lying bathed in blood, being gored in several places, and the bison was couched beside him, apparently waiting to renew the attack had he shewn any signs of life. Mr. M'Donald recovered from the immediate effects of the injuries he received, but died a few months afterwards. Many other instances might be mentioned of the tenaciousness with which this animal pursues its revenge, and I have been told of a hunter having been detained for many hours in a tree by an old bull which had taken its post below to watch him. When it contends with a dog, it strikes violently with its fore-feet, and in that way proves more than

2 o

a match for an English bull-dog. The favourite Indian method of killing the bison is by riding up to the fattest of the herd on horseback, and shooting it with an arrow. When a large party of hunters are engaged in this way on an extensive plain, the spectacle is very imposing, and the young men have many opportunities of displaying their skill and agility. The horses appear to enjoy the sport as much as their riders, and are very active in eluding the shock of the animal, should it turn on its pursuer. The most generally-practised plan, however, of shooting the bison, is by crawling towards them from to leeward, and in favourable places great numbers are taken in pounds. When the bison runs, it leans very much to, first, one side for a short space of time, and then to the other, and so on alternately.

The flesh of a bison in good condition is very juicy and well flavoured, much resembling that of well-fed beef. The tongue is reckoned a delicacy, and may be cured so as to surpass in flavour the tongue of an English cow. The hump of flesh covering the long spinous processes of the first dorsal vertebræ is much esteemed. It is named *bos* by the Canadian voyagers, and *wig* by the Orkney men in the service of the Hudson's Bay Company. The wig has a fine grain, and when salted and cut transversely it is almost as rich and tender as the tongue. The fine wool which clothes the bison renders its skin when properly dressed an excellent blanket; and they are valued so highly, that a good one sells for three or four pounds in Canada, where they are used as wrappers by those who travel over the snow in carioles. The wool has been manufactured in England into a remarkably fine and beautiful cloth, and in the colony of Osnaboyna, on the Red River, a warm and durable coarse cloth is formed of it. Much of the pemmican used by the voyagers attached to the fur companies is made of bison meat procured at their posts on the Red River and Saskatchewan. One bison cow in good condition furnishes dried meat and fat enough to make a bag of pemmican weighing 90 lbs. The bison which frequent the woody parts of the country form smaller herds than those which roam over the plains, but are said to be individually of a greater size.

DESCRIPTION.

The most remarkable features of the male bison are the enormous size of its head, which is carried low; the great conical hump between the shoulders, small piercing eyes, short black horns, and the great profusion of shaggy hair on the fore-parts, which all contribute to give to the animal a wild and malicious aspect. The hind quarters, being clothed with shorter wool, appear disproportionably weak. The forehead is broad, the cheeks rather lank, and the face, which tapers from the eye towards the nose, has a form approaching to triangular. The horns are small, tapering, and acute, set far apart, and nearly erect, being only slightly curved at the base and tips, which point outwards. The bulk of the hump consists, exclusive of a deposit of fat, which varies much in quantity, of the muscles that are attached to the unusually long spinous processes of the posterior cervical and anterior dorsal vertebræ, and are destined for the support of the head. The hair on the forehead, hump, fore-quarters, under-jaw, and throat, is very long and shaggy, and is mixed with much wool. The back, hind-quarters, belly, and legs, are clothed with shorter, and in many parts curly hair. The general colour of the hair, when the animal has acquired its new coat at the close of the summer, is between dark umber and liver-brown, and it has then a considerable lustre. As the hair lengthens during the winter, its tips become paler, and before it is shed in the summer much of it is of a pale, dull, yellowish-brown colour. The *tail* is covered with short fur for the greatest part of its length, but is terminated by a tuft of long, straight, coarse hair, of a blackish-brown colour. The *legs* are strong.

The bison, when full grown, is said to attain at times a weight of two thousand pounds; but 12 or 14 cwt. is generally considered a full size in the fur countries. Its length, exclusive of the tail, is about eight feet and a half; its height, at the fore-quarters, upwards of six feet, and the length of its tail is twenty inches.

The *cow* has a smaller head, and shorter hair on the fore-parts, than the bull.

There is a variety of the buffalo which is nearly white.

ADDENDUM.

[83.]   2. CONDYLURA MACROURA. (Harlan.) *Thick-Tailed Star-Nose.*

Condylura macroura. HARLAN, *Fauna*, p. 39.

PLATE XXIV.

Since the greater part of the preceding sheets were printed off, Mr. David Douglas has presented me with a specimen of this remarkable animal, procured by him on the banks of the Columbia. Dr. Harlan has described an individual which is preserved in the Philadelphia Museum, but its native locality is not mentioned. Mr. Douglas's specimen possesses all the characters ascribed by Dr. Harlan to his; and I have, therefore, considered it to belong to the same species. I received no information respecting its habits.

DESCRIPTION.

The *head* is remarkably large; the *body* is thick and short, and becomes narrower towards the tail, and the hind legs are consequently nearer to each other than the fore-ones. The *nose* is rather thick, and projects beyond the mouth; it is naked towards its end, is marked with a furrow above, and terminates in a flat surface, which is surrounded by seventeen *cartilaginous processes,* with two more anterior ones situated above the nostrils, and a pair of forked ones immediately below the nostrils. The surfaces of these processes are minutely granulated. Some white *whiskers* spring from the side of the nose, and reach about half the length of the head. There are others not so long on the upper and under lips.

The *fur* on the body is very soft and fine, and has considerable lustre. It is longer than the fur of the other two known species. Its colour on the dorsal aspect is dark umber-brown, approaching to blackish-brown. On the belly it is pale liver-brown. When the fur is blown aside, it exhibits a shining blackish-gray colour towards its roots. It is longer on the hind-head and neck than on the belly.

The *tail* is narrow at its origin, but it suddenly swells to an inch and a half in circumference; it then tapers gradually until it ends in a fine point, formed by a pencil of hairs about half an inch long. It is round, or very slightly compressed, and is covered with scales about as large as those on the feet, and with short, tapering, acute hairs, which do not conceal the scales. The hairs covering the upper surface of the tail are nearly black; those beneath are of a browner hue.

The *extremities* are shaped almost precisely like those of the *condylura longicaudata.* Only the palms and toes of the *fore-feet* project beyond the body. The palms are nearly circular, and

CONDYLURA MACROURA.

Published by John Murray. January 1829.

are protected by a granulated skin, like shagreen. The sides of the feet are furnished with long, white hairs, which curve in over the palms. The (5) toes are very short, equal to each other in length, and, together with the back of the hands, are covered with hexagonal scales. The *fore-claws* are white, nearly straight, broadly linear, and acute, convex above, and flat underneath. The palms turn obliquely outwards, which causes the fourth claw to project rather furthest; but the third one measures as much, the second is shorter, and the first and fifth are equal to each other, and a little shorter than the rest.

The *hind-feet* are also turned obliquely outwards, and are scaly, with a few interspersed hairs above, and granulated underneath. The sides are narrow, and present a conspicuous callous tubercle, posterior to the origin of the inner toe. The hind legs are very short, and are clothed with soft brown hair, a tuft of which curves over the heel. There are no hairs on the sides of the hind-feet, like those which form a margin to the fore-ones. The hind-toes are longer than the fore-ones, and are armed with more slender claws, which are white, awl-shaped, curved, and acute. They have a narrow groove towards their points underneath.

DIMENSIONS.

	Inches.	Lines.		Inches.	Lines.
Length of the head and body	4	3	Length of the longest fore-claw	0	3
„ head	1	6	„ hind leg	0	2
„ tail	2	6	„ „ from the heel to the		
„ „ including the pencil of			roots of the toes	0	6
hair at its extremity	3	3	„ longest hind-toe and claw	0	4½
„ naked part of the nose, ex-			„ „ „ claw alone	0	3
clusive of the awl-shaped processes	0	2½	„ fur on the back	0	5
Breadth of the palm	0	4			

The *condylura longicaudata*, described at page 13, has the scales on the feet, particularly the fore ones, more conspicuous, and with fewer interspersed hairs.

There is a third specimen in the Zoological Museum which was procured at Mr. Brookes's sale, which I suppose to be the *condylura cristata* of DESMAREST, or the radiated mole of PENNANT.

Its colour is between umber and chestnut browns, and its tail is somewhat four-sided, slender, and tapering gradually from the root to the tip, which is terminated by a small pencil of hairs. The tail is obscurely scaly, and is covered with strong, short, tapering hairs. It is similar to the other two species in other respects, and has the same number of processes to its nose with the *condylura macroura*.

DIMENSIONS.

	Inches.	Lines.		Inches.	Lines.
Length of head and body	5	3	Length of the hind-foot, from the heel to		
„ the tail	2	6	the end of the middle-claw	1	0

The native place of the specimen is not recorded.

INDEX OF THE SPECIES.

The Latin specific names are in SMALL CAPITALS, and the Synonyms and Native names in *Italics*.

288 INDEX OF THE SPECIES.

INDEX OF THE SPECIES. 299

PAGE

	PAGE		PAGE
Squenoton	261	*Tampeh*	241
Squinoton	261	*Tarbogan*	147
Squirrel, barking	154, 187	*Tawny American marmot*	164
Squirrel, black	191	Tawny marmot	164
Squirrel, burrowing	151, 154	Tawny lemming	128
,, common	187	*Tawny lemming*	136
,, flying	195	*Taxus Labradoricus*	37
,, four-banded	184	*Techallotl*	179
,, four-lined	184	*Temamaçame*	261
,, ground	174, 181	*Terienniak*	83
Squirrel, ground	158	*Terree-anee-arioo*	83
Squirrel, Hudson's Bay	187	*Terreeya*	46
,, lesser gray	191	*Teuthlal-macame*	261
,, pine	190	*Thæthiay*	158
,, red	187	*Thling*	80
,, Severn river	193	Thick-wood badger	147
,, small brown	190	Tiger cat	104
,, small gray	174	*Tlalcoyotl*	41
Stag	251	*Tooktoo*	238, 241
Stagg	232	*Tree-innæuck-kannortoot*	89
Star-nose, long-tailed	13	*Tschernoburi*	94
,, thick-tailed	285	*Tsinantontonque*	93
Stoat	45	*Tsoutayé*	105
Stoat weasel	46	*Tukta*	238, 241
Stone fox	83		
Striped dormouse	181	*Urson*	214
,, ground squirrel	177	URSUS AMERICANUS	14
Subulate bat	3	*Ursus arctos*	14
Suisse	181	URSUS ARCTOS ? Americanus	21
Suzet	180	*Ursus candescens*	24
Swift fox	98	,, cinereus	24
		URSUS FEROX	24
Tail-less marmot	230	*Ursus horribilis*	24
Talpa longicaudata	13	,, Freti Hudsonis	41
TAMIAS, sub-genus	181	,, Labradoricus	37

London: Printed by W. Clowes, Stamford-street.